PERGAMON INTERNATIONAL LIBRARY
of Science, Technology, Engineering and Social Studies

The 1000-volume original paperback library in aid of education,
industrial training and the enjoyment of leisure

Publisher: Robert Maxwell, M.C.

LIPID ANALYSIS

2nd Edition

THE PERGAMON TEXTBOOK
INSPECTION COPY SERVICE

An inspection copy of any book published in the Pergamon International
Library will gladly be sent to academic staff without obligation for their
consideration for course adoption or recommendation. Copies may be retained
for a period of 60 days from receipt and returned if not suitable. When a
particular title is adopted or recommended for adoption for class use and the
recommendation results in a sale of 12 or more copies, the inspection copy may
be retained with our compliments. If after examination the lecturer decides that
the book is not suitable for adoption but would like to retain it for his personal
library, then a discount of 10% is allowed on the invoiced price. The Publishers
will be pleased to receive suggestions for revised editions and new titles to be
published in this important International Library.

Other Pergamon publications of related interest

BOOKS

BUTLER
The Analysis of Biological Materials

CHRISTIE
Lipid Metabolism in Ruminant Animals

HOFFMAN-OSTENHOF *et al*
Affinity Chromatography

REVIEW JOURNAL

PROGRESS IN LIPID RESEARCH

LIPID ANALYSIS

ISOLATION, SEPARATION, IDENTIFICATION AND STRUCTURAL ANALYSIS OF LIPIDS

2nd Edition

By

WILLIAM W CHRISTIE

Hannah Research Institute, Ayr, Scotland

PERGAMON PRESS

OXFORD · NEW YORK · TORONTO · SYDNEY · PARIS · FRANKFURT

U.K.	Pergamon Press Ltd., Headington Hill Hall, Oxford OX3 0BW, England
U.S.A.	Pergamon Press Inc., Maxwell House, Fairview Park, Elmsford, New York 10523, U.S.A.
CANADA	Pergamon Press Canada Ltd., Suite 104, 150 Consumers Rd., Willowdale, Ontario M2J 1P9, Canada
AUSTRALIA	Pergamon Press (Aust.) Pty. Ltd., P.O. Box 544, Potts Point, N.S.W. 2011, Australia
FRANCE	Pergamon Press SARL, 24 rue des Ecoles, 75240 Paris, Cedex 05, France
FEDERAL REPUBLIC OF GERMANY	Pergamon Press GmbH, 6242 Kronberg-Taunus, Hammerweg 6, Federal Republic of Germany

First edition 1973
Second edition 1982

Library of Congress Cataloging in Publication Data

Christie, William W.
Lipid analysis.
(Pergamon international library of science, technology, engineering, and social studies)
Includes bibliographical references.
1. Lipids — Analysis.   I. Title.   II. Series.
[DNLM: 1. Lipids — Analysis. QU 85 C555L]
QP751.C49 1982   574.1'247   82–491

British Library Cataloguing in Publication Data

Christie, William W.
Lipid analysis. — 2nd ed. — (Pergamon international library)
1. Lipids — Analysis
I. Title
547'.77046   QP751
ISBN 0–08–023791–6 Hard Cover
ISBN 0–08–023792–4 Flexicover

Printed in Great Britain by A. Wheaton & Co. Ltd., Exeter

Preface to the Second Edition

IN THE Preface to the first edition of this book (1973), it was stated that "The development of chromatographic techniques, particularly gas-liquid chromatography and thin-layer chromatography, together with advances in spectroscopy, have led to an explosive growth of interest in lipids, and have revolutionised our knowledge of the role that they play in the structure and function of cell membranes, as essential dietary components and in numerous biological processes". In the intervening years, technical developments have continued apace, with the introduction of new materials and methods for lipid analysis. In particular, high performance liquid chromatography, which was virtually ignored in the first edition, is increasingly making a contribution. Nor has the need for improved methods lessened. It is increasingly being realised that not only the composition of lipids, but also their detailed structures and the physical form of their association with proteins in tissue, are of importance in understanding how lipids function and how they may potentially be involved in various human disease states. It has, therefore, been necessary to add a new chapter on the separation of lipoproteins to this edition. The basic plan of the book is the same as in the first edition but each of the original chapters has been extensively re-written to take account of new developments. I have resisted suggestions that the book should be expanded into an encyclopaedic compendium on the subject, in the hope that it will remain on the laboratory bench, not on the library shelf, guiding both the tyro and the expert through the complexities of the practice of lipid analysis.

Acknowledgements

I AM grateful to the Director of the Hannah Research Institute and to the Department of Agriculture and Fisheries for Scotland for permission to write this book. My thanks are due also to Drs. R. C. Noble and J. H. Shand for reading critically the first draft; their comments led to many improvements. Dr. R. A. Clegg assisted greatly in the preparation of the chapter on lipoprotein separations. Mrs A. E. Brice and Mrs L. E. Brawn carefully and patiently typed two drafts of the complete manuscript. Finally, I gratefully acknowledge the assistance of the authors and journals, cited at various points in the text, for permitting me to reproduce certain figures.

W. W. CHRISTIE

Contents

Summary

Introduction

THE approach of the research worker to the analysis of the lipids in a given sample will depend partly on the amount of material in the sample, partly on the equipment and instrumentation available, but principally on the amount of information required. For example, if large numbers of samples must be screened routinely, it may only be possible to perform the more basic compositional analyses on each one, while in other circumstances a detailed knowledge of the composition and structure of all the lipid components of a single sample may be necessary. Structural features of lipids of importance in analytical work are described in Chapter 1. Throughout the book, it is assumed that apparatus for gas-liquid chromatography (GLC) and thin-layer chromatography (TLC) will be available to the analyst. The principles of these and other procedures are described in Chapter 3.

In all analytical procedures, precautions must be taken to minimise the effects of autoxidation (see Chapter 3); in particular, lipids must be handled in an atmosphere of nitrogen wherever possible and anti-oxidants should be added to TLC sprays or to TLC solvent mixtures. Care should be taken to prevent the introduction of contaminants into samples. Also, many of the solvents and other chemicals used in lipid analysis are potentially hazardous to the operator and they should always be handled circumspectly.

Extraction of Lipids from Tissues

Detailed extraction procedures are described in Chapter 2. The method of choice depends partly on the nature of the sample, e.g. plant tissues are extracted first with *iso*-propanol to prevent enzymatic degradation of lipids, and also on the amount of the sample. For reasons of economy in time and materials, simplified extraction procedures may be preferred for very large samples or for numbers of small samples, although more exhaustive extraction procedures must be used when highly detailed analyses are intended. The analyst must also consider, when choosing a method of removing non-lipid contaminants from extracts, whether the gangliosides are required for analysis. The weight of the tissue and of the lipid obtained must be recorded and, in some circumstances, the amount of dry matter in the sample should be determined. This data can sometimes be obtained from a small representative aliquot of the sample.

Fatty Acid Compositions

Methods of determining the fatty acid compositions of lipid samples are discussed in Chapter 5. The methyl ester derivatives are first prepared as described in Chapter 4 (methods for preparing other lipid derivatives are also described here) and the method chosen will depend on the nature of the sample. If free fatty acids are present, an acidic reagent such as methanolic hydrogen chloride must be used although the milder and more rapid alkaline transesterification reagents are to be preferred for glyceride-bound fatty acids (special procedures may be required for short-chain, amide-bound or unusual fatty acids). Infrared and ultraviolet spectroscopy may be of assistance in indicating the presence of some of the less common functional groups in fatty acid chains, and adsorption chromatography (especially TLC) is of value in indicating whether these are polar in nature. Gas-liquid chromatography is the chief method of determining the fatty acid composition of lipids and analyses should preferably be performed on more than one type of polyester liquid phase. Components can be identified provisionally by their retention times relative to authentic standards, by their equivalent chain length values, on the basis of possible biosynthetic relationships or by their behaviour in other ancillary chromatographic techniques, such as silver nitrate TLC or high performance liquid chromatography (HPLC). For unequivocal identification of a component, however, it is necessary to isolate it by some chromatographic procedure or combination of procedures, and establish its structure by definitive chemical and/or spectroscopic techniques. The GLC quantification method selected should be checked and calibrated regularly with standard mixtures of known composition similar in nature to the samples to be analysed. Results for each component fatty acid in a lipid class or mixture of lipids are generally expressed as "weight per cent" of the total but they must be converted to "mol per cent" for structural studies of lipids. In some circumstances, it is necessary to express results as "weight of each fatty acid per unit of tissue", but provided the weight per cent of lipid in the tissue is recorded, the methods of reporting data are interconvertible.

The Analysis of Simple Lipids

This topic is discussed in detail in Chapter 6. Thin-layer chromatography is usually the method chosen for the analysis of simple lipids and it may be performed on the microgram to 100 mg scale. Layers of silica gel G 0.5 mm thick are preferred for larger amounts and those 0.25 mm thick for micro-amounts; hexane-diethyl ether-formic acid (80:20:2 by vol.) is by far the most widely used developing solvent. A few simple lipids can be identified by specific spray reagents (cholesterol and derivatives, free fatty acids or ester bonds), but normally authentic standards are run alongside the samples under investigation. With very small amounts of material or with large numbers of similar samples, fluorometry or charring followed by photodensitometry are generally the methods of choice for lipid quantification; this can be performed while the samples are on the plate or after they have been eluted from the adsorbent. With large amounts of lipid (i.e. 2 mg or more), specific chemical techniques can be applied to quantify individual components, or gas chromatography of the fatty acid constituents of each lipid class in the presence of a known amount of a suitable internal standard may be used so that the fatty acid composition and the amount of each fraction are determined in the same analysis. When larger amounts of particular simple lipids are required for structural analyses say, they can be isolated by column chromatography on silicic acid or Florisil although, because of the difficulty of monitoring columns, it may often be simpler to fractionate samples by preparative TLC on several identical TLC plates with pooling of corresponding fractions.

Alkyldiacylglycerols and neutral plasmalogens are sometimes found with triacylglycerols in lipid samples and can be separated from each other with care by TLC. They can be determined by hydrogenolysis with lithium aluminium hydride and TLC separation of the products. Aldehydes, liberated from plasmalogens by acidic conditions, may be chromatographed as such or in the form of more stable derivatives on similar GLC columns as are used for the analysis of methyl esters of fatty acids.

Results obtained by the above quantification procedures are most often expressed for each component in terms of "weight per cent of the total lipid" although sometimes "weight per unit of tissue" is preferred. Provided that the total weight of lipid per unit of tissue is recorded, however, either method is suitable.

The Analysis of Complex Lipids

Procedures for the analysis of complex lipids are discussed in Chapter 7. Complex lipids can be separated from simple lipids for analysis by rapid column chromatographic procedures on virtually any scale and if glycolipids are also major components of the sample, they can also be isolated at this stage. With small amounts of material, the complex lipids can be obtained by preparative TLC using the solvent system described in the previous section.

The more common phospholipids from animal tissues can be separated by TLC on silica gel H layers in a single dimension (on up to the 10 mg scale) with chloroform-methanol-acetic acid-water (25:15:4:2 by vol.) as developing solvent. Components are identified by their chromatographic behaviour relative to authentic standards or by means of specific spray reagents. Lipids can be determined by phosphorus analysis or charring-densitometry, or by GLC of their constituent fatty acids with a known amount of an internal standard so that the fatty acid composition and amount of each lipid class are determined in the same analysis. Other TLC systems in one direction are available for the analysis of the acidic lipids, which are generally minor components, but two dimensional TLC systems are a valuable alternative as they permit the separation of complex lipids into many more fractions on one TLC plate than is possible with one-dimensional systems. Related procedures are available for plant and bacterial lipids.

Glycosphingolipids can be separated from phosphatides by several chromatographic or chemical procedures. They can be fractionated into single components differing in the number and nature of the hexose units by one dimensional TLC and estimated by determining the amounts of the nitrogenous bases by chemical procedures. HPLC of benzoylated glycosphingolipids also provides excellent separations. Chromatographic methods are available for determining the fatty acid composition, the long-chain base composition and the hexose composition of individual glycosphingolipids.

When detailed analyses of all the minor components are required, more material must be used and it is advisable to convert all the components to the same salt form before commencing the separation process. Combinations of techniques such as DEAE or TEAE cellulose column chromatography in conjunction with preparative TLC or other column procedures must then be used to separate all the lipid components.

Alkyl- and alkenyl-forms of individual phosphatides cannot be separated from the diacyl- form of the phospholipid in the natural state but they can be determined by the hydrogenolysis procedure used for the analogous simple lipids. If the polar phosphorus group is removed or modified chemically, however, separation of the various forms of the phosphatide is sometimes possible.

Results of analyses are expressed in a variety of ways but generally reflect the method used for quantification. For example, "mol per cent phosphorus" or "weight of phospholipid per unit of tissue" are often used for each component when phosphorus analysis has been performed although "mol per cent phospholipid" can also be calculated from this data. "Mol per cent phospholipid" is normally the method of choice when GLC internal standard procedures are used for quantification. "Mol per cent of the total glycolipids" is the simplest method of expressing the results of glycosphingolipid analyses as these generally depend on determinations of the molar amounts of the nitrogenous bases. For the sake of uniformity, the author would prefer to see all results converted to "mol per cent of the total complex lipids."

Structural Analysis of Lipids

Methods of separating lipids into molecular species are described in Chapter 8, and for determining the positional distribution of fatty acids in lipids in Chapter 9. Lipids can be separated into simpler molecular species according to the combined properties of all the fatty acid constituents by silver nitrate and/or reverse phase TLC and often by high temperature GLC; HPLC in a reverse-phase mode is increasingly finding favour with lipid analysts. Combinations of two of these procedures should preferably be used wherever this is feasible. It is usually necessary to convert phospholipids to less polar derivatives prior to analysis in this manner, either by removing the phosphorus group entirely by means of phospholipase C hydrolysis, when the resulting diacylglycerols or a derivative thereof can be subjected to high temperature GLC in addition to the TLC separations, or by rendering the phosphate group less polar by enzymic and chemical means, when compounds labelled with ^{32}P can be studied. GLC of the component fatty acids with an added internal standard or related procedures are generally the methods chosen for determining molecular species separated in this way.

Complicated stereospecific analysis procedures have been developed for determining the distribution of fatty acids in positions 1, 2 and 3 of L-glycerol in triacylglycerols but the composition of position 2 alone is obtained relatively easily by means of hydrolysis with pancreatic lipase. Phospholipase A_2 can be used to establish the distribution of fatty acids between positions 1 and 2 of glycerophosphatides.

In all lipid structural studies, results for fatty acid compositions and molecular species should always be expressed in terms of "mol per cent" of the total.

Other Analyses

Isotopically-labelled lipids can be assayed by essentially two approaches, i.e. either continuously or discontinuously. In the former, radioactivity and mass are monitored at the same time on chromatograms. In the latter, separate aliquots of lipids are isolated for mass determination or for liquid scintillation counting. The advantages and disadvantages of various methods are discussed in Chapter 10.

The techniques used for the separation of lipoprotein fractions are more the province of the protein than the lipid analyst. Refined procedures are available to the specialist, but some simplified procedures for the more important fractions are described in Chapter 11.

CHAPTER 1

The Structure, Chemistry and Occurrence of Lipids

A. INTRODUCTION

The term LIPID has traditionally been used to describe a wide variety of natural products including fatty acids and their derivatives, steroids terpenes, carotenoids and bile acids, which have in common a ready solubility in organic solvents such as diethyl ether, hexane, benzene, chloroform or methanol. A more specific definition is to be preferred, but has yet to be agreed internationally, and the term is nowadays frequently restricted to fatty acids and their naturally-occurring derivatives (esters or amides) and to compounds closely related biosynthetically to fatty acids. It is in this sense that the term is used in this book.

The principal lipid classes consist of fatty acid (long-chain aliphatic monocarboxylic acid) moieties linked by an ester bond to an alcohol, principally the trihydric alcohol, glycerol, or by amide bonds to long-chain bases ("sphingoids" or "sphingoid bases"). Also, they may contain phosphoric acid, organic bases, sugars and more complex components that can be liberated by various hydrolytic procedures. Lipids may be subdivided into two broad classes — "simple", which can be hydrolysed to give one or two different types of product per mol, and "complex", which contain three or more hydrolysis products per mol. The terms "neutral" and "polar" respectively are used more frequently to define these classes, but are less precise and may occasionally be ambiguous; for example, un-esterified fatty acids are normally classed as neutral lipids despite the presence of the free carboxyl group.

A complete analysis of the lipids from a given source, therefore, involves separation of the lipid mixture into simpler types according to the number and nature of the various constituent parts, the identification and estimation of each of these and eventual determination of the absolute amount of each lipid type. Before progressing to this, however, a knowledge of the structure, chemistry and occurrence of the principal lipid classes and their constituents is necessary.

B. THE FATTY ACIDS

The common fatty acids of plant and animal origin contain even numbers of carbon atoms (4–24) in straight chains with a terminal carboxyl group and may be fully saturated or contain one, two or more (up to six) double bonds, which generally but not always have a cis-configuration. Fatty acids of animal origin are comparatively simple in structure and can be subdivided into well-defined families. Plant fatty acids, on the other hand, may be more complex and can contain a variety of other functional groups, including acetylenic bonds, epoxy-, hydroxy- or keto-groups and cyclopropene rings. Bacterial fatty acids usually consist of simpler saturated and monoenoic components, but may also contain odd-numbered, branched-chain and cyclo-propane acids. Very complex high molecular weight acids, the mycolic acids, have been found in certain bacterial species.

1. Saturated fatty acids

The commonest saturated fatty acids are straight-chain even-numbered acids containing 14–20 carbon atoms, although all the possible odd and even-numbered homologues with 2–30 or more carbon atoms have been found in nature. They are named systematically from the saturated hydrocarbon with the same number of carbon atoms, the final -*e* being changed to -*oic*. For example, the acid with sixteen carbon atoms and structural formula

$$CH_3.(CH_2)_{14}.COOH$$

is correctly termed hexadecanoic acid, although it also has a trivial name hallowed by common usage, i.e. *palmitic acid.* To simplify presentation and discussion of fatty acid compositions, shorthand nomenclatures also exist. In the simplest form, fatty acids are designated solely by the number of carbon atoms they possess, e.g. palmitic acid is a C_{16} acid. This compound can be defined more accurately, however, by listing both the number of carbon atoms in the acid and the number of double bonds, separating the two figures by a colon, i.e. taking the above example — 16:0. Table 1.1 contains a list of common saturated fatty acids together with their trivial and systematic names and shorthand designations.

TABLE 1.1. SATURATED ACIDS OF GENERAL
FORMULA $CH_3.(CH_2)_n.COOH$

Systematic name	Trivial name	Shorthand designation
ethanoic	acetic	2:0
propanoic	propionic	3:0
butanoic	butyric	4:0
pentanoic	valeric	5:0
hexanoic	caproic	6:0
heptanoic	enanthic	7:0
octanoic	caprylic	8:0
nonanoic	pelargonic	9:0
decanoic	capric	10:0
hendecanoic	—	11:0
dodecanoic	lauric	12:0
tridecanoic	—	13:0
tetradecanoic	myristic	14:0
pentadecanoic	—	15:0
hexadecanoic	palmitic	16:0
heptadecanoic	margaric	17:0
octadecanoic	stearic	18:0
nonadecanoic	—	19:0
eicosanoic	arachidic	20:0
heneicosanoic	—	21:0
docosanoic	behenic	22:0
tetracosanoic	lignoceric	24:0

Acetic acid is only rarely found in association with higher molecular weight fatty acids in esterified form, although it has been found esterified to glycerol and to hydroxy-fatty acids in some seed oils. C_4 to C_{12} acids are found mainly in milk fats, although the C_{10} and C_{12} acids have also been found in quantity in certain seed oils. *Myristic acid* (14:0) is a minor component of most animal lipids, but is present in major amounts in seed oils of the family Myristicaceae. Palmitic acid is probably the commonest saturated fatty acid and is found in virtually all animal and plant fats and oils. *Stearic acid* (18:0) is also relatively common and may on occasion be more abundant than palmitic acid, especially in complex lipids. Longer chain saturated acids occur less frequently, but are often major components of waxes. C_{15} to C_{19} odd-chain acids can be found in trace amounts in most animal lipids, but can occur in larger quantities in certain species of fish or in bacterial lipids.

Decanoic and higher saturated fatty acids are solids at room temperature. Because of the lack of functional groups other than the carboxyl group, they are comparatively inert chemically, and natural or synthetic lipids containing only these acids can be subjected to more vigorous chemical conditions than those containing polyunsaturated fatty acids.

2. Monoenic fatty acids

Straight-chain even-numbered fatty acids of 10 to 30 carbon atoms containing one double bond of the *cis*-configuration have been characterised from natural sources. Monoenoic acids with double bonds in the *trans*-configuration are also known but are found comparatively rarely. Fatty acids of a given chain length may have the double bond in a number of different positions and a full description of any acid must specify the position and configuration of the double bond (in the recommended numbering system, the carboxyl carbon is C-1). For example, by far the commonest monoenoic acid is *cis*-9-octadecenoic acid, less accurately $\triangle^9$-octadecenoic acid, which has the trivial name *oleic acid* and structure

$$CH_3.(CH_2)_7.CH = CH.(CH_2)_7.COOH$$

cis

In the shorthand nomenclature, the acid is designated 18:1 or, indicating the position of the double bond, 18:1(9).[339] In addition, the position of the double bond can be denoted in the form $(n-x)$, where n is the chain length of the acid and x the number of carbon atoms from the last double bond to the terminal methyl group, i.e. oleic acid is 18:1(n-9) or in the older literature 18:1ω9. IUPAC–IUB Commissions [338, 339] have reluctantly agreed to the first form of this nomenclature because of its convenience to biochemists with interests in fatty acid metabolism in animals, as will become apparent later. Table 1.2 contains a list of the common monoenoic acids together with their systematic and trivial names and their shorthand designations.

Oleic acid is probably the most abundant fatty acid of all and is found in virtually all lipids of animal and plant origin. Various positional isomers exist; for example, *petroselinic acid* (*cis*-6-octadecenoic acid) occurs in seed oils of the family Umbelliferae and *cis*-vaccenic acid (*cis*-11-octadecenoic acid) is the major unsaturated acid in many bacterial species. The *trans*-isomer of oleic acid *(elaidic acid)* is rarely found, but *trans*-vaccenic acid, which is a by-product of the biohydrogenation of polyunsaturated fatty acids in the rumen, is found in small amounts in the lipids of ruminant animals. Many different positional isomers of monoenoic fatty acids may be present in a single natural lipid; for example, five different *cis*-octadecenoic acids and eleven different *trans*-octadecenoic acids have been found in bovine milk triacylglycerols.[290]

Palmitoleic acid is a component of most animal fats and may be present in somewhat greater amounts in fish and some seed oils. Shorter-chain monoenoic acids occur in milk fats, but are rarely found in significant amounts in other tissues. C_{20} and C_{22} monoenoic acids are minor components of most animal lipids, but are found in appreciable quantities in certain seed oils (e.g. rape seed oil) and in fish oils. Odd-chain monoenoic acids are sometimes found as minor components of animal lipids, but may be present in larger amounts in some fish oils and in bacterial lipids.

Animal lipids frequently contain families of monoenoic fatty acids with different chain lengths but with similar terminal structures. Here biosynthetic relationships are obvious as several components may arise by chain elongation or chain shortening of a common precursor as illustrated in Fig 1.1a.

cis-Monoenoic fatty acids of eighteen carbon atoms or less are all low melting compounds. *trans*-Isomers generally have slightly higher melting points than the corresponding *cis*-compounds. Because of the presence of the double bond in the aliphatic chain, monoenoic acids and lipids containing such acids are more susceptible to chemical attack, particularly by oxidising agents, than the corresponding saturated compounds. They are fairly resistant to autoxidation but will succumb to this under very vigorous conditions.

3. Non-conjugated polyunsaturated fatty acids

Non-conjugated fatty acids (often abbreviated to PUFA) of animal and plant origin can be subdivided into several simple families according to their biosynthetic derivation from single specific fatty acid precursors.[490] The acids in each family contain two or more *cis*-double bonds, generally separated by a single methylene group (methylene-interrupted unsaturation), and have the same terminal structure. Table 1.3 contains a list of the more important of these acids.

Linoleic acid (*cis*-9, *cis*-12-octadecadienoic acid) is the commonest and simplest fatty acid of this type, and is found in most plant and animal tissues. Using

TABLE 1.2. THE MORE IMPORTANT MONOENOIC ACIDS OF GENERAL FORMULA
$CH_3.(CH_2)_m.CH=CH.(CH_2)_n.COOH$

Systematic name	Trivial name	Shorthand designation
cis-9-dodecenoic	lauroleic	12:1 (n-3)
cis-9-tetradecenoic	myristoleic	14:1 (n-5)
trans-3-hexadecenoic	—	16:1*
cis-9-hexadecenoic	palmitoleic	16:1 (n-7)
cis-6-octadecenoic	petroselinic	18:1 (n-12)
cis-9-octadecenoic	oleic	18:1 (n-9)
trans-9-octadecenoic	elaidic	18:1*
cis-11-octadecenoic	*cis*-vaccenic	18:1 (n-7)
trans-11-octadecenoic	*trans*-vaccenic	18:1*
cis-9-eicosenoic	gadoleic	20:1 (n-11)
cis-11-eicosenoic	gondoic	20:1 (n-9)
cis-13-docosenoic	erucic	22:1 (n-9)
cis-15-tetracosenoic	nervonic	24:1 (n-9)

* *The $(n-x)$ nomenclature is only used with fatty acids containing cis double bonds.*

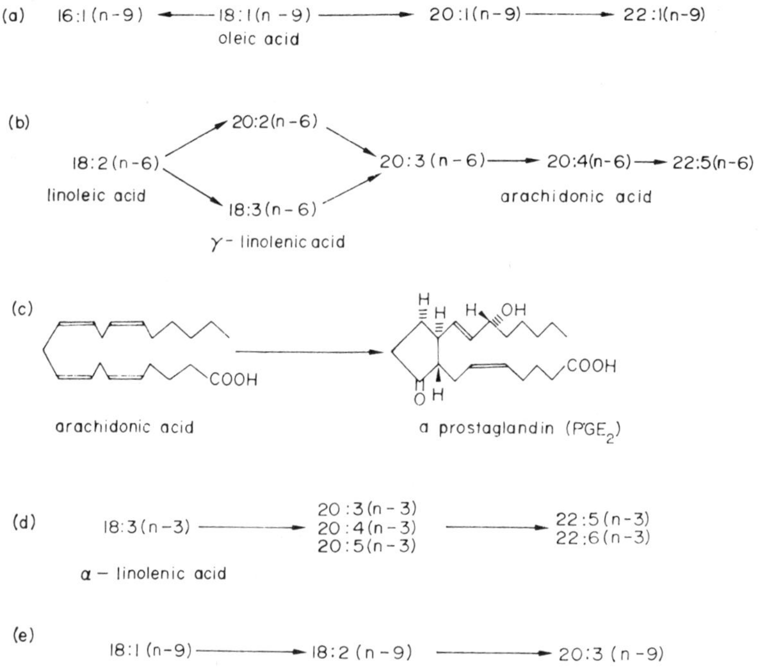

(a) $16:1(n-9)$ ⟵——— $18:1(n-9)$ ————▶ $20:1(n-9)$ ————▶ $22:1(n-9)$
                      oleic acid

(b)
                  $20:2(n-6)$

$18:2(n-6)$ $20:3(n-6)$ ——▶ $20:4(n-6)$ ——▶ $22:5(n-6)$
linoleic acid   arachidonic acid

                  $18:3(n-6)$

                γ - linolenic acid

(c)

       arachidonic acid   a prostaglandin (PGE_2)

(d) $20:3(n-3)$
       $18:3(n-3)$ ————▶ $20:4(n-3)$ ————▶ $22:5(n-3)$
                      $20:5(n-3)$ $22:6(n-3)$

      α - linolenic acid

(e) $18:1(n-9)$ ————————▶ $18:2(n-9)$ ————————▶ $20:3(n-9)$

FIG. 1.1. Biosynthetic relationships between unsaturated fatty acids. a, Elongation and retroconversion of oleic acid. b, Elongation and desaturation of linoleic acid. c, Biosynthesis of prostaglandin E_2 from arachidonic acid. d, Elongation and desaturation of α-linolenic acid. e, Elongation and desaturation of oleic acid.

TABLE 1.3. THE MORE IMPORTANT NON-CONJUGATED
POLYUNSATURATED FATTY ACIDS OF GENERAL FORMULA
$CH_3.(CH_2)_m.(CH=CH.CH_2)_x.(CH_2)_n.COOH$

Systematic name	Trivial name	Shorthand designation
9,12-octadecadienoic*	linoleic	18:2 $(n-6)$
6,9,12-octadecatrienoic	γ-linolenic	18:3 $(n-6)$
8,11,14-eicosatrienoic	homo-γ-linolenic	20:3 $(n-6)$
5,8,11,14-eicosatetraenoic	arachidonic	20:4 $(n-6)$
4,7,10,13,16-docosapentaenoic	—	20:5 $(n-6)$
9,12,15-octadecatrienoic	α-linolenic	18:3 $(n-3)$
5,8,11,14,17-eicosapentaenoic	—	20:5 $(n-3)$
4,7,10,13,16,19-docosahexaenoic	—	22:6 $(n-3)$
5,8,11-eicosatrienoic	—	20:3 $(n-9)$

* The double-bond configuration in each instance is *cis*.

the same shorthand nomenclature as before, it can be designated 18:2(n-6) (it is assumed that there is a methylene-interrupted double bond system). It is an *essential* fatty acid (often abbreviated to EFA) in animal diets, as it cannot be synthesised by the animal yet is required for growth, reproduction and healthy development.[315] In animal tissues, it is the precursor of a family of other fatty acids which are produced from it by desaturation and chain elongation as illustrated in Fig. 1.1b. All have the (n-6) terminal structure and can also function as essential fatty acids. The enzymes in mammalian systems are only able to insert double bonds between the carboxyl group and the first double bond already present in the fatty acid while plant enzyme systems can only insert new double bonds between the last double bond and the terminal portion of the fatty acid.

Arachidonic acid (cis-5, cis-8, cis-11, cis-14-eico-satetraenoic acid) is the most important of the linoleic acid metabolites and is a major constituent of the complex lipids of animal tissues, but is rarely found in plants. It is one of the principal precursors of a highly important series of hormone-like compounds known as prostaglandins[371,612] as shown in Fig. 1.1c. These compounds have profound pharmacological activities and are the subject of intensive study.

λ-Linolenic acid (cis-6, cis-9, cis-12-octadeca-trienoic acid), an important intermediate in the biosynthesis of arachidonic acid, occurs only in minor amounts in animal tissues but is found in appreciable quantities in some seed oils.

Linolenic acid (sometimes termed α-linolenic acid) or *cis*-9, *cis*-12, *cis*-15-octadecatrienoic acid ($18:3(n-3)$) is a major component of plant lipids, particularly of the photosynthetic tissues, but it is rarely a significant constituent of animal lipids. It is none the less extremely important as the primary precursor of another important family of polyunsaturated fatty acids (Fig. 1.1d). Linolenic acid and/or polyunsaturated fatty acids of the (n-3) family are essential fatty acids in fish,[700] but it is still a matter of debate whether they have an essential role in mammals. 5,8,11,14,17-Eicosapentaenoic and 4,7,10,13,16,19-docosahexaenoic acids, in particular, are found in many animal tissues as major components of the complex lipids and they are also found in large amounts in fish oils.

Oleic acid can also be the primary precursor of a family of polyunsaturated fatty acids (Fig. 1.1e). 5,8,11-Eicosatrienoic acid is normally a minor component of animal lipids but can assume significance in the complex lipids of animals deficient in essential fatty acids. There is in addition a family of polyunsaturated fatty acids derived from palmitoleic acid (n-7).

Polyunsaturated acids with more than one methylene group between the double bonds have been found in the lipids from some bacteria, plants and marine organisms but are rarely found in animals, while odd-chain polyunsaturated fatty acids have been found in some fish oils. Isomers of linoleic acid, in which one or more double bonds have a *trans*-configuration, have been isolated from certain seed oils.

Polyunsaturated fatty acids all have very low melting points. The more double bonds they possess, the greater their susceptibility to oxidative deterioration ("autoxidation"). If the acids or their derivatives are subjected to too high temperatures or to alkaline hydrolysis under conditions which are too vigorous, migration of stereomutation of double bonds can occur.

4. Branched-chain and cyclopropane fatty acids

Branched-chain fatty acids are common constituents of bacterial lipids,[583] but can enter the food-chain and appear in animal tissues. Those found most frequently have a single methyl group on the penultimate (*iso*) or antepenultimate (anteiso) carbon atoms (Fig. 1.2), although simple methyl-branched acids with the substituent elsewhere in the chain are occasionally found (e.g. D-(−)-10-methyl-stearic acid or "tuberculostearic acid"). Shorter-chain multibranched fatty acids are major constituents of avian preen glands. *Phytanic acid* (3,7,11,15-tetramethyl-hexadecanoic acid, Fig. 1.2), a metabolite of phytol, is present in trace amounts in animal adipose tissue lipids but assumes major proportions in Refsum's syndrome, a rare condition in which a deficiency in the fatty acid α-oxidation enzyme system prevents the catabolism of the acid causing it to accumulate (reviewed by Lough[454]).

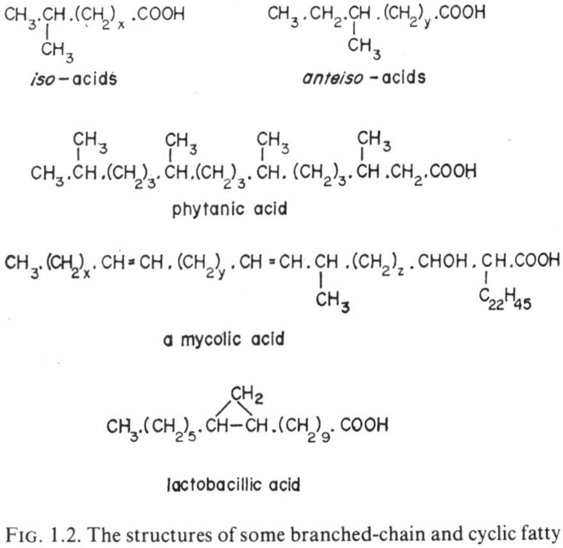

$$CH_3.CH.(CH_2)_x.COOH$$
$$\quad\quad |$$
$$\quad CH_3$$
iso–acids

$$CH_3.CH_2.CH.(CH_2)_y.COOH$$
$$\quad\quad\quad\quad |$$
$$\quad\quad\quad CH_3$$
anteiso–acids

$$CH_3.CH.(CH_2)_3.CH.(CH_2)_3.CH.(CH_2)_3.CH.CH_2.COOH$$
with CH_3 branches
phytanic acid

$$CH_3.(CH_2)_x.CH=CH.(CH_2)_y.CH=CH.CH.(CH_2)_z.CHOH.CH.COOH$$
with CH_3 and $C_{22}H_{45}$ branches
a mycolic acid

$$CH_3.(CH_2)_5.CH-CH.(CH_2)_9.COOH$$
with CH_2 bridge
lactobacillic acid

FIG. 1.2. The structures of some branched-chain and cyclic fatty acids.

Very high molecular weight branched-chain fatty acids having up to eighty or more carbon atoms, the *mycolic* and related acids (Fig. 1.2), have been found in lipids of certain bacteria.[583] Branched-chain acids have been found in only one plant species *(Antirrhinum majus),* but they are often found in small amounts in the depot fats of ruminant animals. Sebum and animal waxes, e.g. wool-wax, may contain significant amounts of branched-chain acids and isovaleric acid is a major component of dolphin triacylglycerols. As they are generally fully saturated compounds, branched-chain acids are comparatively resistant to chemical degradation. None the less, their structures can readily be determined by physical means such as mass spectrometry.

Cyclopropane fatty acids,[127] for example *lactobacillic acid* (11,12-methyleneoctadecanoic acid, Fig. 1.2), are found in bacterial lipids, particularly those of a number of gram-negative and a few gram-positive families of the order Eubacteriales. They are also found, generally in small amounts, in certain seed oils where they may be biosynthetic precursors of cyclopropene fatty acids.

In such compounds, the cyclopropane ring has some of the properties of a double bond although it may sometimes behave as a saturated entity also; while it is resistant to most oxidising agents, it may be disrupted by strong acids or by halogens.

5. Some unusual fatty acids of plant and animal origin

In addition to the more common saturated, monoenoic and C_{18} polyunsaturated fatty acids, plant lipids may contain a wide variety of unusual fatty acids not found in the animal kingdom. The structures of these have been comprehensively reviewed by Smith.[715] So many have been found, indeed, that it is not possible to discuss them all in a brief space, but a list of some of the more important functional groups found in the fatty acid chain will show the potential complexity of the problem of analysing lipids containing such acids. They include acetylenic bonds, conjugated acetylenic and ethylenic bonds, allenic groups, cyclopropane, cyclo-

TABLE 1.4. SOME UNUSUAL FATTY ACIDS FOUND IN PLANTS

Structure	Trivial name	Source

$CH_3.(CH_2)_5.CH.CH_2.CH=CH.(CH_2)_7.COOH$
    $\vert$
   OH   *cis*
	ricinoleic acid	castor oil

$CH_3.(CH_2)_4.C\equiv C.CH_2.CH=CH.(CH_2)_7.COOH$ *cis*	crepenynic acid	*Crepis foetida*
CH_2 $\triangle$ $CH_3.(CH_2)_7.C=C.(CH_2)_7.COOH$	sterculic acid	Sterculiaceae Malvaceae
$CH_3.(CH_2)_3.CH=CH.CH=CH.CH=CH.(CH_2)_7.COOH$ *trans* *trans* *cis*	α-eleostearic acid	tung oil

$CH_3.(CH_2)_4.CH-CH.CH_2CH=CH(CH_2)_7COOH$
            $\searrow\!\!\!\nearrow$
             O
	(+)-vernolic acid	*Vernonia anthelmintica*

cis *cis* —$(CH_2)_{12}COOH$	chaulmoogric acid	*Hydnocarpus* species
$CH_3.(CH_2)_7.CH=CH.(CH_2)_6 CH=CH.(CH_2)_3.COOH$ *cis* *cis*	—	*Limnanthes douglasii* seed oil
$CH_3.(CH_2)_{10}.CH=C=CH.(CH_2)_3.COOH$	laballenic acid	*Leonotis nepetaefolia* seed oil

propene, cyclopentene and furan rings, epoxy-, hydroxy- and keto-groups and double bonds of both the *cis*- and *trans*-configurations and separated by more than one methylene group. Two or more of these functional groups may on occasion be found in a single fatty acid. A brief list of some of the more important of the acids is contained in Table 1.4. Certain of the functional groups may be destroyed by some chemical techniques used widely in the analysis of animal lipids. For example, epoxy-groups and cyclopropene rings are disrupted by acidic conditions such as those used in some transesterification procedures.

Most of these acids occur in obscure seed oils, but a few have commercial importance or are important as biosynthetic precursors of other acids. *Ricinoleic acid* (D-(+)-12-hydroxy-octadec-*cis*-9-enoic acid) is the major component of castor oil and can be pyrolysed to give a number of products of industrial importance. *Crepenynic acid* (octadec-*cis*-9-en-12-ynoic acid) is important as a precursor of a complex series of poly-ynoic metabolites. The commercial value of cotton seed oil is reduced as it contains cyclopropene fatty acids in amounts which are small but sufficient to produce profound physiological effects in animals by inhibiting fatty acid desaturation, when the oil is incorporated into animal feedstuffs.[127]

Although fatty acids with these uncommon functional groups are most often found in plant lipids, they may on occasion be detected in animal tissues after ingestion. Increasing numbers of fatty acids with unusual structural features are being found in fish and other marine organisms. For example, furanoid fatty acids can occur in appreciable quantities in the reproductive tissues of fish in some circumstances.

C. SIMPLE LIPIDS

The simple lipids contain only fatty acid and alcohol components. The alcohol is usually glycerol but may also be a long-chain alcohol or a sterol. Esters of C_2, C_3 and C_4 diols are also known, but are rarely found in greater than trace amounts.[62] Unesterified or free fatty acids, which occur in small amounts in most animal and plant tissues, and free sterols, the most abundant of which is cholesterol, are also classed as simple lipids

1. Triacylglycerols and partial glycerides

In triacylglycerols (also known as "triglycerides"),[338, 339] each hydroxyl group of the trihydric alcohol, glycerol, is esterified to a fatty acid. If the two primary positions contain different fatty acids, a centre of asymmetry is created and the triacylglycerols may exist in different enantiomeric forms; a system of nomenclature must be adopted which takes account of this fact. Several such nomenclatures based on the conventional D/L or R/A systems have been proposed, but a "stereospecific numbering system" is now favoured.[338, 339] In a Fischer projection of a natural L-glycerol derivative (Fig. 1.3), the secondary hydroxyl group is shown to the left of C-2; the carbon atom above this then becomes C-1, that below becomes C-3 and the prefix *sn* is placed before the stem name of the compound. When the stereochemistry of the molecule is not specified, the primary hydroxyl groups are often termed the α- and α¹-positions and the secondary, the β-position. The topic of glyceride chirality has been discussed in detail by Smith.[717] Triacylglycerols are by far the most abundant single lipid class and virtually all the commercially important fats and oils of animal or plant origin and most animal depot fats consist almost entirely of this lipid.

Certain fungal and seed lipids have been found with triacylglycerol components that contain hydroxy-fatty acids, the hydroxl group of which is esterified to an additional fatty acid. Such lipids have been termed *estolides*.

Diacylglyerols (also known as "diglycerides") and monoacylglycerols (or "monoglycerides") contain two mol or one mol of fatty acid per mol of glycerol respectively (Fig. 1.3); collectively they are frequently termed "partial glycerides". They are rarely present in more than trace amounts in fresh animal and plant tissues but 1,2-diacyl-*sn*-glycerols have particular importance biosynthetically as precursors of triacylglycerols and complex lipids. The IUPAC-IUB Commissions on Biochemical Nomenclature[338, 339] have recommended that the names triacyl-, diacyl- and monoacylglycerol

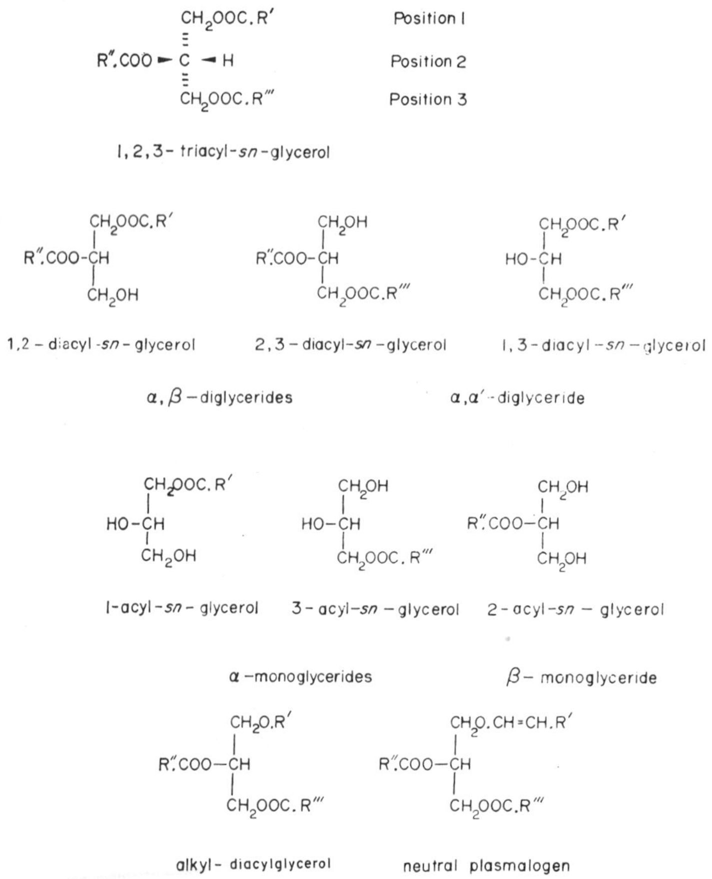

$$CH_2OOC.R'$$ Position I

$$R''.COO \leftarrow C \rightarrow H$$ Position 2

$$CH_2OOC.R'''$$ Position 3

1,2,3- triacyl-*sn*-glycerol

$CH_2OOC.R'$	CH_2OH	$CH_2OOC.R'$
$R''.COO-CH$	$R''.COO-CH$	$HO-CH$
CH_2OH	$CH_2OOC.R'''$	$CH_2OOC.R'''$
1,2 - diacyl-*sn*-glycerol	2,3 - diacyl-*sn*-glycerol	1,3 - diacyl-*sn*-glycerol

α, β – diglycerides α, α' – diglyceride

$CH_2OOC.R'$	CH_2OH	CH_2OH
$HO-CH$	$HO-CH$	$R''.COO-CH$
CH_2OH	$CH_2OOC.R'''$	CH_2OH
1-acyl-*sn*-glycerol	3- acyl-*sn*-glycerol	2-acyl-*sn*-glycerol

α –monoglycerides β– monoglyceride

$CH_2O.R'$	$CH_2.O.CH=CH.R'$
$R''.COO-CH$	$R''.COO-CH$
$CH_2OOC.R'''$	$CH_2OOC.R'''$
alkyl- diacylglycerol	neutral plasmalogen

Fig. 1.3. The structures of simple glycerides.

should replace the terms triglyceride, diglyceride and monoglyceride respectively as the former define the structures of the compounds more precisely.

A full analysis of triacylglycerols and partial glycerides requires that not only the total fatty acid composition of the lipid be determined but also the distribution of fatty acids in each position and the proportions of individual molecular species. The composition and properties of natural triacyl-glycerols have been reviewed comprehensively.[449]

Partial glycerides undergo acyl migration very readily in alcoholic solvents, on heating, in acidic or basic media and even when standing for long periods in the crystalline state, so procedures for separation and analysis should be such as to minimise this occurrence if a knowledge of the structures of these compounds is required.

2. Alkyl-diacylglycerols and neutral plasmalogens

Alkyl-diacylglycerols (or alkyl diglycerides)[720, 721] are lipid components in which a long-chain alkyl group is joined by an ether linkage to position 1 of L-glycerol; positions 2 and 3 are esterified with conventional fatty acids (Fig. 1.3). The alkyl groups of these 1-alkyl-2,3-diacyl-*sn*-glycerols are generally saturated or *cis*-monoenoic even-numbered components with sixteen to twenty carbon atoms. On hydrolysis, fatty acids (2 mol) and 1-alkylglycerols (1 mol) are obtained. The trivial names "chimyl", "batyl" and "selachyl" alcohol are used for 1-hexadecyl-, 1-octadecyl- and 1-octadec-9'-enyl-glycerol respectively. Alkyl ethers occur in small amounts in many animal tissues and are occasionally found in large quantities in the lipids of marine animals. The ether linkage is stable to acidic or basic conditions, although the ester bonds are readily hydrolysed as in more conventional lipids. The alkyl moieties are usually analysed in the form of the 1-alkylglycerol or as volatile non-polar derivatives of this compound.

Neutral plasmalogens are related compounds in which position 1 of L-glycerol is linked by a vinyl ether bond (the double bond is of the *cis*-configuration) to an alkyl group (Fig. 1.3). Such 1-alk-1'-en-1'-yl-2,3-diacyl-*sn*-glycerols have been detected in small amounts only in certain animal tissues. Although the vinyl ether linkage is unaffected by basic hydrolysis conditions, it is disrupted by strong acids yielding one mol of an aldehyde.

$$R.O.CH=CH.R' \xrightarrow{\text{H}_3\text{O}^+} ROH+OHC.CH_2.R'$$

The principal aldehydes usually have sixteen or eighteen carbon atoms and are fully saturated or contain one double bond. Their properties have been reviewed by Mahadevan.[475, 477]

3. Cholesterol and cholesterol esters

Cholesterol is by far the commonest member of a group of steroids with a tetracyclic ring system. In animal tissues, it occurs in the free state in intimate association with esterified lipids in cell membranes and in serum lipoproteins. It is also found esterified to fatty acids *(cholesterol esters)* in animal tissues, especially in the liver, adrenals and plasma. Cholesterol esters are hydrolysed or transesterified much more slowly than most O-acyl lipids. (The correct generic term is cholesterol esters but individual components are designated cholesteryl palmitate, etc.).[338, 339]

Plant tissues contain related sterols, e.g. β-sitosterol, ergosterol and stigmasterol, but only trace amounts of cholesterol. Steroidal hormones also occur in very small quantities in all animal tissues, but these are not lipids in the sense defined earlier. For the same reason, bile acids, which are important metabolites of cholesterol, cannot be discussed in detail here.

4. Wax esters and other simple lipids

Commonly, wax esters consist of fatty acids esterified to long-chain alcohols. The fatty acids are usually straight-chain saturated or monoenoic compounds containing up to thirty carbon atoms, but branched-chain and α- or ω-hydroxy acids are also found on occasion. Saturated long-chain primary alcohols usually predominate, but mono-enoic, branched-chain and secondary alcohols are sometimes present as are dihydroxy fatty alcohols. Wax esters are found in animal (e.g. skin lipids, uropygial glands) and insect (e.g. beeswax, cuticular lipids) secretions. They also occur as protective coatings on plant leaves and fruits, in the lipids of algae, fungi and bacteria, and in marine animals they are sometimes major components of the depot fats and sonar apparatus

In addition to wax esters, waxes may contain free long-chain alcohols, the properties of which have been reviewed,[476, 477] hydrocarbons, aliphatic aldehydes, ketones,[477] hydroxy-ketones and β-dike-tones, and triterpenes. The triterpenoic hydrocarbon, *squalene*, is a major component of some fish lipids, but is found in trace quantities only in animal or plant tissues; none the less, it is important as a biosynthetic precursor of cholesterol. The composition, analysis and biochemistry of waxes and their components have been reviewed in comprehensive monograph.[403] Individual lipid classes generally considered as wax components may be found in trace amounts in most animal tissues.

D. COMPLEX LIPIDS

The complex lipids can be subdivided into three main classes: *glycerophospholipids* (also known as "phosphoglycerides"), which on hydrolysis yield glycerol, fatty acids, inorganic phosphate and an organic base or polyhydroxy compound; *glyco-glycerolipids* (sometimes termed "glycosyldiglycerides"), which on hydrolysis yield glycerol, fatty acids and carbohydrates; *sphingolipids,* which contain a long-chain base, fatty acids and inorganic phosphate, carbohydrates or other complex organic compounds. In glycerophospholipids and glyco-glycerolipids, the non-fatty acid components are linked almost invariably to position *sn*-3 of the glycerol moiety and the stereospecific numbering system described earlier for simple glycerol derivatives is applied. The term glycolipid is used to describe any compound containing one or more mono-saccharide residues linked by a glycosyl linkage to a lipid part and so encompasses glycoglycerolipids and certain of the sphingolipids. The term *phospholipid* denotes any lipid containing phosphoric acid as a mono or diester and so includes glycerophospholipids and the sphingolipid, sphingomyelin. Glycerophospholipids are found in all plant and animal tissues and in micro-organisms; glycoglycerolipids are almost entirely plant and bacterial lipids, although trace amounts have been found in brain. Sphingolipids are important animal lipid constituents, although they have also been found in plants and in some microorganisms. Although they are soluble in most polar organic solvents, glycerophospholipids are generally inso-

luble in acetone and this property is often used in separating them from glycolipids and simple lipids. The occurrence and properties of phospholipids have been reviewed.[740]

1. Glycerophospholipids

(I) PHOSPHATIDIC ACID

Phosphatidic acid or 1,2-diacyl-*sn*-glycerol-3-phosphate (Fig. 1.4) occurs in very small amounts in animal and plant tissues, yet it is extremely important biosynthetically as it is the precursor of all other glycerophospholipids and of triacylglycerols. The fatty acid residues can be removed by mild hydrolytic procedures leaving *sn*-glycerol-3-phosphate (L-α-glycerophosphate), which can be further hydrolysed under more vigorous acidic conditions to glycerol and phosphoric acid. *sn*-Glycerol-3-

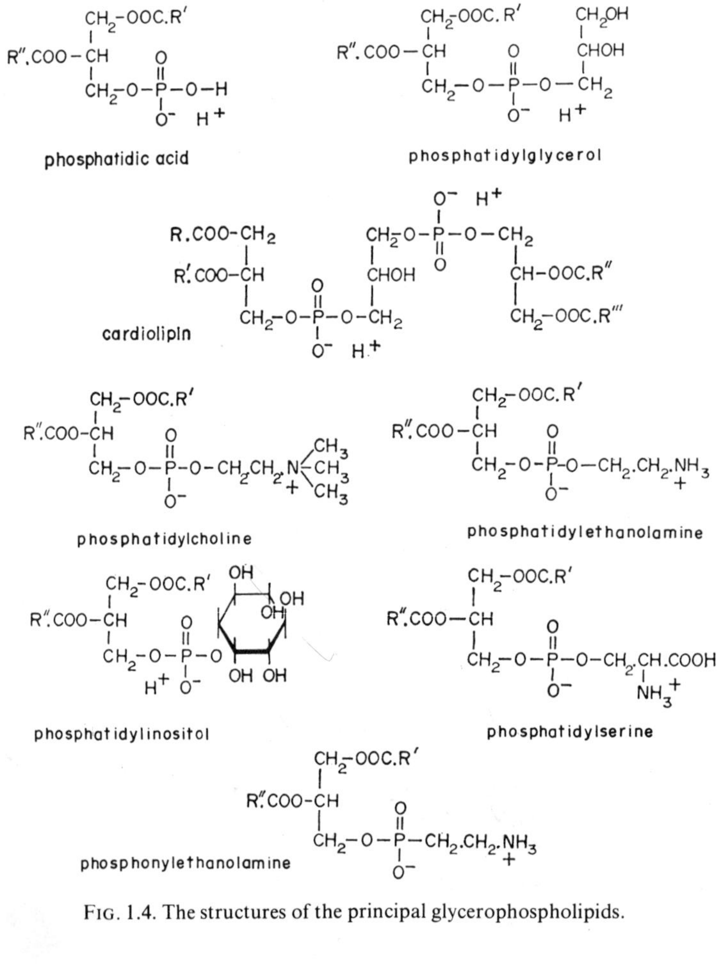

FIG. 1.4. The structures of the principal glycerophospholipids.

phosphate is the biosynthetic precursor of phosphatidic acid, although dihydroxyacetone phosphate may be important in some circumstances. Phosphatidic acid is strongly acidic and is often isolated from tissues in the form of a mixed salt, so that it may be necessary to prepare a single salt form to facilitate its purification.

In common with most glycerophospholipids in animal tissues, position *sn*-1 tends to be occupied by saturated fatty acids, although some monoenoic fatty acids are usually present also, while position *sn*-2 characteristically contains the polyunsaturated fatty acids.

(II) PHOSPHATIDYLGLYCEROL AND POLYGLYCEROPHOSPHOLIPIDS

Phosphatidylglycerol or 1,2-diacyl-*sn*-glycero-3-phosphoryl-1-'*sn*-glycerol (Fig. 1.4) is present in small quantities in many animal tissues, especially in cell mitochondria, and is also found in plant chloroplasts. Polyglycerophosphatides, in particular cardiolipin (or diphosphatidylglycerol), are major components of the lipids of mitochondria, especially in heart muscle. The occurrence and properties of cardiolipin have recently been reviewed.[335] Cardiolipins from some animal species contain very high proportions of linoleic acid and indeed molecular species containing 4 mol of this acid have been isolated. Phosphatidylglycerol and related compounds are acidic lipids, and on hydrolysis yield only glycerol, fatty acids and phosphoric acid in various molar proportions.

Bacterial lipids may contain *lipoamino acids*,[472] O-amino acid esters of phosphatidylglycerol, in which position 3' of the non-acylated glycerol moiety is linked to an amino acid. Usually only one amino acid (commonly lysine, ornithine or alanine) derivative is found in each bacterial species.

(III) PHOSPHATIDYLCHOLINE

Phosphatidylcholine (Fig. 1.4), commonly termed "lecithin", is more often than not the most abundant glycerophospholipid in animal tissues, and is often a major lipid component of plant tissues and of microorganisms. Position *sn*-1 in Phosphatidyl-cholines of animal origin is almost invariably occupied largely by saturated fatty acids while position *sn*-2 contains most of the C_{18}, C_{20} and C_{22} polyunsaturated fatty acids. It is a non-acidic lipid and on hydrolysis yields choline together with the normal glycerophospholipid components.

Small amounts of related compounds with alkyl ether or vinyl ether-containing residues in position *sn*-1 and fatty acids in position *sn*-2 are known. The former compound may now be termed *plasmanylcholine* and the latter *plasmenylcholine*[339] (for which the term "phosphatidalcholine" has also been used). They occur with phosphatidylcholine in some animal tissues and they have been detected in trace amounts in plants.

Lysophosphatidylcholines, in which only one of the two available positions of glycerol is esterified to a fatty acid, are often found in small amounts in tissues when phosphatidylcholine is also present. It is generally believed that position *sn*-1 is esterified in most instances but, as the acyl group migrates very readily, this has not been easy to confirm and there is some evidence that both possible isomers may exist naturally in the same tissue.

(IV) PHOSPHATIDYLETHANOLAMINE

Phosphatidylethanolamine (Fig. 1.4), once given the trivial name "cephalin", is generally the second most abundant class of glycerophospholipid and is present in large quantities in animal and plant tissues; it is frequently the major lipid class in bacteria. Phosphatidylethanolamines of animal origin usually contain more polyunsaturated fatty acids than do the phosphatidylcholines from the same tissue, and these acids are concentrated in position *sn*-2. It is a non-acidic lipid and on hydrolysis yields ethanolamine (1 mol), glycerol (1 mol), fatty acid (2 mol) and phosphoric acid (1 mol).

Alkyl and alkenyl analogues (with the anomalous group in position *sn*-1) are occasionally found in the same tissue with the diacyl component.

Lysophosphatidylethanolamine, in which only one of the two possible positions in the glycerol moiety is esterified, is found infrequently in some animal and plant tissues. Again, because of the ease with which the acyl group migrates, it is not easy to confirm which position is esterified in the natural state.

N-Methyl and NN-dimethyl derivatives of phosphatidylethanolamine are common constituents of the lipids of microorganisms. N-Acyl-phosphatidylethanolamines are occasionally found as minor components of plant tissues and they have been detected in some mammalian tissues.

(V) PHOSPHATIDYLINOSITIDES

Phosphatidylinositol (Fig. 1.4) is a common constituent of animal, plant and bacterial lipids. All compounds of this type are derivatives of the optically inactive form of inositol, myoinositol. Position sn-1 of phosphatidylinositol of animal origin usually contains a high proportion of stearic acid, and position sn-2 a high proportion of arachidonic acid.

Phosphatidylinositol is accompanied by large amounts of di- and triphosphoinositides in brain lipids. In the former, position 4′ of the myoinositol residue is linked to phosphoric acid and in the latter, positions 4′ and 5′ are linked to phosphoric acid. More complex polyphosphoinositides exist. The chemistry and biochemistry of these compounds have been reviewed.[288] Phosphoinositides are strongly acidic and are usually isolated in association with cations, such as those of magnesium or calcium.

(VI) PHOSPHATIDYLSERINE

Phosphatidylserine (Fig. 1.4) is a major component of brain and erythrocyte lipids and is found in small quantities in most animal and plant tissues, and in bacteria. It is a weakly acidic lipid and on hydrolysis yields L-serine together with the normal glycerophospholipid components. It is usually isolated as the potassium salt but may also be associated with calcium, sodium, or magnesium ions. N-Acyl phosphatidylserine has been found in sheep erythrocyte lipids.

(VII) PHOSPHONOLIPIDS

Glycerophospholipids which contain a phosphonic acid esterified to glycerol and a carbon-phosphorus bond to the nitrogenous base have been found in marine invertebrates and in protozoa.[395] The commonest of these phosphonolipids is related to phosphatidylethanolamine, and sometimes termed "phosphonylethanolamine". The phosphorus–carbon bond is extremely resistant to acid hydrolysis and the main products of this reaction are glycerol (1 mol), fatty acids (2 mol) and 2-aminoethylphosphonic acid (1 mol).

Phosphonolipid analogues of phosphatidylcholine, N-methyl- and NN-dimethyl-phosphatidylethanolamine and phosphatidylserine have also been found, in addition to the ceramide derivatives discussed below.

2. Glycoglycerolipids and related compounds

(I) PLANT GLYCOGLYCEROLIPIDS

Plant tissues contain lipids in which 1,2-diacyl-sn-glycerols are joined by a glycosidic linkage through position sn-3 to carbohydrate moieties.[306, 658] The principal components are mono- and digalactosyldiacylglycerols (as illustrated in Fig. 1.5), the hydrolysis products of which are fatty acids (2 mol),

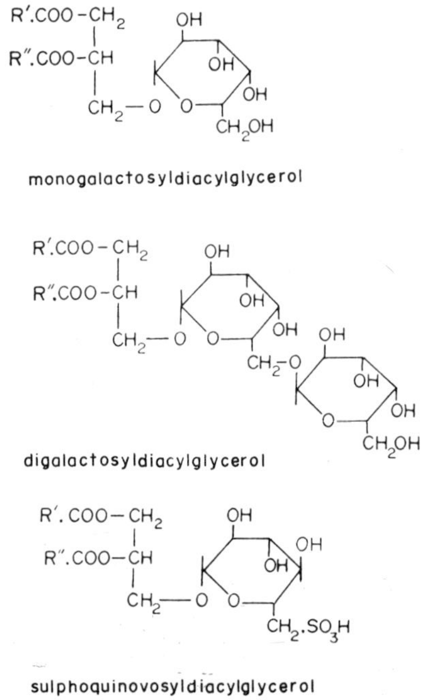

monogalactosyldiacylglycerol

digalactosyldiacylglycerol

sulphoquinovosyldiacylglycerol

FIG. 1.5. The structures of glycosyldiacylglycerols.

glycerol (1 mol) and galactose (1 or 2 mol, respectively). 6-0-Acyl-monogalactosyldiacylglycer-ls are also found in plants, and although they were originally thought to be artefacts, it is now thought that they may sometimes be normal constituents of the tissue. In addition, small amounts of tri- and tetragalactosyldiacylglycerols and glycoglycerolipids containing glucose rather than galactose in one of the monosaccharide residues may be present in some plant tissues.

Plant tissues contain a unique lipid, the *plant sulpholipid* or sulphoquinovosyldiacylglycerol, which is found exclusively in the chloroplasts. This consists of a monoglycosyldiacylglycerol in which position 6 of the monosaccharide moiety is linked by a carbon-sulphur bond to sulphonic acid (Fig. 1.5).

Glycolipids are soluble in acetone and are frequently separated from phospholipids by chromatographic systems in which this solvent is used.

Sterol glycosides, for example, 6-0-acyl-β-D-glucopyranosyl-(1-3')-β-sitosterol, are found in plant tissues (the sterol component may vary).[197]

(II) BACTERIAL GLYCOLIPIDS

A large number of glycosyldiacylglycerols have been isolated from bacterial species. These are 1,2-diacyl-*sn*-glycerols in which position *sn*-3 is linked by an ether bond to a carbohydrate moiety. Mono-, di-, tri- and tetraglycosyl derivatives have been found in which glucose, galactose, mannose, rhamnose or glucuronic acid, in various molar proportions, are the monosaccharide units. One or more fatty acid residues may also be esterified to the sugar components. The structures and occurrence of this complex group of lipids have been reviewed in greater detail than is possible here.[688]

Acylated sugar derivatives have been found in bacteria.[688] These lipids do not contain glycerol but consist of one or more fatty acids esterified to carbohydrates or to inositol.

Yeasts secrete complex extracellular lipids known as *sophorosides* in which a hydroxyl group of a mono-, di- or trihydroxy fatty acid is linked glycosidically to a carbohydrate moiety.[736, 780]

(III) MAMMALIAN GLYCOGLYCEROLIPIDS

Monogalactosyldiacylglycerols are known to be present in small amounts in brain and nervous tissue only of some mammalian species.[658, 757] A related compound in which position *sn*-1 contains an alkyl rather than an acyl group has also been found. A sulphato-galactoglycerolipid, the principal component of which is 1-0-hexadecyl-2-0-hexadecanoyl-3-0-(3'-sulpho-β-D-galactopyranosyl)-*sn*-glycerol, is present as the main glycolipid in testis and sperm cells and it is probably also present in brain; it is sometimes term "seminolipid".[527, 757] Recently, a range of complex glycoglycerolipids have been isolated and characterised from parts of the digestive system and lungs of mammals.

3. Sphingolipids

(I) LONG-CHAIN BASES AND CERAMIDES

Ceramides are amides of fatty acids with long-chain di- or trihydroxy bases, containing twelve to twenty-two carbon atoms in the aliphatic chain, of which the commonest is sphingosine (*trans*-4-sphingenine or (2S, 3R, 4E)-2-amino-4-octadecene-1,3-diol), as shown in Fig. 1.6.

More than sixty long-chain bases have been found in animals, plants and microorganisms,[375, 376] and many of these may be found in a single tissue. The monounsaturated dihydroxy base is found most often in animal tissues, but saturated, diunsaturated and branched-chain dihydroxy bases have also been found. In addition, saturated and monounsaturated straight-chain and branched-chain trihydroxy bases occur. For example, the commonest plant long-chain base is phytosphingosine (4-D-hydroxy-sphinganine or (2S, 3S, 4R)-2-amino-1,3,4-octa-decanetriol, Fig. 1.6). Such bases are frequently represented by a shorthand nomenclature similar to that used for fatty acids; the chain length and number of double bonds are denoted in the same manner with the prefix *d* or *t* to designate di- and trihydroxy bases respectively, e.g. sphingosine is d18:1 and phytosphingosine is t18:0.

The acyl groups of ceramides are generally long-chain (up to C_{26}) saturated or monoenoic fatty acids and long-chain 2-D-hydroxy fatty acids. Tetra-acetylsphingosines have been isolated from the extracellular lipids of yeasts.[736] Free ceramides have been found in small amounts in plant and animal tissues, but generally, they form the basic structural unit of the more complex sphingolipids and are

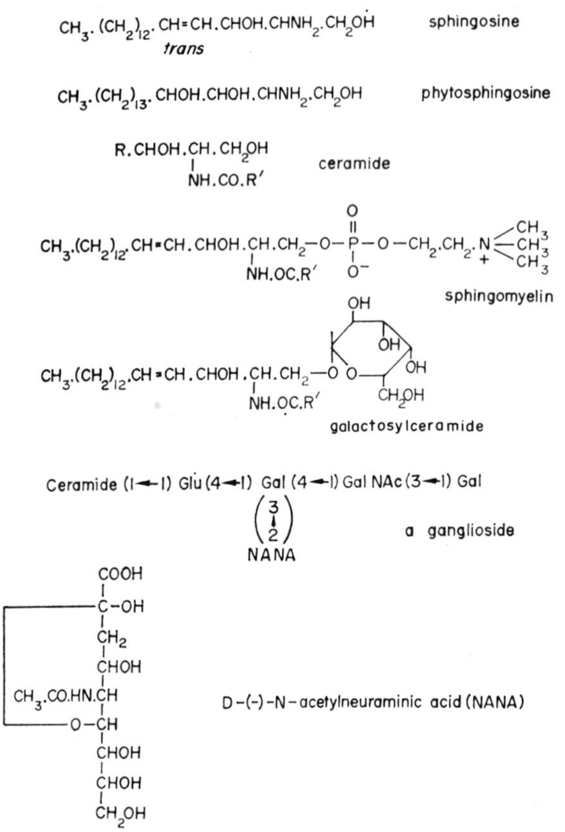

CH$_3$.(CH$_2$)$_{12}$.CH=CH.CHOH.CHNH$_2$.CH$_2$OH   sphingosine
trans

CH$_3$.(CH$_2$)$_{13}$.CHOH.CHOH.CHNH$_2$.CH$_2$OH   phytosphingosine

R.CHOH.CH.CH$_2$OH
|
NH.CO.R′   ceramide

CH$_3$.(CH$_2$)$_{12}$.CH=CH.CHOH.CH.CH$_2$—O—P—O—CH$_2$.CH$_2$.N<CH$_3$
| O$^-$
NH.OC.R′   sphingomyelin

CH$_3$.(CH$_2$)$_{12}$.CH=CH.CHOH.CH.CH$_2$—O
|
NH.OC.R′

galactosylceramide

Ceramide (1←1) Glu (4←1) Gal (4←1) Gal NAc (3←1) Gal
(3
↑
2)
NANA   a ganglioside

COOH
|
C—OH
|
CH$_2$
|
CHOH
|
CH$_3$.CO.HN.CH   D-(-)-N-acetylneuraminic acid (NANA)
|
O—CH
|
CHOH
|
CHOH
|
CH$_2$OH

FIG. 1.6. The structures of sphingolipids and their component parts.

linked through the primary hydroxyl group to phosphate or to complex carbohydrate moieties. The amide linkage is comparatively resistant to hydrolysis but is disrupted by prolonged heating with aqueous alkaline or acidic reagents.

(II) SPHINGOMYELIN

In sphingomyelin, position 1 of the ceramide unit is esterified to phosphorylcholine. It is found as a major component of the complex lipids in nearly all animal tissues but is not found in plants or micro-organisms. Sphingosine is usually the most abundant long-chain base component and it is normally accompanied by sphinganine and C$_{20}$ homologues. The fatty acid constituents are generally non-hydroxy long-chain (C$_{16}$ to C$_{26}$) saturated and *cis*-monoenoic compounds including some of the odd-numbered homologues (23:0, in particular, is often a major component). Sphingomyelin accumulates in the tissues of patients with Niemann–Pick disease.

A sphingolipid analogue of phosphatidylethanolamine (ceramide phosphorylethanolamine) has been found in insects and some freshwater invertebrates and may be present in minute amounts in some animal tissues; ceramide-2-aminoethyl-phosphonic acid is probably the most abundant phosphonolipid and was found originally in the lipids of the sea anemone,[395] but it is also a major component of protozoan lipids.

(III) GLYCOSYLCERAMIDES

Sphingolipids in which the basic ceramide unit is linked through position 1 of the long-chain base by a glycosidic link to glucose or galactose or to a polysaccharide unit are known as *monoglycosylceramides* ("cerebrosides") or *oligoglycosylceramides* respectively (the term "cerebroside" is occasionally used incorrectly to embrace the latter compounds). The occurrence, chemistry and biochemistry of these compounds have been reviewed.[757, 828, 834] As the name suggests, cerebrosides were first found in brain lipids, although they are minor components of most animal tissues and they have also been found in plants (where the hexose is glucose). The major brain cerebroside is a monogalactosylceramide (Fig. 1.6). Glucosylceramides are more common components of other animal tissues, particularly kidney, liver, plasma and spleen. They are also found in plants and in this instance, the major long-chain base is most commonly phytosphingosine.

Di, tri- and tetrahexosylceramides normally accompany cerebrosides in animal tissues. The first of these may contain two galactose units (found in kidney only) or one galactose and one glucose unit (lactosylceramide). The tri- and tetraglycosyl-ceramides are related in structure to these compounds and contain, for example, further galactose or N-acetylgalactosamine units. Other related complex glycosphingolipids contain fucose as the terminal sugar and/or as a branch of one of the internal glycosyl moieties (*fucolipids*). The trivial name "globoside" is occasionally used to denote tetraglycosylceramides. Oligoglycosylceramides containing from five to over twenty residues have been isolated from mammalian tissues but many of these have yet to be fully characterised. The fatty acid residues tend to be mainly 2-hydroxy saturated or monoenic components.

Sulphate esters of glycosylceramides with the sulphate group linked to position 3 of a galactosyl moiety are major components of brain lipids and are found in trace amounts in other tissues. They are known as *sulphoglycosylsphingolipids* (formerly "sulphatides") and their chemistry and bio-chemistry have been reviewed.[263] The main components are derivatives of galactosylceramide and galactosylglucosylceramide.

Mono- and oligoglycosylceramides have been intensively studied as they have a tendency to accumulate in certain rare diseases such as Gaucher's disease (accumulation of glucosyl-ceramide), Fabry's disease (triglycosylceramide) and metachromatic leukodystrophy (sulphomono-glycosylceramide).

(IV) Gangliosides

Gangliosides are highly complex types of ceramide polyhexosides that contain one or more sialic acid group (N-acyl derivatives of neuraminic acid, often abbreviated to NANA) (see Fig. 1.6).[757, 828, 834] They are termed mono-, di-, tri- and tetrasialogangliosides according to the number of sialic acid residues in the molecule. In addition, they contain glucose, galactose and N-acetylgalacto-samine units. These compounds were given the name "gangliosides" as they were first found in high concentration in the ganglion cells of the central nervous system, but they occur, albeit in much lower concentrations, in all tissues, although the N-acyl group of the sialic acid residue may differ from tissue to tissue or from species to species. For example, N-glycolyl-neuraminic acid derivatives are often detected in tissues. The major monosialoganglioside of human brain has the structure shown in Fig. 1.6.

Sphingosine and its C_{20} homologue, with smaller amounts of the related dihydro-compounds, are the main long-chain bases in brain gangliosides. A very high proportion of the fatty acid components of gangliosides from brain tissue is stearic acid while no hydroxy acids are found. In other tissues, the fatty acid pattern is similar to that of the other

glycosphingolipids. Gangliosides are soluble both in polar organic solvents and in water so must be extracted from tissues with care. In Tay–Sachs disease, there is a marked increase in the proportion of specific gangliosides in the brain lipids.

(V) Phytoglycolipids

These complex lipids, which have been found in a variety of plant seeds, consist of ceramides, in which phytosphingosine is the main long-chain base, linked through phosphorylinositol to complex oligosaccharides (inositol, hexuronic acid, hexo-samine, mannose, galactose, arabinose and fucose have been identified as components). For example, full characterisation of a tetrasaccharide phyto-glycolipid has been achieved.[740]

E. STRUCTURAL FEATURES OF LIPIDS IMPORTANT IN ANALYSES

The depth of the approach of any research worker to the analysis of lipids will depend on his own specific requirements. For many, for example in simple nutritional experiments, it may only be necessary to isolate and identify the fatty acid components and to determine the overall fatty acid composition of the total lipids. Biochemists, on the other hand, may have more rigorous requirements and may hope to isolate, identify and estimate all the various lipid classes in a tissue, and to further estimate the composition of the fatty acid and other components of these lipid classes. The ultimate aim may be to determine the composition of fatty acids esterified in each position of all the glycerol-containing lipids and to isolate and determine single molecular species, that contain one specific fatty acid in each position, of these lipids.

The first step in any analysis is the isolation of lipids from tissues by extraction with organic solvents, and the removal of non-lipid contaminants from these extracts. Procedures have been developed to ensure that there is no loss of the more water-soluble lipids such as gangliosides.

If the fatty acid composition of the total lipids is required, the lipids are generally converted to the methyl ester derivatives of the fatty acids by an appropriate procedure for gas chromatographic analysis. With animal lipids, this is usually a straightforward process, though care must be taken to minimise autoxidation of polyunsaturated fatty acids. It may be necessary to isolate individual components at times to determine the position of the double bonds. Plant lipids must be handled with a little more circumspection, however, and it may be necessary to examine these by adsorption chromatography and by spectroscopic procedures to detect unusual functional groups in the fatty acid components which might be destroyed under certain esterification conditions.

Single classes of lipid can be isolated from mixtures by combinations of chromatographic procedures which differentiate lipids according to the number, polarity and acidity or basicity of their constituent parts. They can often be provisionally identified by their chromatographic behaviour relative to that of authentic standards, and by spectroscopic techniques and specific spray reagents. However, unambiguous identification may require that the various products of hydrolysis be isolated, characterised and estimated.

Enzymatic hydrolysis procedures have been developed for determining the positional distribution of the fatty acid components of a number of simple and complex glycerol-containing lipids. Finally, single or at least simpler molecular species of individual lipid classes can be isolated by combinations of procedures according to the degree of unsaturation and/or the combined chain lengths of their aliphatic components. Enzymatic hydrolysis may be required to obtain a complete knowledge of the structures of these simpler species.

CHAPTER 2

The Isolation of Lipids from Tissues

QUANTITATIVE isolation of lipids free of non-lipid contaminants from tissues must ideally be achieved before the lipid analysis itself is begun. Carelessness at this preliminary stage may result in the loss of specific components or in the production of artefacts. For example, the presence of excessive amounts of free fatty acid, phosphatidic acid or N-acylphosphatidylethanolamine in tissues is usually indicative of faulty storage or extraction. Lipids can be extracted from tissues by a number of organic solvents, but special precautions are necessary to ensure that enzymes are deactivated and that the recovery is complete. Non-lipid contaminants must then be eliminated from the extract by washing or by column chromatography procedures, before the sample is ready for analysis. Precautions must be taken at each stage to minimise the risk of autoxidation of polyunsaturated fatty acids or hydrolysis of lipids.

A. GENERAL PRINCIPLES OF SOLVENT EXTRACTION PROCEDURES

1. Storage of tissues

Ideally, animal, plant or bacterial tissues should be extracted as soon as possible after removal from the living organism, so that there is little opportunity for changes to occur in the lipid components. This is especially important with brain lipids. When this is not feasible the tissue should be frozen as rapidly as possible, for example, on dry ice, and stored in sealed glass containers at –20°C in an atmosphere of nitrogen. The freezing process may permanently damage the tissue as the osmotic shock disrupts the cell membranes so that the original environment of the tissue lipids is altered and lipids may come in contact with enzymes from which they are normally

protected. In particular, lipolytic enzymes are released that may hydrolyse the lipids on prolonged standing, (even at–20°C) or on thawing, and contact with organic solvents can aid the process. For example, appreciable deacylation of phospholipids was observed in bacterial extracts frozen solid at –16°C; the rate at–10°C was greater than at+39°C in some circumstances[292]. Similar phenomena have been observed in plant and animal tissues on storage. The presence of large amounts of unesterified fatty acids in animal or plant tissues may be an indication that some irreversible damage to the tissues and subsequently to the lipids has occurred. Also, in plant tissues in particular, phospholipase D may be released to attack phospholipids so that there is an accumulation of phosphatidic acid and related compounds. Other alterations to lipids can occur that are more subtle and less readily discerned. The lipases in small samples of plant[287] or animal[232] tissue can be deactivated by plunging them into boiling water for brief periods, and the shelf-life of material treated in this way is considerably prolonged. Boiling with dilute acetic acid solution may have a similar effect.[577, 578] It is occasionally recommended that tissues be stored in saline solution, but Holman[314] has advised that they be stored under chloroform in all-glass containers or in bottles with teflon-lined caps at –20°C. Eventually, the tissue samples should be homogenised and extracted with solvent without being allowed to thaw.

2. The solubility of lipids in organic solvents

Pure single lipid classes are soluble in a wide variety of organic solvents, but many of these are not suitable for extracting lipids from tissues as they are not sufficiently polar to overcome the strong forces of association between tissue lipids and the other cellular constituents, such as proteins. None the less, polar complex lipids, which do not normally dissolve readily in non-polar solvents, may on occasion be extracted by these when they are in the presence of large amounts of simple lipids such as tri-acylglycerols, so the precise behaviour of any given solvent as a lipid extractant can not always be predicted. The ideal solvent or solvent mixture for extracting lipids from tissues should be sufficiently polar to remove all lipids from their association with cell membranes or with lipoproteins, but should not react chemically with those lipids. At the same time, it should not be so polar that triacylglycerols and other non-polar simple lipids do not dissolve and are left adhering to the tissues. The extracting solvent may also have a function in preventing any enzymatic hydrolysis of lipids if chosen carefully, otherwise it may actually stimulate side reactions. Increasingly, attention is being given to the potential toxicity of solvents to the operator.

Factors affecting the solubility of lipids in solvents have been comprehensively reviewed by Zahler and Niggli[867]. The two main features of lipids which affect their solubility in organic solvents are the non-polar hydrocarbon chains of the fatty acid or other aliphatic moieties and any polar functional groups such as phosphate or sugar residues. Lipids which contain no markedly polar groups, for example triacylglycerols or cholesterol esters, are highly soluble in hydrocarbon solvents such as hexane, benzene or cyclohexane and also in slightly more polar solvents such as chloroform or diethyl ether; they are rather insoluble, however, in polar solvents such as methanol. Their solubility in alcoholic solvents increases with the chain length of the hydrocarbon moiety of the alcohol so they are generally more soluble in ethanol and completely soluble in n-butanol. Similarly, the shorter the chain length of the fatty acid residues, the greater the solubility of the lipid in more polar solvents; tributyrin is completely soluble in methanol, while tripalmitin is virtually insoluble in this solvent. Polar lipids, on the other hand, may be only sparingly soluble in hydrocarbon solvents unless solubilised by association with other lipids, but they dissolve readily in more polar solvents such as methanol, ethanol or chloroform. It is important to recognise that the water in tissues or that used to wash lipid

extracts can alter markedly the properties of organic solvents.

It is generally believed that no single pure solvent is suitable as a general purpose lipid extractant, although Lucas and Ridout[459] have presented evidence that 20 volumes (ml per g of tissue) of ethanol will extract essentially all the lipids from liver homogenates in 5 min under reflux. Unfortunately they do not appear to have extended their work to other tissues. Ethanol–diethyl ether (3:1 or 9:1 v/v) mixtures are used more frequently, particularly to remove lipids from lipoprotein fractions. n-Butanol saturated with water has been recommended for the extraction of cereals or wheat-flour and may have wider uses.[492, 521] Diethyl ether or chloroform alone are good solvents for lipids but poor extractants of complex lipids from tissues. On the other hand, they are very useful solvents for removing non-polar lipids from triacylglycerol-rich tissues such as adipose tissue or oil seeds as they do not co-extract significant amounts of non-lipid contaminants. Unfortunately these solvents may also promote the action of phospholipase D when used to extract plant tissues[382] as does n-butanol.[145] n-and iso-Propanol inhibit this reaction and the latter isomer, which has the lower boiling point of the two, is frequently used especially as a preliminary extractant with plant tissues. An isopropanol-hexane mixture has also been recommended for the extraction of lipids from animal tissues as it is a comparatively low-toxicity mixture[279]; it does not extract gangliosides quantitatively, however.

Acetone is a poor solvent for phospholipids and is often used to precipitate such compounds from solution in other solvents. The favoured technique is to take up the lipid mixture in diethyl ether and then to add 4 volumes of cold anhydrous acetone when much of the phospholipid precipitates.[276] This procedure is useful for preparing a phospholipid-rich fraction in bulk, but such preparations are of little use in lipid analysis because the phospholipids obtained do not accurately represent those in the original lipid mixture; the solvent partition method described in Chapter 7 is better for the purpose. Tissue water and mutual solubility effects with other lipid components may permit acetone to extract more phospholipids from animal or plant tissues than might be predicted from a knowledge of the solubility of pure lipids in the dry solvent. Acetone may also react chemically with certain lipids (see below). Glycolipids are soluble in acetone and chromatographic solvents, in which this solvent is employed, are frequently used in the separation of glycolipids from phospholipids.

Although there are limitations to its use and alternatives are frequently being suggested (c.f. references 671–673), most workers in the field appear to accept that a mixture of chloroform and methanol in the ratio of 2:1 (v/v) will extract lipid more exhaustively fron animal, plant or bacterial tissues than most other simple solvent systems (the water in the tissues should perhaps be considered as a ternary component of this system). The tissue may be homogenised initially in the presence of both solvents, but better results are often obtained if the methanol (10 volumes; ml/g of tissue) is added first, followed, after brief blending, by 20 volumes of chloroform. More than one extraction may be necessary, but with most tissues the lipids are removed almost completely after two or three treatments. Generally there is no need to heat the solvents with the tissue homogenate, but this may on occasion be necessary with wet bacterial cells.[810] However, the extractability of tissue is variable and depends both on the nature of the tissue and of the lipids. For example, the extractability of gangliosides from brain is reduced if the concentration of monovalent cations in the tissue is reduced by dialysis or by washing and if the pH of the tissue is lowered.[728] It may be necessary in some instances to employ more stringent extraction procedures; Ways and Hanahan[817] recommend separate re-extraction with chloroform on its own followed by methanol alone, while others[640] advise that a five-stage extraction procedure, using both acidic and basic solvent systems, may be more successful.

Water is a poor solvent for all lipids but water–methanol mixtures, such as are obtained on washing chloroform–methanol extracts to remove non-lipid contaminants, may dissolve significant amounts of the more polar lipids, especially gangliosides.

The more common simple and complex lipids are in general readily extracted with chloroform–methanol mixtures but there are exceptions, for example, polyphosphoinositides or lysophospholipids. With polyphosphoinositides, it is necessary to ensure that tissues are stored in a manner such that enzymatic degradation is minimised, and to extract with the solvents, initially in the presence of calcium chloride, and subsequently after acidification.[285] When lysophospholipids are major components of the tissue extracts, it has been recommended that acid or inorganic salts be added during extraction with chloroform–methanol, or better that water-saturated n-butanol be used to extract the lipid[68] (acid conditions may be harmful to plasmalogens).

3. Removal of non-lipid contaminants

Most polar organic solvents used to extract lipids from tissues also extract significant amounts of non-lipid contaminants such as sugars, urea, amino acids and salts. When the polar solvents have been removed by evaporation or distillation, the lipids may be taken up in a small volume of a relatively non-polar solvent such as hexane:chloroform (3:1 v/v) leaving many of the extraneous non-lipid substances behind.[459] Separations of this kind are rarely complete so the procedure is little used nowadays, although it should not be discounted where large numbers of samples have to be purified for routine analysis, particularly for some of the simple lipid components. Other procedures which have been tried but with only limited success include dialysis, adsorption and cellulose column chromatography, electrodialysis and electrophoresis. It has recently been demonstrated that a pre-extraction of tissues with 0.25% acetic acid will remove all potential contaminants of lipid extracts, while simultaneously deactivating the lipolytic enzymes.[577, 578] This procedure has to date been applied only to brain tissue and to soybeans but may repay further investigation.

Most of the contaminating compounds can be removed from chloroform–methanol (2:1 v/v) mixtures simply by shaking the combined solvents with one quarter their total volume of water, or even better with a quarter of the total volume of a dilute salt solution (e.g. 0.88 per cent potassium chloride solution).[213] The solvents partition into a lower phase of composition, chloroform–methanol–water in the ratio 86:14:1 (v/v/v) and an upper phase in which the proportions are 3:48:47 (v/v/v) respectively. The lower phase, which comprises about 60 per cent of the total volume, contains the purified lipid and the upper phase contains the non-lipid contaminants together with any gangliosides which may have been present (varying amounts of other glycolipids may also be in this layer on occasion). If these are minor components or are not required for further analysis, then a simple washing procedure of this kind yields satisfactory lipid samples. Indeed, gangliosides can be recovered from the upper phase following dialysis and lyophilisation[368] (see also Chapter 7). Although there has been some debate as to the best method for quantitative recovery of gangliosides, most workers in the field still appear to favour this or closely-related methods.[97, 111, 753, 757]

It is important that the proportions of chloroform–methanol–water in the combined phases should be reasonably close to 8:4:3 (v/v/v). If it is necessary to wash the lower phase again, methanol–water (1:1 v/v), i.e. a mixture similar in composition to the upper phase, should be used to maintain these proportions otherwise losses of polar lipids may be greater than expected. Bligh and Dyer[73] have developed a related procedure which is more suitable for large samples as it uses smaller volumes of chloroform and methanol and the water already present in the samples is taken into account when adding further water in the washing step. With any washing procedure, it should be noted that centrifugation may be necessary or of assistance in ensuring complete separation of the layers.

A more elegant and complete, though more time-consuming, method of removing non-lipid contaminants is to carry out the washing procedure by liquid/liquid partition chromatography on a column. The aqueous washing phase is immobilised on a column of a dextran gel such as Sephadex G–25, while the organic lower phase is passed through the column. This type of lipid purification procedure was first developed by Wells and Dittmer[820] and has proved useful in a number of laboratories, but a modification of their method described by Wuthier[855] is simpler and more suitable for large numbers of samples. Chloroform, methanol and water in the ratio 8:4:3 (v/v/v) are partitioned as in a conventional "Folch" wash. The column of Sephadex G–25 is packed in the upper phase and the crude lipid extract is applied to the column in a small volume of the lower phase. Lipids free of contaminants are eluted rapidly from the column by further lower phase. Gangliosides and non-lipids

can be recovered from the column by washing with upper phase, and the column regenerated for further use. The procedure can also be used to remove acid and alkali from lipid samples. Siakotos and Rouser[696] have described a more complicated Sephadex column procedure based on a similar principle in which larger amounts of lipids can be purified and the gangliosides are obtained in a discrete fraction free of non-lipids. The procedure appears to be particularly suited to the analysis of bile lipids, as the various bile acid components are obtained in distinct fractions separate from the conventional lipids.[640] It is very time-consuming, however, and conditioning and regeneration of the columns is a rather lengthy process so that it is not suitable where large numbers of samples have to be purified routinely. When such column procedures are used to purify the lipids, it is no longer necessary to stick to the ratio of chloroform–methanol of 2:1 (v/v) in the extracting solvent, particularly as in many instances chloroform–methanol (1:1 v/v) may be a better extractant.

4. Artefacts of extraction procedures

If chloroform-methanol, or indeed any alcoholic extracts which contain lipids, are refluxed or stored for long periods in the presence of very small amounts of tissue sodium carbonate or bicarbonate, transesterification of many of the lipids may occur and large amounts of methyl esters are found in the extracts.[455] The problem can be largely circumvented by adjusting the pH of the aqueous medium to 4–5. Similar findings are regularly reported by other workers and it is possible that both acidic and basic non-lipid contaminants may catalyse the reaction. However, as small amounts of methyl esters may occur naturally in tissues, confirmation should be obtained, when they are detected, as to whether they are natural or artefacts of extraction or storage. This can be done simply by extracting the tissues with solvents which do not contain any alcohol, such as diethyl ether,[175] hexane[386] or acetone-chloroform,[440] and repeating the analysis for methyl esters on this material. Some rearrangement of plasmalogens may occur when they are stored for long periods in methanol.[804]

Acetone should not be used to extract brain lipids,

as it causes rapid dephosphorylation of poly-phosphoinositides.[168, 821] Acetone extraction of freeze-dried tissue may result in the production *in vitro* of an acetone derivative (imine) of phos-phatidylethanolamine.[34, 298]

Although 6-O-acyl-galactosyldiacylglycerols are known to be natural components of plant tissues,[297] it is possible that they are formed as artefacts by acyl transfer from other lipids when the cells are disrupted, as they are found in much smaller amounts when the tissues are homogenised in the presence of the extracting solvent.[296] Similar difficulties may be encountered with bacterial homogenates.[688]

The problem of enzymatic modification of lipids during extraction of plant tissues, especially, is discussed above.

5. Some practical considerations

All solvents contain small amounts of potential lipid contaminants and should be distilled before use. Plastic containers or apparatus (other than that made of Teflon®, should be avoided at all costs, as plasticisers (usually diesters of phthalic acid) are leached out surprisingly easily and may appear as spurious peaks on chromatograms and affect UV spectra. Wet animal tissues alone in contact with plastic have been reported to extract small amounts of these compounds and organic solvents will extract very large amounts. Other potential contaminants are discussed in Chapter 3.

Polyunsaturated fatty acids will autoxidise very rapidly if left unprotected in air. Although natural tissue antioxidants such as tocopherols may afford some protection, it is advisable to add an additional antioxidant such as BHT ("butylated hydroxy toluene" or 2,6-di-*tert*-butyl-*p*-cresol) at a level of 50–100 mg per litre to the solvents.[853] This need not interfere with later chromatographic analyses (see Chapter 5). Whenever possible, extraction pro-cedures should be carried out in an atmosphere of nitrogen, and both tissues and tissue extracts should be stored at −20° under nitrogen. It is helpful to deaerate solvents by flushing them with nitrogen before use. Further discussion of the problem will be found in Chapter 3.

Tissues should be homogenised with solvents in a Waring blender or a similar instrument in which the drive to the knives or grinders is from above, so that there is no contact between solvent and any washers or greased bearings. With difficult tissues, clean sand may be added to aid the grinding process. Lyophilised tissues are particularly difficult to extract and it may be necessary to rehydrate them before extraction to ensure quantitative recovery of lipids.

Solvents should be removed from lipid extracts under vacuum in a rotary film evaporator at or near room temperature. When a large amount of solvent must be evaporated, it should be concentrated to a small volume and then transferred to as small a flask as is convenient so that the lipids do not dry out as a thin film over a large area of glass. There is no need to bleed in nitrogen continuously during the evapora-tion process as the solvent vapours effectively displace any air, but the vacuum should eventually be broken with nitrogen. Lipids should not be left for any time in the dry state but should be taken up and stored in an inert non-alcoholic solvent such as chloroform. Last traces of water may be removed by codistillation with chloroform or ethanol.

It should always be remembered that chloroform and methanol are highly toxic and that mixtures are powerful irritants, if they come in contact with the skin. They should only be used in well-ventilated areas. (Filtration is probably the operation most likely to introduce appreciable amounts of vapour into the atmosphere).

It may sometimes be advantageous to extract a small sample of the tissue separately to obtain the weight of lipid per g of wet tissue, and also in order that the amount of dry matter in the tissue be determined from the residue. The weight of lipid recovered from a given amount of tissue should always be recorded.

B. RECOMMENDED PROCEDURES

In the following procedures, it is assumed that all the precautions mentioned above (Section A4 and 5) will be followed and that tissues will have been stored in an appropriate manner.

1. Extraction of large amounts of tissue

Where large amounts of tissue have to be extracted and a complete recovery of lipids is not essential, the procedure of Bligh and Dyer[73] offers some advantages as it does not use as large volumes of solvent as other methods. The following differs from that originally proposed as it has now been shown that it is advisable to filter the monophasic system before adding water, to prevent loss of acidic phospholipids such as phosphatidylinositol.[569] The yield of lipids is generally 95 per cent or better.

"It is assumed that 100 g of the wet tissue to be extracted contains 80 g of water. 100 g of the tissue is homogenised for 4 min in a Waring blender with a solvent mixture consisting of 100 ml of chloroform and 200 ml of methanol. If the mixture has two liquid phases, more chloroform–methanol should be added until a single phase is achieved. The mixture is filtered through a sintered glass funnel and the tissue residue is rehomogenised with 100 ml of chloroform and filtered once more. The two filtrates are combined, transferred to a 1 litre graduated measuring cylinder, 100 ml of 0.88 per cent potassium chloride in distilled water is added and the mixture shaken thoroughly, before being allowed to settle. The mixture should now be biphasic (further aqueous solution may be added to ensure this). The upper layer with any interfacial material is removed by aspiration. The lower phase contains the purified lipid and is filtered before the solvent is removed on a rotary evaporator. The lipid is stored in a small volume of chloroform at –20°C for further analysis."

Care should be taken to ensure that the ratio of chloroform–methanol–water is close to 5:10:4 in the monophasic system and 10:10:9 in the biphasic system. A related procedure has been described for tissues that contain a large amount of water, e.g. invertebrate tissue.[791]

Tissues that are very rich in lipid, such as adipose tissue or oil seeds, may be extracted first with diethyl ether or chloroform which will not remove significant amounts of non-lipid contaminants. Last traces of lipid remaining in the tissue can then be recovered by the above procedure.

2. Chloroform–methanol (2:1, v/v) extraction and "Folch" wash

The procedure of Folch, Lees and Stanley[213] is by far the most frequently quoted method for preparing lipid samples from tissues. It has not proved suitable in all circumstances but its modification, proposed by Ways and Hanahan,[817] is generally satisfactory. This extraction yields approximately a 95–99 per cent recovery of lipids but gangliosides and occasionally some of the glycolipids may be lost in the washing step unless the aqueous phase is specifically retained so that they can be recovered.[368] The following procedure is suitable for animal tissues.

"1 g of tissue is homogenised for 1 min with 10 ml of methanol, then 20 ml of chloroform is added and the process continued for a further 2 min. The mixture is filtered and the solid residue resuspended in chloroform–methanol (2:1 v/v, 30 ml) and homogenised for 3 min. After filtering, the solid is washed once more with chloroform (20 ml) and once with methanol (10 ml). The combined filtrates are transferred to a measuring cylinder and one quarter of the total volume of the filtrate of 0.88 per cent potassium chloride in water is added; the mixture is shaken thoroughly and allowed to settle. The upper layer is removed by aspiration, one quarter of the volume of the lower layer of water–methanol (1:1) is added and the washing procedure repeated. The bottom layer contains the purified lipid which can be recovered as above."

More extensive, and therefore more time-consuming, extraction procedures will guarantee very little loss, but individual requirements must determine whether the additional effort is worth while. For example, Rouser et al.[640] recommend a five-stage extraction procedure in which acidic and basic solvent systems are used. As mentioned earlier, special precautions are necessary for the complete extraction of polyphosphoinositides or lysophospholipids from tissues.

Plant tissues must be extracted first with a solvent that inhibits the action of lipases; isopropanol is used most frequently for the purpose. The following procedure has been recommended by Nichols.[539, 540]

"The plant tissues are macerated with 100 parts by weight of isopropanol. The mixture is filtered, the solid is extracted again is a similar manner and finally is shaken overnight with 199 parts of chloroform–isopropanol (1:1, v/v). The combined filtrates are taken almost to dryness, then are taken up in chloroform–methanol (2:1 v/v) and given a "Folch" wash as above. The purified lipids are recovered from the lower layer as before."

3. Sephadex G-25 columns for removing non-lipid contaminants

The procedure originally described by Wells and Dittmer[820] has been widely used and gives excellent results. Wuthier's modified procedure below[855] is somewhat simpler, gives equally satisfactory results and is more suitable for the purification of large numbers of samples. Both methods give good yields of the simpler glycolipids although gangliosides remain on the column to be eluted with the non-lipid contaminants. The crude lipid extracts to be purified are obtained simply by removing the unwashed solvent from the monophasic filtrate after extraction by any of the above procedures.

"Chloroform, methanol and water in the ratio 8:4:3 (v/v/v) are mixed and partitioned. The upper phase (hereafter referred to as UP) and lower phase (LP) are separated and retained. 25 g of Sephadex G-25 (Pharmacia Fine Chemicals, Uppsala, Sweden) are soaked overnight in 100 ml of UP, then are washed with 4×100 ml of UP. Columns (1 cm i.d. $\times$ 10 cm high) are packed with a slurry of this material. The column is capped with a filter paper disk and rinsed with 10 ml of UP followed by 10 ml of LP. The crude unwashed lipid extract (200 mg), obtained as described above, is taken up in 2–5 ml of LP, filtered to remove any precipitated non-lipid and is applied to the column, which is eluted with 25–30 ml of LP at a flow rate of up to 1 ml per minute. The pure lipid is eluted in this solvent and is recovered as before (any UP which leaks through with the eluate does not contain any impurity). Gangliosides and non-lipid contaminants can be eluted with 50 ml of UP and the column is regenerated for further use by washing with a further 20 ml of LP."

The complex procedure devised by Siakotos and Rouser[696] can only be recommended in special circumstances, for example in the analysis of bile lipids[640] or for the isolation of gangliosides.[696]

A variety of alternative procedures for the extraction of particular membrane preparations have been described by Zahler and Niggli.[867]

CHAPTER 3

Chromatographic and Spectroscopic Analysis of Lipids. General Principles

A. A STATEMENT OF THE PROBLEM

LIPID samples obtained from tissues by the methods described in Chapter 2 are complex mixtures of individual lipid classes and some means must be devised to obtain each of them in a pure state. Often, no single procedure will achieve the desired separations, and combinations of techniques must be used until the required pure lipid classes are obtained. Adsorption chromatographic procedures are generally used to separate each of the various simple lipid classes from the complex lipids. The latter may be further fractionated by adsorption chromatography or by ion-exchange chromatography, or by combinations of both, until the necessary separations are obtained. Lipid classes may be identified by their reactions with specific chemical reagents, by various spectroscopic techniques, or by their chromatographic behaviour relative to that of authentic standards, and the amounts of each determined by appropriate methods. The separation of single lipid classes into simpler molecular species is neither necessary nor desirable at this stage. The fatty acid composition of each lipid class can then be determined by gas–liquid chromatography of the methyl ester derivatives, prepared by transesterification of each, while other hydrolysis products can be estimated if necessary.

Ultimately, simpler molecular species of lipids may be isolated by partition chromatography, by gas chromatography or by chromatography on adsorbents specifically impregnated with reagents which form complexes with certain functional groups, such as double bonds, in the fatty acid or other alkyl moieties.

In common with all fields of research, the instruments available for lipid analysis are becoming increasingly complex and, therefore, increasingly expensive. None the less, much excellent work can be done with some comparatively simple apparatus. A gas chromatograph does not fall into the latter category, unfortunately, and there is unlikely to be any dispute that little good lipid research can be carried out without access to such an instrument. Indeed, the vast explosion of information on the chemistry and biochemistry of lipids over the last 20 years has been due principally to the development and exploitation of this technique. Apparatus for column or thin-layer chromatography can be comparatively inexpensive and versatile. Infrared and ultraviolet spectrophotometers are likely to be available to most research workers, and mass and nuclear magnetic resonance spectrometers, fluorometers and photodensitometers are becoming more commonplace in laboratories. All these refined spectroscopic and other techniques are valued weapons in the lipid analyst's armoury but are not always indispensable to good work; the gas chromatograph is essential.

In this book, greatest attention is given to procedures that utilise gas chromatography and basic column and thin-layer chromatography equipment, although techniques which require more complicated instrumentation are also described. Although paper and silicic acid-impregnated paper chromatography have their uses and their devotees, it is apparent that they offer relatively few lipid separations that cannot be achieved by other methods and are not considered further here. In the following sections of this chapter, the basic principles of various chromatographic procedures and of spectroscopic aids to analysis are discussed;

specific applications of these are dealt with later in the appropriate chapters.

B. CHROMATOGRAPHIC PROCEDURES

1. Gas–liquid chromatography

(I) PRINCIPLES

Gas–liquid chromatography (often abbreviated to GLC or GC) is a form of partition chromatography in which the compounds to be separated are volatilised and passed in a stream of inert gas (the mobile phase) through a column in which a high boiling-point liquid (the stationary phase) is coated onto a solid supporting material. The substances are separated according to their partition coefficients, which are dependent on their volatilities and on their relative solubilities in the liquid phase. They emerge from the column as peaks of concentration, ideally exhibiting a Poisson distribution. These peaks are detected by some means which converts the concentration of the component in the gas phase into an electrical signal, which is amplified and passed to a continuous recorder so that a tracing is obtained with an individual peak for each component as it is eluted. With a suitable detector, the areas under the peaks bear a direct linear relationship to the mass of the components present.

The theory of gas–liquid chromatography has been the subject of a number of excellent textbooks and need not be discussed in detail here, but certain relationships or definitions are particularly useful to the lipid analyst and are worth repeating. Figure 3.1

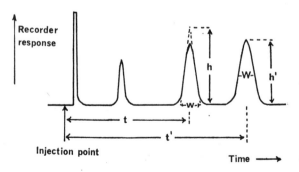

FIG. 3.1. Schematic diagram of a gas chromatographic recorder trace.

illustrates a schematic gas chromatography tracing. Shortly after introducing the sample (injection point), there is often a small air peak followed by a peak for the solvent in which the sample was dissolved. The base line should soon stabilise and peaks for the various components begin to emerge. The time from the point of the injection to that when the maximum amount of each component is emerging (i.e. to when the peak has reached its maximum height) is known as the *retention time* of the substance. The efficiency of a column is usually expressed in terms of the concept of numbers of *theoretical plates*, originally devised for distillation columns, which can be calculated using the simple formula:

$$n = 16 \times \left(\frac{t}{w}\right)^2$$

where n is the number of theoretical plates, w the width of the base (distance between two tangents drawn from the side of the peak to the base line) and t the retention time (or distance) of the component. The various column parameters such as gas flow rate, length of column and amount of liquid phase are chosen to optimise this figure in order to obtain the maximum resolution of peaks. Although in theory the longer the columns, the better the resolution that should be achieved, this is not always so and very long columns are not always practicable. The desired separations may then be better obtained by varying the nature of the stationary phase.

One further theoretical relationship is worth mentioning, i.e. for a homologous series of compounds.

Chain length = a constant × log (retention time)
As most lipids occur in nature in such series, this simple formula can be a useful means of provisionally identifying lipid components (see Chapter 5).

(II) APPARATUS

A large number of manufacturers now produce gas chromatographs commercially. Individual instruments vary greatly in design and in versatility and the choice of a particular make is dependent on the needs of the individual. All have certain features in common, however. If they are considered in order of contact with the sample, the first of these essential requirements is some means of applying the sample to the column. Usually this is done by injecting the sample, dissolved in an appropriate volatile solvent (1–5 μl), by syringe through a silicone rubber septum directly into the packing material of the column ("on-column injection"), so that the flow of gas is not interrupted. If the sample is not very volatile, a flash-heater may be used to ensure that it enters the column entirely as a vapour. The immediate contact with the liquid phase, which is achieved with on-column injection, affords some protection to the more labile components and this method is now preferred in most circumstances. Some commercial instruments are equipped with an accessory which will automatically inject samples — a useful feature where large numbers of similar samples must be analysed.

The columns on which the components are separated are the key to good analysis, and are responsible for much of the versatility of gas chromatography and are discussed in greater detail below. In many instruments, there are two identical columns and two detectors which are balanced electrically against each other to minimise background variations. The columns are held in an oven, the temperature of which is maintained accurately at the required point within 0.05–0.1°C. It is frequently less important that the precise temperature of the oven be known than that it be maintained constantly at the set temperature.

Temperature programming is a useful facility on many instruments. If the sample to be analysed contains components differing widely in volatility, it is advantageous to start the analysis with the column at a low temperature so that the more volatile components are separated as single coherent peaks, and then to raise the temperature at a fixed reproducible rate so that the less volatile components are eluted in a reasonable time. The sample emerges from the column into the detector, which is maintained, by a separate temperature control, at a slightly higher temperature than that eventually reached by the column, so ensuring that there is no change in the response of the detector during temperature programming.

The flame ionisation detector is chosen most frequently for lipid analysis. Eluted components are burned in a flame of hydrogen and air forming ions that are detected and measured by an electrical

system. Although the mechanism of the ionisation process is not fully understood, this detector is very sensitive, has a good signal-to-noise ratio, is rugged and does not deteriorate significantly with prolonged use. Provided gas flow rates are constant during the analysis, the response of the detector to homologous series of compounds is linearly dependent on the mass of each component over a wide range of molecular weights. The flow rates of hydrogen and air together with that of the carrier gas must be adjusted to the optima (instructions are generally supplied by manufacturers) for maximum sensitivity. The detector does not respond to some simple carbon compounds such as formic acid or carbon disulphide and the latter is often used as a solvent for samples to be analysed.

Argon ionisation detectors and thermal conductivity detectors are now little used; the latter is simple and robust though not particularly sensitive and may still have some advantages for preparative gas chromatography as the sample passes through the detector unchanged.

There are of course many other features of importance to consider in choosing a gas chromatograph, including sensitive gas flow controllers and the quality of the amplifier and recorder supplied with the instrument. Only the highest purity gases should be used. Helium is probably the best carrier gas but is prohibitively expensive outside the U.S.A., and nitrogen (containing less than 5 ppm of oxygen) is most often favoured as an alternative. Argon is more expensive than nitrogen but should contain much less oxygen so has advantages for the chromatography of more labile compounds. Containers of molecular sieves inserted into the gas lines to remove impurities improve the signal-to-noise ratio on the gas chromatographic traces.

(III) COLUMNS

The heart of the gas chromatograph is the column on which the substances are separated. There are three basic types: packed, wall-coated open-tubular (capillary or WCOT) and support-coated open-tubular (often abbreviated to SCOT).

Packed columns are the work horses of gas chromatography. They consist of glass or metal tubes, generally 4 mm or 2 mm in diameter (i.d.) and

1.5–2.5 m in length, coiled to fit the oven unit and filled with an inert solid support coated with the liquid phase. Stainless steel has been a popular choice of material for the columns themselves because of its durability and because it is comparatively chemically-inert to lipids. Aluminium and nickel are occasionally used also for the purpose. All metal columns must be thoroughly cleaned with solvent before use to remove any grease, and it may be advantageous to remove any active sites by silanisation. Glass columns are more fragile but otherwise have a number of advantages over metal columns; the surface is almost completely inert, it is easy to see how efficiently the column has been packed and whether any breaks in the packing appear during use, while deterioration of the top of the column packing at the inlet end is immediately apparent. Also, they are more easily emptied for re-use. Columns with good glass to metal seals are commercially available and are especially popular where the columns remain *in situ* in the gas chromatograph for long periods, as breakages are most likely to occur on removing or replacing columns in the oven unit.

The list of liquid phases available to the analyst is almost endless but, in practice, certain polyester liquid phases have proved themselves most useful for fatty acid analysis and a few silicone elastomers have advantages for higher molecular weight components. These will be discussed in greater detail in later chapters where analyses of specific lipid classes are described. The solid supporting materials for the liquid phases are generally diatomaceous earths, graded so that the particles are of uniform size, and deactivated by acid washing to prolong the life of the liquid phases and by silylation to minimise any adsorptive effects on the solutes.

It is claimed[328] that the most uniform coating of the liquid phase on the support is obtained if a solution is filtered through a bed of the support and the whole spread out as a layer to dry; the amount of liquid phase that remains on the support must be determined by experiment and depends somewhat on the nature of the support. It is also possible to achieve satisfactory coatings by evaporating a solution of the liquid phase in the presence of the support in an indented flask (to stir the solid material) on a rotary evaporator, but great care must be taken with this method to ensure that the support

is not damaged. Precoated supports are commercially available at prices close to those of the starting materials. A column is packed by adding the coated support in small amounts via a funnel while tapping gently and applying a vacuum to the exit end of the column. When it is filled, a glass-wool plug (acid-washed and silanised) is placed on top of the packing at the inlet end to consolidate it. If the columns are too loosely packed, the separations are poor; if too tightly packed, they may block or the injection syringes may plug easily. The column must be conditioned for up to 48 hr at a temperature just above that at which it is to be operated, before being used. 1.5–2 m Columns containing a support with 10–15 per cent (w/w) liquid phase should have an efficiency of 3000–5000 theoretical plates. Such columns normally have a working life of well over a year and this can periodically be prolonged by repacking the top 3–5 cm with fresh packing material.

Open-tubular or capillary columns, the theory of which has been comprehensively reviewed by Ettre,[205] consist of lengths of up to 100 m of narrow bore stainless steel or, increasingly, glass tubing (normally 0.25 mm in internal diameter), the inner wall of which is coated with the liquid phase. Such columns are highly efficient, 20,000–100,000 theoretical plates overall, and give quite remarkable separations; for example, resolution of positional isomers of monoenoic fatty acids is possible. They are particularly useful when coupled directly to the inlet of a mass spectrometer as each peak is likely to contain only a single component, and more positive identifications can be made. Some of the early capillary columns, especially those made of stainless steel, had a number of disadvantages; lengthy analysis times were necessary and the prolonged exposure to the metal surface resulted in decomposition of polyunsaturated components; coating the columns with liquid phase was difficult and not always reproducible, the working life of the columns was very short, often only a few weeks and they were difficult to repack. The increasing commercial availability of equipment for preparing glass capillary columns, which have less reactive surfaces, and the development of more thermally-stable stationary phases have greatly reduced the magnitude of these problems. In addition to their value in resolving mixtures of great complexity, glass capillary columns offer advantages in permitting quantitative analyses at the limit of detectability or, by means of short columns, very rapid analyses at moderate resolution. Developments in this area have recently been reviewed.[357, 679]

Before capillary columns can be used, the gas chromatographs should be equipped with suitable connectors to the inlet and exit ports. Various types of injection system are available commercially to ensure that the column is not overloaded and to minimise the amount of air and solvent injected onto the column with the sample.

Support-coated open-tubular columns[206, 601] consist of wider bore tubing than the capillary columns (0.5 to 1.25 mm in internal diameter), and contain a finely powdered solid support coated with the liquid phase. They are usually shorter in length than the capillary columns (10–15 m) and the efficiencies are a little less (10,000–15,000 theoretical plates). They can, however, take much larger loads of sample than the capillary columns, the analysis time is shorter, the operating temperature is lower and the useful working life can be longer. Such columns cannot easily be repacked, however, and would generally be considered too expensive for routine use. Although they enjoyed a brief vogue, they have largely been superseded by glass-capillary gas chromatography.

(IV) QUANTIFICATION OF COMPONENTS

It appears to be generally accepted that the most accurate routine method of estimating the amount of material in a gas chromatographic peak is by electronic digital integration of the amplified signal from the detector, and this is especially true for temperature-programmed analyses. A number of sophisticated electronic integrators are commercially available specifically for the purpose but the detector output can also be fed into a digital voltmeter and then via suitable software into a computer that has been programmed for the analysis.[81, 114] Such equipment is much less expensive than before and is increasingly available to analysts.

The alternative is to measure the area of the peaks or some parameter proportional to this, manually. Planimetry or measuring the area of the triangle enclosed by tangents drawn to the sides of the peaks to the base line have been used, but are not particularly accurate methods, i.e.

$$\text{Area } \alpha\, h \times w \text{ (Fig. 3.1)}$$

A better procedure is to multiply the height of the peak by the width at half-height.

$$\text{Area } \alpha\, h' \times W \text{ (Fig. 3.1)}$$

If the peaks obey a pure symmetrical Poisson distribution, the height of the peak multiplied by its retention time is also proportional to the area.[55] Conditions closest to the ideal are obtained in isothermal analyses.

$$\text{Area } \alpha\, h \times t' \text{ (Fig. 3.1)}$$

Although peaks are never completely symmetrical, there have been such marked improvements in the design of gas chromatographs and in the quality of the inert supports and liquid phases over the last few years, that this method of quantification is more accurate for isothermal analyses than reports in the earlier literature might lead one to believe. It has the advantages that two large distances are measured so that the measuring errors are smaller, it gives reasonably accurate results with incompletely resolved peaks and it is the simplest and most rapid method. A correction factor can be applied to increase the accuracy of the method to compensate for the fact that the sample is applied to the column as a finite band rather than as a point source,[82] i.e. the widths of the peaks at half-height are plotted against the retention times of components and the straight line through these intercepts the base line at a point in front of the point of the actual injection. (The errors of extrapolation are such that the point can perhaps better be found by trial and error with chromatographic traces of simple mixtures of known composition). If retention distances are measured from this point, more accurate results can be obtained, particularly for early running components. This procedure cannot be used with temperature programmed analyses, however. Usually the areas of the peaks, determined by one of the above methods, are summed and the area of each is expressed as a percentage of the total, so that the relative proportion of each component in the mixture is obtained.

All methods of peak area measurement should be calibrated with standard mixtures of known composition, similar to the samples to be analysed, and the calibration should be checked at regular intervals. It may then be necessary to adjust the area measurements by incorporating response factors obtained from the calibrations into the calculations. Detectors usually respond to the weight of a given component present and weight percentages may have to be converted to molar percentages by multiplying the results by suitable arithmetic factors. If the above precautions are taken, an absolute error of less than 1 per cent of major components (i.e. one that is 10 per cent or more of the total) of any mixture should easily be achieved.

(V) RANGE OF APPLICATIONS

Gas-liquid chromatography is used to analyse most substances that can be volatilised without decomposition. The highest molecular weight compounds that have been subjected to the technique are triacylglycerols containing over sixty carbon atoms, which can be eluted from highly thermostable silicone liquid phases at temperatures approaching 350°C. Although this is close to the thermal cracking temperature of such glycerides, the liquid phase appears to offer some protection. It is unlikely that compounds of very much higher molecular weight will ever be successfully subjected to conventional gas chromatography.

The most useful application of gas chromatography in lipid analysis is in the determination of the fatty acid compositions of lipids and, by a judicious choice of liquid phases, a complete analysis of fatty acids separated both by chain-length and degree of unsaturation in a given sample can be achieved. Indeed, positional isomers of unsaturated fatty acids can be separated on open-tubular columns. Volatile derivatives of lipids must often be prepared for GLC analyses; for example, fatty acids are customarily converted to methyl esters and the free hydroxyl groups of partial glycerides are acetylated. In addition to fatty acid analysis, gas chromatography can be used to determine cholesterol, glycerol, inositol, carbohydrates and many other compounds, released on hydrolysis of lipids in the form of volatile derivatives.

Although it is primarily an analytical tool, gas chromatography may be used preparatively, by inserting a stream splitter between the end of the column and the detector and collecting the effluent in a suitable trap (see Chapter 5). Gas chromatographs can also be operated with the detectors in series or in parallel with appropriate counting equipment, so that mass and radioactivity measurements on compounds eluted from the columns are determined together in a single analysis (see Chapter 10).

2. Adsorption column chromatography

In adsorption chromatography, differences in the degree to which lipid components are adsorbed on to a solid support, relative to their solubility in an appropriate solvent, are utilised as a means of separation. Lipids are held by the adsorbents in a variety of ways including hydrogen bonding, van der Waals' forces and ionic bonding. The more polar the functional groups contained in the lipid, the more strongly is it adsorbed; fatty acid chains are non-polar and have little effect in relation to polar functional groups such as free hydroxyl or keto groups or the polar head-groups of complex lipids. Lipids are released from the adsorbent by passing solvents of increasing polarity through it, and are therefore separated by such procedures according to the number (but generally not type unless they contain polar functional groups) of fatty acids in the molecule and the number and type of other polar functional groups. The adsorbent may be held in glass columns and eluted continuously with solvents (column chromatography), or it may be supported as thin layers on glass plates and the solvent allowed to pass through the adsorbent by capillary attraction (thin-layer chromatography, often abbreviated to TLC). Column chromatography can be used as a large-scale preparative procedure of comparatively low resolution or as a high resolution analytical or small scale preparative procedure (high performance liquid chromatography or HPLC sometimes also termed "high pressure" or "high speed" liquid chromatography). The principles of adsorption chromatography have been reviewed in some detail by Stein and Slawson.[732]

(I) PREPARATIVE-SCALE COLUMN CHROMATOGRAPHY ON SILICIC ACID

By far the most widely used adsorbent is silicic acid, a partially hydrated silicon dioxide (also often termed silica or silica gel) and it is available from a number of manufacturers. Different brands and even different batches of the same brand may vary in such properties as particle size, degree of hydration, surface area and in the proportions of trace organic and inorganic contaminants. In general, the finer the particles of silicic acid, the greater the surface area available for adsorption and the better the separations that can be obtained. If the particles are too small, however, the flow rate through the column may be rather slow and, in practice, a compromise must be sought which allows reasonable flow rates without unduly prejudicing the quality of the separations. 200-Mesh silicic acid has been commonly used but some manufacturers now market products specially prepared to have larger particle sizes without sacrificing surface area, and these permit more rapid flow rates to be attained.

The degree of hydration of the silicic acid has an important influence on the quality of the separations. If there is too little water, lipids are strongly held and are not eluted from the column in sharp bands but tend to "tail". If the adsorbent contains too much water, lipids are not adsorbed sufficiently strongly and the separations are poorer. The nature of the separation may also be changed through partition effects between bound water and polar lipids. Although the optimum degree of hydration must be found by trial and error, it is generally around 5 per cent. Water can be removed from the adsorbent by heating it at 110°C for several hours or by prewashing the column with dehydrating solvents such as acetone[305]. Water can be added uniformly to an adsorbent by adding the desired amount to the dry powder in a stoppered flask, turning this slowly until any large lumps break up and leaving for about 2 hr to allow the water to distribute evenly.

Glass columns with a sintered disc at the bottom to support the adsorbent and a tap below this, so that the flow of solvent can be controlled, are suitable for column chromatography (Teflon taps are preferable when lengthy separations are involved) but the dead volume should be as small as possible. A filter disc on top of the sintered disc minimises blocking of the latter. Highly sophisticated columns with very small dead spaces at the top and bottom are commercially available but these are more elaborate and expensive than is necessary for most routine lipid separations. The adsorbent is packed in a slurry of the first solvent

to be used in the separation and is allowed to settle with gentle tapping until it forms a bed at least ten times greater in height than in diameter (long thin columns give better separations than short wide ones). Columns should never be allowed to run dry as channels are formed that cause uneven elution of lipids.

The amount of lipid that can be applied to a column is variable and depends on the magnitude of the differences in polarity between the various components to be separated. In general, 30 mg of lipid per g of absorbent is a reasonable load but this can be varied with circumstance. The sample should be applied to the column in as small a volume as possible of the least polar eluting solvent, washed on to the bed of adsorbent carefully until no lipid remains above the surface, and then the main solvent reservoir can be attached.

Components are removed from the column by eluting with solvents of gradually increasing polarity. Wren[852] has described a modification of Trappe's eluotropic series of solvents. This is, in order of increasing polarity, petroleum ether (hexane) < cyclohexane < carbon tetrachloride < benzene < chloroform (containing no ethanol as stabiliser) < dichloromethane < diethyl ether < chloroform stabilised with 1 per cent ethanol < ethyl acetate < acetone < acetonitrile < methanol < acetic acid < water. Eluant composition can be changed in a discontinuous manner (stepwise elution), which is simple in practice and suited to many of the more common lipid separations, or it can be changed continuously (gradient elution) by means of a simple device.[305] Although the merits of the two types of elution procedures have been widely debated, no clear favourite has emerged. It appears that the commoner phospholipid classes may be better separated by stepwise elution, but that gradient elution may give better results with cerebrosides and sulphatides. Both methods may have to be tried with difficult samples.

The quality of the separations is also dependent on the rate of flow of the eluant and again the optimum may have to be determined by trial and error, though flows of 1–3 ml/min usually give satisfactory results. With very tightly packed columns, it is necessary to apply pressure to the top from a suitable pump or from a nitrogen cylinder in order to obtain satisfactory flow rates. This is preferable to applying suction from below which causes the column to dry up at the bottom.

Column chromatography with silicic acid offers advantages in some circumstances over TLC as the lipids are probably better protected against autoxidation and larger quantities can be separated. Partial glycerides will isomerise to some extent on silicic acid, but most other lipids are unaltered in its presence.

Lipids in general do not possess chromophores or functional groups which can be readily detected spectrophotometrically so that, until recently, it has not been possible to monitor the effluents of liquid chromatographic columns continuously. Detectors developed for HPLC (see below) can be applied to preparative-scale separations also. In the absence of such equipment, the eluate from columns is collected in small volumes (10–20 ml) by means of a fraction collector and aliquots of constant volume are taken from each of these (or from every 2nd or 4th), so that the amount of lipid can be determined by some means. For example, the solvent can be removed and the lipid weighed on a microbalance. Usually the solution is placed on a small aluminium pan weighting 5–10 mg which is dried on a hot plate at 60°C for a few minutes then in a vacuum desiccator to constant weight. The pans should never be touched by hand as sufficient grease may be deposited to alter the weight significantly. As the lipid autoxidises during the process, the pans are discarded after weighing. Such gravimetric procedures are best reserved for when large quantities of material are being chromatographed.

In an alternative method,[25, 215, 702] the solvent is removed from an aliquot of each fraction of the eluate and 2 ml of a reagent, prepared by dissolving 2.5 g of potassium dichromate in 1 litre of 18M sulphuric acid, is added. The whole is heated at 100°C for 45 min with shaking, cooled and the decrease in the absorbance of potassium dichromate ($Cr_2O_7=$) at 350 nm is determined. The lipid is oxidised and the dichromate reduced by a proportionate amount that can be determined from the absorbance measurements. Appropriate blanks must be run and the procedure calibrated by authentic standards. The procedure is useful for neutral lipids, but less so for phospholipids which are best estimated by a direct phosphorus determination by the method described in Chapter 7. It is also possible to apply aliquots from discrete fractions of the eluate as spots to TLC plates, and to use procedures for locating TLC samples (e.g. iodine or charring, see below) to estimate the amount of material present roughly by eye or more accurately by photodensitometry.

TLC with microplates may be useful for identifying and estimating the purity of fractions from columns. When the adsorption characteristics and the lipid elution profile of a silicic acid column have been determined, it may no longer be necessary to monitor fractions so precisely.

(II) PREPARATIVE-SCALE COLUMN CHROMATOGRAPHY ON ADSORBENTS OTHER THAN SILICIC ACID

Florisil, which is the trade name for a coprecipitated mixture of magnesia and silicic acid produced by the Floridin Co. (Pittsburg, U.S.A.), offers some advantages over silicic acid in the preparative-scale separation of simple lipids.[105] It is supplied as a coarse mesh powder with a high adsorptive capacity for lipids so that large amounts of these may be separated at higher solvent flow rates than are normally permissible. While the separations that can be achieved are similar in many ways to those obtained with silicic acid, acidic lipids are very strongly adsorbed and must be eluted with acidic solvent systems. Recoveries of phospholipids are generally poor, especially of phosphatidylcholine, and as some magnesium silicate is eluted with polar solvents, it may be necessary to purify complex lipids obtained in this way by a "Folch" wash (see Chapter 2).

When Florisil is washed with concentrated hydrochloric acid, much of the magnesium is removed and the product is in effect a coarse mesh silicic acid with a similar adsorptive capacity to the original Florisil, but with identical chromatographic properties to silicic acid.[106] The high flow rates obtained with acid-washed Florisil render it suitable for rapid routine separations. It is prepared as follows:

"Florisil (300 g; 60–100 mesh) is mixed with concentrated hydrochloric acid (900 ml) and heated on a steam bath for several hours. The hot supernatant liquid is decanted and the adsorbent is washed with a little more acid and heated overnight with a further 900 ml of acid. The product is filtered and washed till neutral with distilled water, and the acid and washing treatment repeated. The neutral residue is washed with approximately 400 ml each of methanol, chloroform–methanol (1:1, v/v), chloroform and diethyl ether. The product is first air-dried and then activated by heating overnight at 110–120°C."

Alumina has also been used as an adsorbent for column chromatography but there have been reports of extensive alteration to lipids when it is used, including autoxidation of double bonds, isomerisation of partial glycerides and hydrolysis of lipids. Different batches of alumina appear to vary more in properties than does silicic acid so for most purposes, the latter is now preferred.

(III) HIGH PERFORMANCE LIQUID CHROMATOGRAPHY (ADSORPTION)

In HPLC, modern technology has been applied to each stage of the column chromatographic process from injection of the sample, to maintaining the composition and rate of flow of the solvent, to the properties of the adsorbent and to detection of the separated components, in order to optimise the quality of the separations that can be achieved. A wide variety of commercial equipment is available. Several general textbooks on the subject have been published[269, 725] and Aitzetmuller[19] has comprehensively reviewed applications in lipid analysis.

The adsorbent generally favoured is silicic acid in the form of microspherules, manufactured commercially to closely-controlled sizes (commonly 5–30 μm), shape, porosity and degree of hydration; preparations are available under various trade names. The columns are generally of stainless steel, 20–50 cm long and 2–4 mm in internal diameter, and are tightly packed with the adsorbent.[269 725] After an analysis, the absorbent can be regenerated by solvent-washing procedures for re-use but it will slowly deteriorate, especially if aqueous solvents are used in analyses.

Because of the narrow bore of the columns and the tight-packing of the very uniform adsorbent, very high pressures are necessary to drive an adequate flow of solvent through the columns and it is necessary to use special pumps, operating on the

piston or diaphragm principle, for the purpose. Gradient elution tends to be favoured for many lipid separations and solvent-mixing devices can be connected, either before the pump i.e. only one costly pump is needed, or between several solvent pumps and the column. Commercial equipment is readily available but similar devices can be constructed in a good workshop.[19]

A considerable effort has been expended in the search for a lipid-sensitive detector for monitoring the eluate from columns, but with only limited success. Differential refractometers, which sense minute differences in the refractive index of an eluant brought about by material eluting from the columns, are readily available and are fairly sensitive but can, in practice, only be used with an eluant of constant composition, so limiting the range of applications. There have been attempts to overcome this drawback by using balanced columns, one for the chromatography of the sample and the other for reference purposes, or by using mixtures of solvents with very similar refractive indices but with different polarities, but such procedures also have their limitations. Universal detectors, in which the eluate from HPLC columns is coated onto a moving wire, which is dried to remove solvent then passed through a flame ionisation detector for quantification of any substances present, have been described by many research workers.[19]. For example, that constructed by Stolyhwo et al.[738] appeared particularly promising. Commercial instruments were apparently not sufficiently successful and are no longer available. Ultraviolet (UV) detectors are easily obtained but are of little use for monitoring column separations, as few lipids have functional groups that absorb in the appropriate region of the spectrum. However, isolated double bonds in unsaturated lipids do exhibit a specific absorbance at 200-206 nm, and some variable wavelength detectors can be purchased that can be utilised in this part of the spectrum. Traces of peroxides in the lipids or of aromatic compounds in the eluting solvent can swamp this effect because of their much higher extinction coefficients. Lipid analysts have demonstrated considerable ingenuity in converting particular lipids to derivatives that contain aromatic groupings which can be monitored by UV detection in specific separations. When HPLC is used as a quantitative analytical procedure with most of these detectors, care is necessary in calibration and it is essential that standards very similar to the samples to be analysed are used.

HPLC in the absorption mode would find innumerable applications if a suitable universal detector for lipids were to become available, but the cynical observer will note that many more papers are appearing describing model separations than those describing analyses of real samples.

3. Thin-layer adsorption chromatography

(I) PROCEDURES

In thin-layer chromatography (TLC), the adsorbent is held on glass plates in a thin layer, as the name suggests. A very fine grade silica gel is by far the most common adsorbent used for the purpose and this may contain calcium sulphate as a binder to ensure adhesion of the layer to the plate. Such mixtures are commonly termed "silica gel G" although, strictly speaking, this is the trade name of a commercial product. The adsorbent is applied to the plate in the form of an aqueous slurry (2 ml of water per g of adsorbent) by means of a suitably-designed spreader, so that even layers of a predetermined thickness are obtained. The plates are air-dried briefly then activated by heating in an oven at 110–120°C for 2 hr and stored in an airtight box or in a desiccator. A number of manufacturers supply complete kits of equipment for preparing TLC plates, but if it is intended to work with silver nitrate-impregnated layers (see later section), the spreader should be made of anodised or silver-plated aluminium or some other inert material so that it is impervious to the reagent. While it is possible to purchase pre-coated TLC plates ready for immediate use, they would generally be considered too expensive for routine applications. Small TLC plates suitable for simple separations can be made by dipping microscope slides in a slurry of silica gel in chloroform (0.25 g per ml) and allowing them to dry in the air,[575] or alternatively micro-spreading equipment is available. In addition, it may be practical to cut up commercial pre-coated layers (on plastic backing) into small pieces.

Samples are applied as discrete spots or as narrow streaks, 1.5–2 cm from the bottom of the plate, in a solvent (frequently chloroform) by means of a syringe or with a sample applicator made specifically for the purpose (many are available commercially), and the plate is then placed in a tank containing the eluting solvent. Lining the tanks with filter paper, to saturate the atmosphere inside with solvent vapour, speeds up the analysis, especially with polar solvents, and may occasionally improve the resolution. The solvent moves up the plate by capillary action taking the various components with it at differing rates, according to the extent to which they are held by the adsorbent. When the solvent nears the top of the plate, the plate is removed from the tank, dried in air or in a stream of nitrogen (depending on whether it is intended to recover the lipids for further analysis) to remove the solvent and sprayed with a reagent that renders the lipids visible. These should appear as a line of discrete spots or bands.

The spray may be a chemical reagent, which is specific for certain types of lipid or for certain functional groups (see later chapters), or it may be a non-specific reagent that renders all lipids visible. A 0.1 per cent (w/v) solution of 2',7'-dichlorofluorescein in 95 per cent methanol is most frequently used for the latter purpose and causes lipids to show up as yellow spots under UV light. Alternatively, an aqueous solution of Rhodamine 6G (0.01 per cent, w/v) may serve the purpose, in which case lipids appear as pink spots under UV light. Rhodamine 6G is particularly useful when alkaline solvent systems have been used and 2',7'-dichlorofluorescein is to be preferred with acidic solvents. Both sprays are non-destructive and the lipids can be recovered from the plates for further analysis. (It has been claimed that 0.1% aqueous 8-anilino-1-naphthalene sulphonate (ANS) as the ammonium salt is a more sensitive spray reagent).[800] Water can be used as a non-destructive spray, when large amounts of lipids are separated preparatively; they show up as white spots on a translucent background. Lipids become visible as brown spots, if the plate is left for a few minutes in a tank of iodine vapour, but the iodine reacts to some extent with polyunsaturated fatty acids which cannot then be recovered for analysis in other ways. Alternatively, the plates may be sprayed with a solution of 50 per cent sulphuric acid and the lipids made visible as a black deposit of carbon by heating the plates at 180°C for an hour or so. 20 per cent (w/v) ammonium bisulphate in water has also been used to char lipids,[872] the advantage being that the vapours given off during the heating process are less corrosive than when concentrated acids are used. 3% Cupric acetate in 8% orthophosphoric acid has similar advantages.[209] Although charring procedures have the obvious disadvantage that they completely destroy the lipids, they are very sensitive and as little as 1 μg of lipid can be detected by this means. Sterols give a red-purple colour in a few minutes with charring reagents before blackening, and this is a useful diagnostic guide.

For analytical purposes, layers of adsorbent 0.25 mm thick or less give maximum resolution, but in preparative applications, thicker layers are necessary to take heavier loads of lipids. The thicker the layer, the poorer the resolution and, in practice, a compromise must be sought and layers 0.5–1.0 mm thick are usually preferred; indeed with thicker layers, there may be mechanical difficulties in keeping the adsorbent adhering to the plate. The amount of lipid that can be applied to a preparative TLC plate varies with the ease of separation of the components of the mixture. For example, 25–50 mg of simple lipids can often be applied to a 20 × 20 cm plate with a layer 0.5 mm thick of silica gel G. On the other hand, only 4 mg or so of phospholipids can be separated effectively on such plates.

Lipids separated by TLC can be recovered after they have been detected by an appropriate non-destructive method; by scraping the adsorbent band into a small chromatographic column or sintered disc funnel and eluting with solvents of appropriate polarity. Chloroform or diethyl ether containing 1–2 per cent methanol (by volume) will elute most simple lipids, although partial glycerides may isomerise in alcoholic media, and chloroform–methanol–water (5:5:1 by volume) will quantitatively elute most polar lipids. Lipids can also be recovered by repeatedly mixing the adsorbent with solvent in a test tube, centrifuging and decanting the supernatant liquid. The dye used to detect the lipid is eluted by these procedures along with the lipid and can be removed from non-polar lipids by washing them in diethyl ether or chloroform through a short column of Florisil. 2′,7′-Dichlorofluorescein can be removed from non-acidic polar lipids by dissolving them in chloroform–methanol (2:1 by volume) and giving them a rapid "Folch" wash with a solution of tris buffer (0.05M) of pH 9 or with dilute ammonia. The dye does not necessarily interfere with subsequent analyses; for example, if the fatty acid components are transesterified for GLC analysis, the dye is not eluted from the GLC column.

Complicated lipid mixtures cannot always be separated by thin-layer chromatography in one direction but can often be resolved by rechromatography in a second direction (two dimensional TLC). In this method, the sample is applied to the plate as a spot in the bottom left-hand corner of a square TLC plate and the plate run normally in a selected solvent system. When the solvent has run close to the top of the plate the latter is removed from the tank and dried thoroughly in a desiccator under vacuum so that atmospheric moisture is not permitted to deactivate the adsorbent. The plate is then turned anti-clockwise through 90° and redeveloped with a second solvent system (some workers apply the lipid to the bottom right-hand corner and turn the plate clockwise).

As in column chromatography, the quality of the separations and the distances that individual lipid classes will migrate is influenced by the degree of hydration of the adsorbent. This is affected by the time and temperature of activation of the plates, by the storage conditions and, as plates are inevitably exposed to the air during the application of samples and during development, by the relative humidity of the atmosphere. Different proprietary brands of silica gel and different batches of the same brand have different particle sizes and surface areas and, therefore, different adsorptive properties and this also affects the nature of the separations. If these factors are very rigorously controlled, reproducible Rf values for different lipid classes can be obtained, but this is rarely worth the effort involved, especially with two-dimensional separations. A much more common practice is to apply a mixture of authentic lipid standards, that migrate in a known order, to the plate alongside the unknown mixture so that direct comparison is possible.

TLC offers a number of advantages over column chromatography. It is more rapid and sensitive, gives better resolution and the nature of the separation is very soon apparent. This last feature is particularly useful in preparative applications because if the desired separation is not achieved with one development, the plate can be given a second or third development in the same direction with the same or a different solvent, often resulting in an improved resolution. Lipids appear to be more stable to autoxidation on thin-layer adsorbents than has generally been believed.[706, 707] None the less, it is advantageous to add antioxidants such as BHT at a level of 0.01% to the sample, to the eluting solvent or to the spray reagents (BHT migrates with the solvent front even in non-polar systems) to protect the lipids during subsequent analyses. Although larger amounts of lipid can be purified or separated in a single operation by column chromatography, because of the difficulty of monitoring column eluates, it may often be quicker and require less effort to separate lipids preparatively by chromatography on several identical TLC plates and combine the corresponding fractions.

Silicic acid-impregnated paper[207, 383] is occasionally used as an alternative to TLC, and similar though not identical types of separation are obtained with the two procedures. It is claimed that the impregnated papers are superior in that a greater variety of spot tests can be applied to a given chromatographic paper, developed and stained papers are more easily stored than TLC plates and autoradiography is simpler. On the other hand, TLC is a more versatile technique and in particular, sensitive charring reagents can be used to detect lipids and much greater amounts of material can be separated preparatively.

In recent years, high performance TLC (HP-TLC) procedures have been developed i.e. giving separation efficiencies of approximately 5,000 (or three times normal) or more theoretical plates.[874] In one form, a special apparatus is used in which the solvent front is continuously evaporated and renewed, resulting in greatly improved separation efficiencies, but few applications to lipid analyses appear to have been described. A more accessible form of the technique consists of use of commercially-prepared plates, precoated with silica gel of a smaller and more uniform particle size than is generally employed. With such plates, both the resolution and speed of separations are improved and there are indications that more accurate quantifications can be achieved. Disadvantages are that only very small amounts of sample can be applied to the plate without overloading and that commercial precoated plates are costly.

(ii) Quantification

Charring followed by photodensitometry is probably the most popular method of quantifying components separated by TLC, although it leaves much to be desired in terms of convenience and accuracy. In one procedure, bands of lipid on the adsorbent are scraped into tubes to which the dichromate reagent, described earlier, is added. The tubes are heated as before, centrifuged to precipitate the silica gel and the absorbance of the solution is measured at 350 nm.[25, 215, 702] Such charring procedures may give high blank values in the presence of silica gel, so reducing the accuracy of estimating small amounts of lipid or minor components of a given mixture, but the problem can be lessened by thoroughly extracting the adsorbent with solvents before making up the TLC plates. For example, silica without binder can be washed in a Buchner funnel with 8 volumes of chloroform–methanol–formic acid (2:1:1 by volume) followed by 4 volumes of distilled water. The adsorbent is finally dried in an oven at 110°C for 48 hr with occasional stirring.[572] Plates can also be developed to the top with a solvent such as diethyl ether–methanol (1:1 v/v) then used again in the normal way after reactivation (they must of course be used in the same direction). In an alternative procedure, the plate is sprayed with chromic acid solution[637] or 3 per cent cupric acetate in 8 per cent phosphoric acid solution[209] and the amount of charred material, obtained after heating the plate at 180°C for 25 min, measured by means of a scanning photodensitometer as illustrated in Figure 3.2. The areas of the

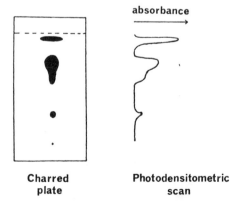

Fig. 3.2. Charred TLC plate and corresponding photodensitometric scan.

peaks on the recorder trace are proportional to the amount of lipid originally present, although the relationship is not a simple one.[113, 814] The procedure has a number of disadvantages; the sample is destroyed, the yield of carbon is variable and affected by degree of unsaturation, authentic standards are necessary for calibration but are not always available, and constant updating of the calibration is necessary. In addition, a scanning photodensitometer is an expensive specialised piece of equipment and is not to be found in all laboratories.

A modified charring procedure has been described by Shand and Noble[685] that promises to overcome many of the difficulties associated with the more traditional method. Lipids are separated by TLC in the conventional manner before being charred by spraying with 3% cupric acetate and 8% orthophosphoric acid[209] and heating at 180° in a forced-draught oven for 15 min. On cooling, the bands are scraped into a scintillation vial and are suspended in an emulsion with water and an emulsifier-scintillator, chosen according to whether the liquid scintillation counter has a cooling unit or not.[550] The amount of quenching due to the presence of the charred material is measured by counting the vials in the external standard mode of a liquid scintillation counter, and is proportional to the amount of lipid present. The results are interpolated into previously derived curves of mass of lipid against the external standard channels ratio. A different calibration curve is required for each lipid class, but these are virtually independent of the degree of unsaturation of the fatty acid moieties and are not affected by the amount of silica gel present. The calibration appears to be remarkably stable and does not have to be checked at frequent intervals. A number of advantages are claimed for the method; it is rapid, very reproducible and has a wide dynamic range. Liquid scintillation counters are freely available in biochemical and medical laboratories and do not require to be dedicated to a single type of analysis (unlike photodensitometers). Generally, they are designed for automatic analysis of large numbers of samples, and data acquisition and processing are capable of a high degree of computerisation, so that the method appears to be suited to routine analyses.

A non-destructive method for quantification of lipids of comparable precision to the charring procedures consists of the use of fluoro-metry[303, 547, 631, 661] in which the fluorescence of a dye, produced by the presence of lipid, is measured. The dye can be incorporated into the adsorbent,[631] or sprayed on to the developed plate,[303, 547] and the fluorescence measured by a scanning fluorometer, or the lipids can be recovered from the plates, the dye added in an appropriate solvent and the fluorescence measured by placing the mixture in cuvettes in a fluorometer.[661] The fluorescence is directly proportional to the amount of lipid present, but depends on the nature of the lipid, so calibration curves must be prepared with standards. As double bonds in alkyl chains cause quenching, these standards should be very similar in nature to the samples to be analysed. Standards should also be run on the same plate as the unknowns to compensate for differences in quenching obtained with different batches of adsorbent. Rhodamine 6G (0.01 per cent in water) has been used most often as the dye, but a spray of 1-anilino-8-naphthalene sulphonate (ANS) at the same concentration has certain advantages[303] (N-phenyl-1-naphthylamine (NPN) is more readily removed from phospholipids later if these are required for further analysis).[64] In a typical analysis, spots of the lipids to be analysed and standards are placed in a row on the plate which is developed in the normal way, then dried thoroughly and sprayed evenly with the dye. It is then dried thoroughly again and the individual lanes of lipid on the plate are scanned with the fluorometer; the amount of each lipid is obtained from the calibration curve after making an allowance for quenching. The procedure is non-destructive so that the lipids can be recovered for further analysis if necessary; however, the range of lipids tested is still somewhat limited and the technique has not been used in many laboratories, possibly because of the cost of the equipment necessary.

Phospholipids separated by TLC can be accurately quantified by phosphorus determination (see Chapter 7) in the presence of the adsorbent, although high blank values may be obtained because of impurities in the silica gel. Chemical methods can also be used for other hydrolysis products of lipids e.g. cholesterol, glycerol, etc. (See Chapters 6 and 7).

Gravimetric methods can be unreliable for estimating lipids separated by thin-layer chromatography as small amounts of impurities, including calcium sulphate, silica gel and the indicating dye may also be eluted from the plates and weighed.[404]

As all lipids contain fatty acids, it is possible to determine the amounts of lipid classes separated by chromatographic procedures by determining the amounts of the fatty acids which they contain. Chemical methods are available for the purpose (see Chapter 6), but gas chromatography can also be used. Typically, the fatty acid components of each lipid are converted to methyl esters in the presence of a known amount of the ester of an acid that does not occur naturally in the sample (e.g. an odd-chain compound), and this serves as an internal standard. Transesterification can be performed on pure lipids eluted from TLC plates or it can be carried out in the presence of the adsorbent, but the method chosen should be such that no losses can occur before all the fatty acid components are in the form of methyl esters (see Chapter 4). By means of gas chromatography, the total amount of the fatty acids relative to that of the standard is obtained by dividing the sum of the areas of the relevant peaks on the recorder trace by that of the internal standard (see Fig. 3.3). This method has been widely used to determine the amounts of molecular species of single lipid classes,[256] but can also be used to estimate natural mixtures of lipids;[141] it has the additional merit that both the fatty acid compositions and the amounts of the lipid classes in a given mixture are determined in a single analysis. Related procedures are used in the estimation of intact glycerides, cholesterol esters, cholesterol and many of the hydrolysis products of lipids, such as glycerol or carbohydrates, and these applications are discussed in greater detail in later chapters. Methods of this kind are not often suited to the routine analysis of large numbers of samples, however.

There have been a number of attempts to marry flame ionisation detection systems to thin-layer chromatography as an aid in quantification (reviewed by Mangold and Mukherjee[484]), but the only system that appears to have been taken to successful commercial fruition is that marketed under the trade name "Iatroscan TH 10 Analyser" (Iatron Laboratories of Japan). The TLC medium is a quartz rod (0.9 mm in diameter) to which a layer of silica gel (75 to 100 μ thick) is fused by means of a sintering process. Compounds are separated on these "chromarods" in the conventional manner, excess solvent is removed and up to 10 rods are placed in a rack in the instrument to be fed automatically through a flame ionisation detector. The signal is passed to an amplifier and thence to a recorder, so yielding a trace similar to that from a gas chromatograph, or to an integrator so that the mass of each component can be determined. The act of passing the rod through the flame regenerates it, and it can be reutilised up to 100 times. Provided care is taken in maintaining a correct degree of hydration of the chromarod, excellent separations of lipids can be achieved (see references 137, 701, 789). Unfortunately, despite the high cost of the instrument and of the chromarods, it has been the author's experience that the accuracy and reproducibility of quantification still leaves something to be desired i.e. it is no better than more conventional charring procedures.

4. Liquid–liquid partition chromatography

In partition chromatography, lipids are separated according to differences in their partition coefficients between two immiscible liquids. The procedure is especially useful for separating lipids with homologous or vinologous series of aliphatic residues and in particular for the isolation of fatty acids or their esters. In low-resolution applications, a *cis*-double bond has approximately the same effect on the partition coefficient of a lipid as two

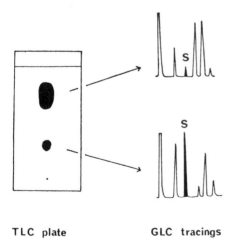

TLC plate   GLC tracings

FIG. 3.3. The use of a fatty acid internal standard to estimate, by means of GLC, lipids previously separated by TLC (peak "S" is the internal standard).

methylene groups; thus palmitic and oleic acids or their derivatives are not easy to separate and are known as a *critical pair*. Also, *cis-* and *trans-* components are not readily separated, but a triple bond is equivalent to more than two double bonds and polar functional groups will drastically alter the partition coefficient of a compound. Where the compounds to be separated differ markedly in their partition coefficients, for example normal fatty acid and hydroxy acid esters or neutral lipids and phospholipids, useful separations can be achieved with a few transfers in separating funnels. For more difficult separations, such as simple fatty acid homologues, several hundred transfers on a specially constructed countercurrent distribution machine are necessary. This is probably the mildest separatory procedure available to the lipid chemist, although it is tedious, time-consuming and the equipment is expensive. It has proved particularly useful for the isolation of labile fatty acids from seed oils.

Chromatographic procedures utilising this phenomenon are more convenient in practice. In conventional chromatography, a polar solvent (the stationary phase) is held on an inert support, in a column or on a TLC plate, and a less polar solvent (the moving or mobile phase) flows continuously past it. Paper chromatography separates compounds by this principle as in this instance the stationary phase is water bound to the cellulose. In *reverse-phase* partition chromatography, the non-polar solvent is held on the inert support while the polar solvent moves. Useful systems utilising both principles have been described for separating fatty acids and esters, and for separating molecular species of triacylglycerols and phospholipids, but reverse phase systems are of greater utility.

In low-resolution or preparative-scale reverse-phase column chromatography, the non-polar stationary phase is held on an inert support such as silanised celite[332] or Kieselguhr while the mobile phase, in which the sample is applied, is percolated through the column. The stationary phase is generally a hydrocarbon such as heptane or a high-boiling silicone liquid, while the mobile phase may be acetonitrile, glacial acetic acid or nitroethane containing various proportions of methanol, acetone or water. Other more exotic stationary phases, which have been used but never widely adopted are polymeric materials, which do not require an inert support and do not "bleed" material into the eluant; they include rubber, Factice, polyethylene and Teflon (Fluon). More recently, alkylated Sephadex derivatives have been described that have advantages in terms of uniformity and general accessibility.[199, 200, 446, 557] When the eluant is constant in composition throughout the analysis, differential refractometers can be used to detect and estimate lipids eluting from the columns. Otherwise, the problems of quantification are similar to those in preparative-scale adsorption column chromatography.

The technology of HPLC has also been applied to reverse-phase chromatographic separations. By far the most widely used stationary phases in this instance consist of hydrocarbonaceous material chemically-bonded onto microparticles of silica gel (5–10 μ in diameter; generally, they are prepared by reaction of octadecyltrichlorosilane with silica gel and are available from several suppliers under various trade names). The properties of such liquid phases have been reviewed by Horvath and Melander.[330] Alkylated Sephadex has proved to be too compressible for HPLC applications. The problems of detection and quantification of compounds separated by HPLC in the reverse-phase mode are similar to those discussed earlier for HPLC in an adsorption mode. The resolution which can be achieved by HPLC is of course very much better than that obtained in preparative-scale applications and in some circumstances is superior to that obtained with gas–liquid chromatography.

Components emerge from liquid–liquid chromatography columns in much the same manner as from GLC columns, and the efficiency of the former can be calculated from the elution profiles of the components in terms of numbers of theoretical plates, using a relationship between their retention and elution volumes (c.f. retention time and width of peak) similar to that described earlier in this chapter for GLC columns.

Thin-layer reverse-phase systems, in which the non-polar stationary phase is held on a layer of inert support material on a glass plate and developed in a tank containing the polar mobile phase, have also been developed. Such plates are impregnated by slowly and very gently immersing plates, coated in the normal manner with a layer of silica gel or

Kieselguhr, in a 5–10 per cent solution of the chosen stationary phase in a volatile solvent. When it is thoroughly soaked, the plate is removed and the volatile solvent is allowed to evaporate off in the atmosphere at room temperature. In a second method, the plate is developed in a tank containing a solution of the stationary phase then is removed and dried off as before. Silanised silica gel has itself been used in reverse-phase TLC but has a very low carrying capacity. Finally, Litchfield[448] makes plates for reverse-phase TLC by preparing a slurry consisting of hexadecane (4 g) in hexane (85 ml) and silanised silicic acid (50 g). After thorough shaking, the mixture is poured into a conventional TLC spreader and the layers are made in the normal way. The plates must be left overnight in a ventilated area to allow the hexane to evaporate before being used. If the stationary phase is a high boiling silicone such as Dow Corning 200 fluid, or a high molecular weight hydrocarbon such as tetradecane or hexadecane, the plates can be stored for long periods in sealed containers (undecane may evaporate off, however). Samples are applied to the plate in a volatile solvent such as pentane which is allowed to evaporate, before the plate is placed in a tank containing the mobile phase, 80–100 per cent saturated with stationary phase, and is allowed to develop in the normal manner.

Theoretical and practical aspects of reverse-phase TLC have been reviewed.[656, 799] The procedure has one important advantage over adsorption TLC in that Rf values are more reproducible. The main drawbacks are that it can be messy and that lipids are difficult to detect, especially in preparative applications, as the stationary phase interferes with non-specific non-destructive spray reagents. None the less, there is a report that 2′,7′-dichloro-fluorescein sprays can be used to detect lipids on Kieselguhr impregnated with liquid paraffin;[827] lipids appear as green spots on a purple background under UV light. In analytical applications, it is possible to heat the plates to a sufficiently high temperature to evaporate off such stationary phases as undecane and then detect the lipids by charring techniques. Most organic compounds can also be detected on hydrophobic layers by iodine vapour or with a spray of 5 per cent phosphomolybdic acid in ethanol (followed by heating at 120°C for 5 min when lipids appear as blue spots); chromic acid solution

may be used on siliconised layers. All these procedures are destructive to lipids so are not suitable for preparative chromatography. For this purpose, it is usually necessary to compromise and destroy part of the sample by exposing one edge of the developed plate to iodine vapour to locate the ends of the bands, which can then be removed from adjacent regions of the plate. Lipids are recovered from these bands in the manner described for adsorption TLC, but it may be necessary subsequently to subject the components isolated to adsorption chromatography to completely eliminate any adhering stationary phase.

Compounds that emerge as critical pairs on reverse-phase chromatography can usually be resolved by chromatography on adsorbents impregnated with silver nitrate (see following section). Indeed, two-dimensional TLC systems in which components are separated in one direction by silver nitrate chromatography and in the second direction by a reverse-phase system have been described.[387] Reverse-phase systems have also been utilised to separate normal fatty acids from those containing polar functional groups, but adsorption chromatography is move convenient and is now generally preferred for the purpose.

5. Chromatography on adsorbents containing complexing agents

(I) SILVER NITRATE CHROMATOGRAPHY

In silver nitrate (or "argentation") chromatography, the property exhibited by silver compounds of forming polar complexes reversibly with the double bonds of the aliphatic moieties of lipids is used as a means of separating them according to the number, configuration (*cis* or *trans*) and to some extent the position of those double bonds. The first procedure in which the principle was applied was a countercurrent distribution system, but it is now more generally used in conjunction with adsorption chromatography. Morris[508] has reviewed the principle and applications of the method.

For silver nitrate thin-layer chromatography, 3 to 5 per cent (larger amounts are helpful in a few applications only) by weight of silver nitrate relative to the weight of adsorbent is incorporated into the slurry used to make the plates. These are activated in the usual way and will retain their activity for a month or so, if stored in the dark in a desiccator. Such plates can also be prepared by immersion (very carefully!) of previously prepared silica gel-coated plates in a solution of 4% silver nitrate in aqueous acetone or methanol (9:1). Silver nitrate plates having improved resolution and better keeping properties are prepared by using 30% ammonia, rather than water, in preparing the slurry with silica gel, but care must be taken to activate the plates in a well-ventilated area.[847] On exposure to light, the plates blacken rapidly and it is important that they be handled and developed in a darkened room or cupboard whenever possible. Lipids on the plate can be visualised under UV light after spraying with 2′,7′-dichlorofluorescein solution, when they appear as yellow spots on a red-purple background, and they can be recovered from the plates as described earlier for adsorption chromatography, although special precautions may be necessary to remove small amounts of silver that are also eluted. Ion-exchange resins may be used to remove silver from phospholipids separated in this way,[429] but washing the extract in hexane:ether (1:1 v/v) with sodium chloride solution and then with 0.05M tris buffer (pH 9) or dilute ammonia solution will remove both the silver and the dye from simple lipids.[21] (See also Chapter 8.)

Silicic acid impregnated with silver nitrate has been used in columns to effect separations of lipids, and acid-washed Florisil so treated[29, 830] is particularly useful as it permits the separation of much larger quantities of lipid. Acid-washed Florisil, prepared as described earlier in this chapter, is impregnated with silver nitrate as follows (basically the procedure of Willner,[830] which is more economic in its use of silver nitrate than those described by others).

"Acid-washed Florisil (35 g) and 14 ml of a 50 per cent aqueous solution of silver nitrate are mixed, and the mixture shaken until free flowing (20 min). It is then left overnight in an oven at 120°C to activate, cooled and stored in the dark."

The adsorbent is packed into columns in the conventional manner, except that the columns are wrapped in black paper or aluminium foil to exclude light. Silver complexes are more stable at lower temperatures so cooling the columns with a water jacket may aid the separations. Improved silver

nitrate–TLC separations at low temperatures have also been recorded.[514] Ion-exchange resins impregnated with silver nitrate have some advantages over adsorption columns as the silver is not eluted by polar solvents.[678] Chloroform should be avoided wherever possible in silver nitrate chromatography, either in applying lipids to plates or as a component of the eluant, as it reacts with silver nitrate with the formation of nitric oxide, causing a blackening of the adsorbent.

Silver nitrate chromatography has a singularly useful application in the separation of *cis*- from *trans*-olefinic compounds, a feat that is not easily attained by gas chromatography, for example. A further important advantage of the method is that traces of silver left with the lipids separated by this technique, afford some protection against autoxidation.[708] On the other hand, small amounts of adventitious peroxides in solvents can react with double bonds in the presence of silver nitrate in thin layers to form epoxides; this can be troublesome with very small samples or those of very high specific radioactivity.[121] In combination with gas chromatography or liquid partition chromatography, silver nitrate chromatography is a powerful method of isolating simpler molecular species of lipids.

(II) GLYCOL-COMPLEXING AGENTS

Chromatographic adsorbents impregnated with sodium borate or boric acid are useful for separating compounds with adjacent free hydroxyl groups. For example, vicinal diols of the *threo*- and *erythro*-configuration form non-polar complexes reversibly with borates and compounds containing these groups, which are not normally separable by adsorption chromatography, can be separated; the *threo*-complexes form more readily and are less polar than the *erythro*-compound so have greater mobility on the impregnated adsorbents.[507] Partial glycerides can also be separated on boric acid impregnated adsorbents where the complexing agent, in addition to improving the separations, stabilises the compounds and prevents acyl migration.[767] Boric acid forms complexes with ceramides having a double bond in position 4 of the long-chain base (although the reason for it is not known), and the property has been used in the separation of molecular species of ceramides (see Chapter 8).[516] Boric acid eluting with the lipids is removed by washing the diethyl ether solution with ice-cold distilled water (the solution should be dried over anhydrous sodium sulphate immediately).

Boric acid-impregnated layers on TLC plates are prepared by incorporating sufficient boric acid in the slurry used to prepare the plates so that it comprises 10 per cent by weight of the final adsorbent,[507] and plates are activated and stored in the usual way. Acid-washed Florisil impregnated with boric acid for column chromatography can be prepared by adding methanol containing sufficient boric acid to amount to 10 per cent by weight of the adsorbent, evaporating off the solvent on a rotary evaporator and activating at 110°C for 1 hr.[684]

Sodium arsenite-impregnated layers also give remarkable separations of diols, including diastereoisomers of these.[507] The free diols must be recovered by alkaline hydrolysis of the stable complex that is formed, after this is eluted from the adsorbent with solvents. The separations obtained with borate and arsenite-impregnated layers differ in nature, so the two systems may be used to complement each other.

6. Ion-exchange cellulose chromatography

DEAE cellulose column chromatography is a useful technique for the separation of complex lipids in comparatively large amounts.[640] The principle of the separation process is not fully understood but is partly ion-exchange chromatography of the ionic moieties of the polar lipids, and partly adsorption of highly polar non-ionic parts of complex lipids, for example the hydroxyl groups of inositol or carbohydrates. The DEAE cellulose packing material is converted to the acetate form, from which charged non-acidic lipids (such as choline or ethanolamine-containing lipids) can be eluted with chloroform containing various proportions of methanol. Weakly acidic lipids are eluted with glacial acetic acid and it is necessary to add an inorganic salt or ammonia to the solvent to elute strongly acidic or highly polar lipids. Glycolipids can also be separated by the procedure, and ceramide polyhexosides are eluted as a single complex fraction which can be further purified by other techniques. DEAE cellulose in the borate form and triethylaminoethyl (TEAE) cellulose[640] afford useful alternative

separations and recently some distinctive separations of phospholipids, especially some of the more acidic ones, have been achieved on carboxymethyl-cellulose in the sodium form.[146] The amount of lipid that can be applied to such a column without overloading depends somewhat on the nature of the components present, but as a rough guide, approximately 300 mg can be applied to a 30×2.5 cm column. The DEAE cellulose, which should be a fibrous grade (fines reduced), is prepared in the acetate form as follows.[638, 640]

"The adsorbent is washed on a Buchner funnel with 1M aqueous hydrochloric acid (3 bed volumes), followed by water to neutral pH and then with 0.1M aqueous potassium hydroxide (3 bed volumes) followed by water. The cycle is repeated three times. It is converted to the acetate form by washing with 3 bed volumes of glacial acetic acid and by leaving overnight in this solvent. The column is packed with the slurry of DEAE cellulose in glacial acetic acid, adding small amounts of material at a time and tamping down gently with a wide glass rod. If the fines generated by the washing procedure are removed by decanting the upper layers of the slurry two or three times before packing the column, better flow rates are obtained. The bed is freed of acid by washing with 5 bed volumes of methanol, 3 bed volumes of chloroform–methanol (1:1 v/v) and 5 bed volumes of chloroform. Non-polar coloured compounds such as azulene can be run through the column to determine the bed volume and to detect imperfections in the packing."

Lipids eluting from the columns may be quantified by the procedures described above for adsorption column chromatography. Although it is generally considered as a larger-scale preparative technique, it can be scaled down for semi-micro isolation purposes.

C. SPECTROSCOPY

Infrared (IR) spectroscopy was the first of the spectroscopic methods to be applied to the analysis of lipids in general and fatty acids in particular and, although there have been comparatively few new developments involving the technique in recent years, it is probably still the spectroscopic method that would be chosen for the preliminary examination of an unknown lipid. Ultraviolet (UV) spectroscopy is now used much less frequently by lipid analysts, but it still has important specific applications. On the other hand, nuclear magnetic resonance (NMR) spectroscopy and mass spectrometry have been widely used for problems of lipid structure determination and new applications are continually being developed. IR, UV and NMR spectroscopy are non-destructive techniques, so that samples may be recovered if necessary for further analysis; mass spectrometry requires so little material that it can often be considered in much the same light.

Most spectroscopic methods are based on empirical collations of vast amounts of data obtained from model compounds of known structure, and in the interpretation of spectra of unknown compounds, a knowledge of these data is required. It will only be possible to reproduce a small part of such information here but further details can be obtained from more specialised publications. In particular, Chapman[117] has surveyed the earlier literature and Klein and Kemp[400] have reviewed more recent developments in the application of spectroscopic techniques to lipid structural analysis in some detail. The general principles and range of applications of these techniques are now discussed. Specific uses will be dealt with later in the appropriate sections.

1. Infrared absorption spectroscopy

Infrared spectra are obtained when energy of light in the infrared region at a given frequency is absorbed by a molecule, thereby increasing the amplitude of the vibrations of specific bonds between atoms in the molecule. The most useful and conveniently measured region of the infrared spectrum (limited by sodium chloride optics) is over a range of wavelength of 2.5–15 μm (equivalent to wave numbers of 4000–667 cm^{-1}), although the *near* infrared region (0.8–2.5 μm or 12,500–4000 cm^{-1}) may also contain features of interest in lipid analysis. Bonds absorb energy and vibrate in mainly two ways; *stretching* vibrations, in which the distance between atoms increases and decreases, and *bending*

vibrations, in which the position of an atom is altered relative to the line of the original bond. Bonds between specific atoms in defined molecular environments absorb energy at certain fundamental frequencies, which are characteristic and can often be used to diagnose the presence of particular bonds in unknown molecules. These fundamental frequencies are affected only slightly by the presence of other atoms in the molecule, but minor shifts of frequency can in themselves sometimes be useful diagnostic aids.

With modern double-beam infrared spectrometers, the spectrum from 2.5 to 15 μm is scanned continuously and the percentage transmission through the sample is recorded directly on a chart, graduated linearly in wave numbers (cm^{-1}) or more commonly in wavelengths (μm). Samples may be in the vapour phase, in solution, or as solids and liquids, and very fast scanning instruments are available so that spectra can be obtained from compounds in the effluent of a gas chromatograph. Carbon tetrachloride and carbon disulphide are chosen most frequently as solvents, as they are transparent over a wide range of wavelengths and can be used safely in cells with windows of rocksalt. Smaller samples or compounds that are insoluble in these solvents are analysed in potassium chloride or potassium bromide pellets. Samples and solvents must be thoroughly dried as water obscures part of the spectrum and may attack the cell windows. Further practical details can be found in the many excellent reviews and textbooks on the subject. Very small samples are best analysed as a smear in the centre of a sodium chloride disc in an instrument fitted with a beam condenser.

A great deal of valuable information about the physical state of lipids can be obtained from their spectra in the solid state but is not within the scope of this text. Solution spectra are much simpler and much more useful for detecting and estimating specific functional groups in lipids. By far the most common use of infrared absorption spectroscopy is for the determination of isolated *trans*-double bonds, which absorb at 10.3 μm (960–970 cm^{-1}), in lipids.[765] The frequencies due to the carbonyl function of most lipids are easily seen and identified, but *cis*-double bonds, unlike those of *trans*-configuration, do not give rise to major absorption bands that lend themselves to characterisation in the usual infrared region (bands in the near infrared spectra can be used for this purpose, however). Many other functional groups, which may occur in fatty acid alkyl chains, give rise to prominent absorption bands, and organic bases in phospholipids absorb at distinct characteristic frequencies which can be used as an aid to identification.[51, 536] The application of infrared spectroscopy to lipid analysis has been frequently reviewed.[117, 118, 216]

2. Ultraviolet absorption spectroscopy

The ultraviolet spectrum of a compound is generally measured over the range 220–400 nm, although the visible range up to 800 nm may also be covered by a single instrument. Quartz cells are transparent over this range and compounds are dissolved in a solvent such as 95 per cent ethanol for analysis. Conjugated double bonds absorb strongly in this region and before the development of gas chromatography, the concentrations of the main classes of polyunsaturated fatty acids were commonly determined by isomerising the double bonds with strong alkali at elevated temperatures, to form conjugated systems and the absorption at the characteristic wavelengths for conjugated dienes, trienes and so forth were determined. More recently ultraviolet spectrophotometers have come on the market which can be used at wavelengths down to 200 nm, and so cover a range in which isolated *cis*-double bonds exhibit a specific absorbance (~ 206 nm).

Nowadays, UV spectroscopy is used principally in the analysis of natural fatty acids containing conjugated double bond systems or for the estimation of conjugated dienes formed by the action of lipoxygenase or as a result of autoxidation. These and other uses of UV spectroscopy have been discussed in greater detail elsewhere.[117, 579]

3. Nuclear magnetic resonance spectroscopy

The nuclei of certain isotopes are continuously spinning with an angular momentum, which can give rise to an associated magnetic field. If a very powerful external magnetic field is applied to the nucleus and made to oscillate in the radio frequency range, the nucleus will resonate between different quantised energy levels at specific frequencies, absorbing some of the applied energy. Such very small changes in energy can be detected, amplified and displayed on a chart. The trace obtained, of the variation in the intensity of the resonance signal with increasing applied magnetic field, is the NMR spectrum. In organic compounds, the isotope of hydrogen, ^1H, displays this phenomenon whereas the main isotopes of carbon, oxygen and nitrogen do not, so the resonance frequencies of hydrogen atoms in molecules are those most often measured and the technique in this instance is referred to as proton magnetic resonance spectroscopy. One of the less naturally-abundant isotopes of carbon, ^{13}C, also exhibits the phenomenon but until comparatively recently, the inherent 6000 fold loss of sensitivity relative to proton magnetic resonance limited the availability of the technique. Developments in instrumentation and data processing have made ^{13}C NMR much more accessible and since all carbon atoms in organic compounds give distinctive signals, whether or not they are linked to protons, a great deal of structural information can be obtained from the spectra. Unfortunately, much larger samples (about 200 mg) are still required to obtain a spectrum than would be considered appropriate for most analytical techniques.

Proton magnetic resonance spectroscopy then, remains the most valuable form of the technique to lipid analysts. The frequency at which any given hydrogen atom in an organic compound resonates is strongly dependent on its precise molecular environment and is subjected to "chemical shifts" by the presence of other adjacent atoms. The fine structure of each absorption band is determined by the weak magnetic forces of neighbouring hydrogen atoms and this feature (known as spin–spin coupling), together with the extent of the chemical shift, may be used to deduce the arrangement of hydrogen atoms and often the complete structure of an unknown compound. As the area under an absorption band is proportional to the number of hydrogen atoms in an identical environment in the molecule responsible for that signal, additional information on the structures of unknown compounds can be obtained by integrating the signals while recording the spectrum. The theory of NMR spectroscopy is complex and the reader should consult one of the excellent textbooks on the subject before tackling the interpretation of spectra.

There are basically two main types of PMR spectroscopy, broad-band and high-resolution. The former can be used to obtain much valuable information on the physical state and environment of lipid molecules, but has limited use as an analytical tool and will not be considered further in this book. High resolution PMR spectroscopy is a valued technique for obtaining information on the structures of unknown lipids and has been widely used to solve lipid structural problems. Instruments operating at 60 or 100 MHz are those most often seen in chemical laboratories, but more powerful instruments (220 MHz) are coming into more general use.

The PMR spectra of aliphatic compounds are not always distinctive as the signals from many of the methylene protons tend to overlap. Some resolution of these can be achieved by the addition of chemical shift reagents (generally lanthanide derivatives), which effect a separation of signals from certain protons. PMR spectra of enantiomers can also be distinguished after the addition of chiral shift reagents.

Compounds must be in solution for analysis and the solvent should preferably not contain the isotope ^1H. Carbon tetrachloride is suitable for non-polar lipids, but deuterochloroform may also be used and deuterated methanol has been added to this to effect solution of phospholipids. Chemical shifts are not measured in absolute units but are recorded as parts per million of the resonance magnetic field. Tetramethylsilane is added to the solvents as an internal standard, and in the conventional system it is given the arbitrary value 10 on the so-called τ (tau) scale. (On the usual charts, values increase from left to right with the increasing strength of the magnetic field). In a less frequently used system (the δ scale), tetramethylsilane is given the value zero. To convert:

$$\delta = 10 - \tau$$

After a spectrum has been obtained, all the material is recoverable for further analysis.

The technique has been widely used for lipid structure determinations, particularly for the identification and location of double bonds in fatty acids, but also to detect and locate other functional groups such as cyclopropane rings, hydroxyl groups, triple bonds and oxirane rings in fatty acids. Glycerides, glyceryl ethers, glycolipids and phospholipids have distinctive spectra which aid in their identification. The application of NMR spectroscopy in the analysis of lipids has been the subject of several detailed reviews.[117, 245, 392, 400]

4. Mass Spectrometry

In the mass spectrometer, organic compounds in the vapour phase are bombarded with electrons and form positively charged ions, which can fragment in a number of different ways to give smaller ionised entities. These ions are propelled through a powerful magnetic field and are separated according to their mass to charge (m/e) ratio. Ions are collected in sequence, as the ratio increases, on a suitable detection system and are displayed as peaks on a chart. The largest peak or base peak in the spectrum is given an arbitrary intensity value of 100 and the intensities of all the other ions are normalised against this, so that data can be presented in a uniform manner. The ion with the highest m/e value is generally (although not always) that of the original ionised molecule and is termed the parent ion (M^+). With low resolution instruments, peaks appear at unit mass numbers, but at higher resolutions the masses of individual ions can be measured with sufficient accuracy for the molecular formula of each to be unequivocally determined.

Molecules do not fragment in an arbitrary manner but tend to split at weaker bonds, such as those adjacent to specific functional groups, or according to certain complex rules which are now fairly well understood and have been formulated from the analysis of model compounds. The mass numbers and molecular formulae of the major ions can then be used to detect the presence and location of specific functional groups in a compound. Indeed, an accurate determination of the molecular weight and molecular formula of a compound can in itself be a valuable guide to its identity.

Mass spectrometry can be used to locate double bonds in aliphatic chains after conversion of the parent compound to an appropriate derivative; other functional groups, such as keto, hydroxyl, epoxyl and some cyclic groups in the fatty acid chains, are more readily identified and their positions confirmed. In particular, methyl branches, which are located only with difficulty by other methods, are comparatively easily identified by mass spectrometry. The technique has been used for the identification of higher molecular weight lipids such as ceramides, long-chain bases and glycerides, but it is most useful for the analysis or identification of pure compounds rather than natural mixtures. It is rarely used as a quantitative technique, although it has been used in this way for certain purposes. Mass spectra of model phospholipids have been obtained, but so far the procedure has been little used as an aid to the identification of natural compounds of this type. It is possible to connect the outlet of a gas chromatograph through a suitable interface to a mass spectrometer, so that individual compounds are identified as they are eluted; this has proved a particularly powerful tool in the hands of lipid analysts. It is advisable to use highly thermostable silicone phases in the GLC columns so that material bleeding from them does not contaminate the samples and/or the GLC–mass spectrometer interface.

An alternative form of the technique, chemical ionisation mass spectrometry, in which the initial ionisation and fragmentation of the molecules is brought about by impact with an ionised reactant gas, is less freely available to lipid analysts but has advantages in certain applications. For example, it does not effect such extensive disruption of the molecules and gives more distinctive molecular ions from sensitive compounds, such as alcohols or their trimethylsilyl ethers.

Although samples are destroyed during mass spectrometry, only very small amounts are necessary for an analysis (<0.1 mg) so that this is not a serious disadvantage. The application of mass spectrometry to lipid structure determinations has been extensively reviewed.[117, 392, 400, 464, 465, 871]

5. Optical rotary dispersion

The determination of the absolute configuration of a compound is a problem not often faced by lipid analysts but is nonetheless of considerable importance. Optical rotary dispersion measurements with modern recording spectropolarimeters

are essential to such a study, but the problem of assignment of configuration is too complex to lend itself to simple rationalisations. Smith has comprehensively reviewed current knowledge of fatty acid and long-chain base asymmetry[716] and glyceride chirality.[717]

D. SOME PRACTICAL CONSIDERATIONS

1. Autoxidation of lipids

The problem of autoxidation of lipids has already been mentioned briefly in Chapter 2. It can give rise to very real difficulties in the chromatography of lipids and the greater the degree of unsaturation of the fatty acid components, the greater the risk. For example, linoleic acid is autoxidised twenty times as readily as oleic acid, and each additional double bond in the molecule increases this rate by a factor of at least two. The mechanism of autoxidation involves attack by free radicals and is enhanced by high energy photons (e.g. ultraviolet light) and certain metal ions; once initiated, the reaction is autocatalytic.[461] With fatty acid moieties containing two or more double bonds, one effect of autoxidation is that the double bonds move to form conjugated systems and exhibit distinctive ultraviolet spectra which can be used to estimate the extent of reaction.[400]

To prevent autoxidation, lipids should be handled whenever possible in an atmosphere of nitrogen, and antioxidants such as BHT[538, 853] (see Chapter 2) should be added at a level of 0.05 per cent to thin-layer chromatographic solvents and spray reagents in particular. Solvents should be flushed with nitrogen to displace any dissolved oxygen. BHT volatilises in a stream of nitrogen and during solvent removal on a rotary evaporator. It is eluted at the solvent front in most TLC systems and so can be removed from lipids easily if this is necessary for any purpose. It is more troublesome in gas chromatography as it chromatographs on many of the more useful polyester columns close to methyl myristate which it may obscure, but if this is not a major fatty acid component, as in many animal or plants lipids, the nuisance value of BHT may be negligible. BHQ (1,4-dihydroxy-2-*tert*-butyl benzene) has been recommended as an alternative to BHT.[538] It has similar elution characteristics to BHT in adsorption chromatography, but is eluted much later on GLC. On occasion, it may appear as little more than a slight elevation of the base line but in other circumstances it may obscure essential components of the sample analysed.

As mentioned in Chapter 2, large volumes of solvents should be removed from lipids in a rotary film evaporator at or near room temperature. Small volumes of solvent can be evaporated by directing a stream of nitrogen onto their surface. It should be recognised that over-vigorous application of such evaporation techniques can cause some loss of lipids which might normally be considered as involatile e.g. methyl esters of medium chain fatty acids up to and occasionally including C_{14}. Purified lipids separated by chromatographic procedures should be stored in non-alcoholic solvents (such as chloroform) in sealed containers flushed with nitrogen at −20°C. If they are to be stored for long periods, it is advisable to seal them in glass vials under vacuum.

2. Solvents

All solvents, including high purity analytical grades, contain traces of impurities, some of which, for example antioxidants, may have been deliberately introduced by manufacturers. As large volumes of solvents are often required for the isolation of very small amounts of lipid, serious contamination can occur. If high quality solvents cannot be purchased directly, those of poorer grades should at least be redistilled before use and, on occasion, further purification may be necessary.

Diethyl ether and other ethers develop peroxides on storage and antioxidants are frequently added by the manufacturer to minimise this. Distillation from potassium hydroxide pellets is effective in removing most impurities (including much of the water and perhaps the antioxidants), but a large volume should not be purified unless it is to be used quickly. Care must be taken to ensure that the distillation flasks are never allowed to boil dry, as serious explosions can occur. Last traces of peroxide can be removed by percolating the solvent down a column of activated alumina. When dry ether is required for any purpose, sodium wire can be freshly extruded from a sodium press into the solvent distilled from potassium

hydroxide. Again, appropriate precautions should be observed.

The chloroform sold by most manufacturers contains 0.25–2 per cent ethanol, which codistils with it, as a stabiliser. In this form, the solvent is much more polar and has quite different elution properties for lipids in adsorption chromatography than has the alcohol-free solvent. If necessary, the ethanol can be removed by washing the chloroform several times with half its volume of distilled water, drying the product over anhydrous sodium sulphate and storing in the dark (to minimise the photochemical formation of phosgene) over anhydrous granular calcium chloride. Alcohol-stabilised chloroform is used for most chromatographic purposes detailed later, unless a statement is made to the contrary. Analysts are increasingly becoming aware of the toxic properties of chloroform, which should only be used in well-ventilated areas.

Hexane and most aliphatic hydrocarbons may contain olefins, which can be removed prior to distillation by washing with concentrated sulphuric acid, followed by 10 per cent sulphuric acid containing potassium permanganate. Some manufacturers now supply hydrogenated solvents.

Benzene usually contains water, most of which distils over in the first small fraction, which is discarded, and it can be maintained in the dry state by storing in the presence of sodium wire. It is not always recognised that benzene is a highly toxic compound, and great care must be taken so that there is little escape of the vapour into the laboratory atmosphere. Although the early literature on lipid analysis frequently specified the use of benzene as a solvent, there are in fact very few applications for which it is essential and for which no adequate substitute exists.

Methanol may contain traces of aldehydes and amines, much of which is removed by distillation from potassium hydroxide pellets. When dry methanol is required, it is prepared as follows:

"10 g of dry magnesium turnings and 30–50 ml of methanol are warmed together with a few crystals of iodine. A vigorous reaction soon starts and when all the magnesium has dissolved, a further 1.5 litres of methanol are added and the mixture is refluxed for 1 hr, after which it is distilled. The first few mls should be discarded but most of the remainder may be collected.''

All solvents, particularly ethers or those containing halogens, should be stored out of direct sunlight in brown bottles. More detailed purification procedures for solvents can be found in textbooks on laboratory practice in organic chemistry, such as that by Vogel.[807]

3. Contaminants

Apart from the potential contaminants in solvents, extraneous lipid-like materials may appear from a variety of sources. Plasticisers (discussed earlier in Chapter 2) are the most common of these and plastic ware of any kind should be avoided in lipid analytical laboratories. It is well known that organic solvents will leach out plasticisers, which can interfere in analysis, but it is not always appreciated that warm water can also have this effect. Other contaminants, e.g. antioxidants, can also be eluted from plastics by both aqueous and organic solvents.[560] Similarly, it is not always recognised that lipids can themselves dissolve in plastics, especially silicone rubber or polyvinyl, and so be lost.[439] All laboratory reagents whatever their nature may from time to time contain troublesome impurities, and vigilance must be exercised continuously to detect and eliminate these. For example, 2',7'-dichloro-fluorescein may contain impurities which give rise to spurious peaks on the GLC recorder trace, unless the reagent is washed with hexane before use.[572] Other contaminants may arise from fingerprints, soaps, hair preparations, tobacco smoke, cosmetics, laboratory greases, the exhausts of vacuum pumps, floor polish and so forth; the list could be endless.

4. Equipping a laboratory for lipid analysis

Most conventional chemical laboratories can be used for lipid analysis, but certain features are particularly desirable. For example, sealed bench tops that do not need waxing are useful as waxes can be picked up on glassware and contaminate lipid samples. Indeed, sealed bench tops and floors are essential if isotopically-labelled lipids are to be analysed. An efficient fume cupboard or hood is necessary for handling toxic reagents, including solvents or isotopically-labelled lipids, and for spraying TLC plates. As many of the solvents required for lipid analysis are highly inflammable, special precautions against accidental fires should be taken.

Conventional glassware with ground-glass joints can be used for chemical manipulations and too wide a range of joint sizes should be avoided for maximum interchangeability. As a great deal of lipid analytical work can be performed on a small scale, a range of semi-micro glassware is valuable. The author makes extensive use of test-tubes, with ground-glass joints, of about 15 ml capacity for carrying out chemical reactions. When it is necessary to extract aqueous solutions with organic solvents in such tubes, the layers are separated by means of disposable Pasteur pipettes. The test-tubes can be used in standard centrifuges and can be equipped with small condensers when solvents must be heated under reflux. An aluminium heating block drilled with holes of the correct size is valuable in the latter circumstance. Rotary evaporators are essential for the removal of large volumes of solvent and these should have greaseless joints and be constructed so that solvent vapours are only permitted to come in contact with glass surfaces.

A supply of purified nitrogen should be available in the laboratory so that an inert atmosphere can be maintained in vessels containing lipids. If a stream of nitrogen is used to remove small volumes of solvents from lipid samples or to dry developed TLC plates, this should be performed in a fume cupboard.

The chromatographic equipment required will depend on the needs of the individual, but some general suggestions can be made. A gas chromato-graph will probably be essential but should be situated in a separate room or as far as possible from benches where inflammable solvents may be handled. The features of the instrument desirable for lipid analysis have been discussed above and are also considered in Chapters 5 and 8. For safety reasons, the cylinders containing the carrier and detector gases and the main regulators should be situated outside the laboratory, although it is convenient to have single stage regulators within the laboratory to control gas flow rates. The use of hydrogen cylinders can be avoided entirely, by making use of hydrogen generators, and air cylinders can be substituted by a pulseless air pump (though drying agents are necessary to prevent condensation occurring on compression). The TLC equipment required will probably comprise an adjustable spreader (ideally made from a material impervious to silver nitrate), plate leveller, glass plates of various sizes, developing tanks and airtight boxes or desiccators for storing activated plates. A range of glass columns will probably be required for column chromatography.

It is impossible to list all the apparatus that might be of value as needs will vary with the nature of the work, but centrifuges, shakers, UV lights, magnetic stirrers, vortex mixers, heating mantles, thermo-statically-controlled water baths and balances are likely to be needed from time to time, and there should be access to spectrophotometers. Similarly, it is not possible to give a definitive list of chemicals that will be used, but a range of common organic solvents, thin-layer and column adsorbents, anhydrous sodium sulphate, potassium bicarbonate, sodium or potassium hydroxide, and a source of distilled water will certainly be needed in most laboratories. A range of lipid standards is required and these, together with lipid samples, should be stored in a refrigerator or deep-freeze.

CHAPTER 4

The Preparation of Derivatives of Lipids

MANY simple and complex lipids are either too polar or of too high a molecular weight to be subjected to some chromatographic procedures, so it is necessary to convert them to volatile and/or non-polar derivatives for further analysis. In addition, it may be necessary to hydrolyse or transesterify lipids to obtain the component parts for a complete analysis. Because of the high sensitivity of thin-layer and gas chromatographic analysis procedures, small amounts of material (certainly less than 10 mg) may be all that is required and most of the procedures described below are on this scale. Any conventional Pyrex glassware with ground-glass joints can be used for the reactions but for many the author has found it convenient to use test-tubes of about 15 ml capacity with a ground-glass joint at the top. Organic and aqueous layers can be separated efficiently in these with the aid of Pasteur pipettes. Precautions should be taken to prevent autoxidation of lipids (see Chapter 3). Methods of preparing derivatives in general have been reviewed.[72,185]

A. HYDROLYSIS (SAPONIFICATION) OF LIPIDS

Lipids may be hydrolysed by heating them under reflux with an excess of dilute aqueous ethanolic alkali and the fatty acids, diethyl ether-soluble non-saponifiable materials and any water-soluble hydrolysis products recovered for further analysis. When the water-soluble components (such as glycerol, glycerophosphoryl choline, etc.) are required, special procedures must be used and these are discussed in later chapters. The free fatty acids and the diethyl ether-soluble non-saponifiable components are separately recovered in the following procedure:

"The lipid sample (100 mg) is hydrolysed by refluxing it with a 1M solution of potassium hydroxide in 95 per cent ethanol (2 ml) for 1 hr. The solution is cooled, water (5 ml) is added and the solution is extracted thoroughly with diethyl ether (3×5 ml). It may be necessary to centrifuge to break any emulsions that form. The solvent extract is washed several times with water, dried over anhydrous sodium sulphate and the non-saponifiable materials are recovered on removal of the solvent in a rotary evaporator. The water washings are added to the aqueous layer, which is acidified with 6M hydrochloric acid and extracted with diethyl ether or hexane (3×5 ml). The free fatty acids are recovered after washing the extract with water, drying it over anhydrous sodium sulphate and removing the solvent in the usual way."

The non-saponifiable layer will contain any hydrocarbons, long-chain alcohols and sterols originally present in the lipid sample in the free or esterified form. If the sample contained any glycerol ethers or plasmalogens, the deacylated residues will also be in this layer. When short-fatty acids are present in the lipids (C_{12} or less), it is necessary to extract the acidified solution much more exhaustively, and even then it may be almost impossible to recover very short-chain fatty acids such as butyric quantitatively. Epoxyl groups and cyclopropene rings in fatty acids are normally disrupted by acid, but with care they will survive the above procedure if the exposure to the acidic conditions is short.

Cholesterol esters are hydrolysed very slowly by most reagents and may not have reacted completely if the above method is used, so if they are major components of the mixture, longer reflux times are necessary. Similarly, N-acyl derivatives of long-chain bases are not so readily saponified by alkali, but hydrolysis is virtually complete when sphingomyelin, for example, is refluxed for 10 hrs in 1M KOH in methanol–water (9:1, v/v).[374]

Non-saponifiable materials and free fatty acids can also be obtained by separating the total acidified extract by thin-layer chromatography, using the solvent systems suggested in Chapter 6, eliminating the step in which the alkaline solution is extracted. The free fatty acids are easily separated from the other products of hydrolysis, which can be individually isolated and identified. As an alternative, acidic and neutral materials can be separated by ion-exchange chromatography using the following procedure.[873]

"One gram of DEAE-Sephadex (Pharmacia Fine Chemicals, Sweden), type A-25 (capacity 3.2 m-equiv per g), is successively washed on a Buchner funnel with small amounts of 1M hydrochloric acid, water, 1M potassium hydroxide and water again (the procedure is repeated three times), the last wash until neutral. It is then washed twice with methanol (25 ml), and with 25 ml of diethyl ether–methanol–water (89:10:1 by volume). It is slurried in the latter solvent mixture, equilibrated overnight and packed into a small column. The sample, containing no more than 100 mg of fatty acids, is washed through with 25 ml of the solvent mixture; the neutral materials are eluted while the acids remain on the column. The latter can be eluted with diethyl ether–methanol (9:1, v/v) saturated with carbon dioxide."

Polyunsaturated fatty acids are not altered by the mild hydrolysis conditions described above, but if the reaction time is prolonged unduly or if too strong an alkali is used, some isomerisation of double bonds can occur.[351]

B. THE PREPARATION OF METHYL AND OTHER ESTERS OF FATTY ACIDS

Before the fatty acid composition of a lipid can be determined by gas chromatography, it is necessary to prepare the comparatively volatile methyl ester derivatives of the fatty acid components. This must be by far the commonest chemical reaction performed by lipid analysts, yet it is often poorly

understood; the topic has been comprehensively reviewed.[129,159,690] There is no need to hydrolyse lipids to obtain the free fatty acids before preparing the esters as most lipids can be transesterified directly. No single reagent will suffice for all purposes, however, and one must be chosen that best fits the circumstances. Esters prepared by any of the following methods can be purified if necessary by preparative TLC (see Chapter 5 for further details). Care should be taken in the evaporation of solvents as appreciable amounts of esters up to C_{14} can be lost if this step is performed carelessly.[80] Esters other than methyl may be required from time to time for specific purposes.

1. Acid-catalysed esterification and transesterification

Free fatty acids are esterified and O-acyl lipids transesterified by heating them with a large excess of anhydrous methanol in the presence of an acidic catalyst. If water is present it may prevent the reaction going to completion. The commonest and mildest reagent used for the purpose is 5 per cent (w/v) anhydrous hydrogen chloride in methanol. It can be prepared by bubbling hydrogen chloride gas (which is commercially available in cylinders or can be prepared by dropping concentrated sulphuric acid slowly on to fused ammonium chloride or into concentrated hydrochloric acid[807]) into dry methanol, but a simpler procedure is to add acetyl chloride (5 ml) slowly to cooled dry methanol (50 ml).[39] Methyl acetate is formed as a by-product, but does not interfere seriously with methylations

$$CH_3OH + CH_3.CO.Cl \rightarrow CH_3.CO_2CH_3 + HCl$$

at this concentration. It is usual to heat the lipid sample in the reagent under reflux for about 2 hrs, but they may also be heated together in a sealed tube at higher temperatures for a shorter period. Alternatively, equally effective esterification is obtained if the reaction mixture is heated in a stoppered tube at 50°C (reducing the glassware requirements) overnight.

Non-polar lipids, such as cholesterol esters or triacylglycerols, are not soluble in reagents composed predominantly of methanol, and will not react in a reasonable time, unless a further solvent is added to effect solution. Benzene has been used most often for the purpose but because of its toxicity, it is advisable to use some other solvent such as toluene, chloroform or tetrahydrofuran.

A solution of 1–2 per cent (v/v) concentrated sulphuric acid in methanol transesterifies lipids in the same manner and at much the same rate as methanolic hydrogen chloride. It is very easy to prepare, but if the reagent is used carelessly, some decomposition of polyunsaturated fatty acids may occur. Boron trifluoride in methanol (12–14 per cent w/v) is also used as a transesterification catalyst and in particular as a rapid esterifying reagent for free fatty acids.[520] The reagent has a limited shelf-life unless refrigerated, and the use of old or too concentrated solutions may result in the production of artefacts and the loss of large amounts of polyunsaturated fatty acids.[222,402,453] In view of the large amount of acid catalyst used in comparison with other reagents and the many known side reactions, it is the author's opinion that boron trifluoride in methanol has been greatly overrated as a transesterifying reagent, although it is undoubtedly of value for the rapid methylation of unesterified fatty acids. Boron trichloride in methanol does not appear to have been much used for esterification, but is almost as effective as boron trifluoride–methanol and does not bring about the same unwanted side reactions.[86,402]

Methanolic hydrogen chloride (5 per cent) is then probably the best general purpose esterifying agent. It methylates free fatty acids very rapidly and can be used to transesterify other O-acyl lipids; it is generally used as follows:

"The lipid sample (up to 50 mg) is dissolved in tetrahydrofuran (1 ml) in a test tube and 5 per cent methanolic hydrogen chloride (2 ml) is added. The mixture is refluxed for 2 hr, then water (5 ml) containing sodium chloride (5 per cent) is added and the required esters are extracted with hexane (2 × 5 ml) using washable or disposable Pasteur pipettes to separate the layers. The hexane layer is washed with water (4 ml) containing potassium bicarbonate (2 per cent) and dried over anhydrous sodium sulphate. The solution is filtered and the solvent removed under reduced pressure in a rotary film evaporator or in a stream of nitrogen."

No additional solvent is necessary if free fatty acids alone are to be methylated (also only 20 min at reflux is required), or if phospholipids are to be transesterified. The reaction can be scaled up considerably although for very large amounts of lipid, methanol containing concentrated sulphuric acid is probably more economical; for example, 50 g of lipid in chloroform (100 ml) can be transesterified with 200 ml of methanol containing 4 ml concentrated sulphuric acid.

The same method is used to prepare dimethylacetals from aliphatic aldehydes or plasmalogens (see below) and when it is used on lipid samples containing such compounds, acetals are formed which may contaminate the methyl esters. N-acyl lipids are transesterified very slowly with these reagents (see below). If acidic reagents are permitted to super-heat in air, some artefact formation is possible.

2. Base-catalysed transesterification

O-Acyl lipids are transesterified very rapidly in anhydrous methanol in the presence of a basic catalyst. Free fatty acids are *not* esterified, however, and care must be taken to exclude water from the reaction medium to prevent their formation as a result of hydrolysis of lipids. 0.5 M Sodium methoxide in anhydrous methanol, prepared simply by dissolving fresh clean sodium in dry methanol, is the most popular reagent but potassium methoxide or hydroxide have also been used as catalysts. The reagent is stable for some months at room temperature, especially if oxygen-free methanol is used in its preparation. The reaction is very rapid; phosphoglycerides, for example, are completely transesterified in a few minutes at room temperature. It is performed as follows:

"The lipid sample (up to 50 mg) is dissolved in tetrahydrofuran (1 ml) in a test-tube, and 0.5 M sodium methoxide in anhydrous methanol (2 ml) is added. The solution is maintained at 50°C for 10 min then glacial acetic acid (0.1 ml) is added, followed by water (5 ml) and the required esters are extracted with hexane (2 × 5 ml), using a Pasteur pipette to separate the layers. The hexane layer is dried over anhydrous sodium sulphate containing 10 per cent solid potassium bi-carbonate and filtered, before the solvent is removed under reduced pressure on a rotary film evaporator."

As with acid-catalysed procedures, an additional solvent is necessary to solubilize non-polar lipids such as cholesterol esters or triacylglycerols, but is not required if they are not present in the sample. Chloroform should probably not be used because dichlorocarbene, which can react with double bonds, is generated by reaction with sodium methoxide. Cholesterol esters are transesterified very slowly and may require twice as long a reaction time as that quoted. The quantities of lipid used can be scaled up considerably; for example, 50 g of lipid is transesterified in toluene (50 ml) and methanol (100 ml) containing fresh sodium (0.5 g) in 10 min at reflux, and a related procedure has been used to transesterify litre quantities of oils.[534] Under the conditions described above, no isomerisation of double bonds in polyunsaturated fatty acids occurs, though prolonged or careless use of basic reagents may cause alterations to fatty acids.

Amide-bound fatty acids, as in sphingolipids, are not affected by alkaline transesterification reagents under such mild conditions, and this fact is sometimes used in their purification. Also, aldehydes are not liberated from plasmalogens with basic reagents but this does occur under acidic conditions.

Although free fatty acids are not esterified under the basic conditions described above, methyl esters can be prepared by exchange with N,N-dimethyl-formamide dimethyl acetal in the presence of pyridine.[766] Alternatively, if 0.2M methanolic (m-trifluoromethylphenyl)trimethylammonium hy-droxide is used as a transesterification reagent, quaternary ammonium salts of free fatty acids form which decompose, yielding methyl esters, when injected onto a heated gas chromatography column.[471]

3. Diazomethane

Diazomethane reacts rapidly with unesterified fatty acids forming methyl esters in the presence of a little methanol, which catalyses the reaction. The reagent is generally prepared as a solution in diethyl ether by the action of alkali on a nitrosamide, e.g. N-methyl-N-nitroso-p-toluene-sulphonamide ("Dia-zald", Aldrich Chemical Co., Milwaukee, U.S.A.), in the presence of an alcohol. Solutions of diazomethane are stable for short periods if stored refrigerated in the dark over potassium hydroxide pellets, but if kept too long polymeric by-products form which may interfere with subsequent gas chromatographic analysis.

Diazomethane is highly toxic and potentially explosive so great care must be exercised in its preparation; in particular strong light and apparatus with ground glass joints must be avoided, and it should only be prepared in an efficient fume cupboard. In addition, the intermediate nitros-amines are among the most potent carcinogens known. Diazomethane, accordingly, should only be used when no other reagent is suitable.

The procedure of Schlenk and Gellerman[663] is particularly convenient for the preparation of small quantities of diazomethane for immediate use. In this instance, there is very little by-product formation and, if sensible precautions are taken, the risk to health is minimal.

"A simple apparatus is required that can be quickly assembled by a glassblower. It consists of three tubes with side arms that are bent downwards and arranged so that the arm of each projects into and is near the bottom of the next tube. A stream of nitrogen is saturated with diethyl ether in the first tube and carries diazomethane, generated in the second tube, into the third tube where it esterifies the acids. The flow of nitrogen through diethyl ether in tube 1 is adjusted to 6 ml per min. Tube 2 contains carbitol (0.7 ml), diethyl ether (0.7 ml) and 1 ml of an aqueous solution of potassium hydroxide (600 g/l). The fatty acids (5–30 mg) are dissolved in diethyl ether–methanol (2 ml; 9:1 by vol) in tube 3. About 2 mmole of N-methyl-N-nitroso-p-toluenesulphonamide per mmole of fatty acid in ether (1 ml) is added to tube 2 and the diazomethane which is formed is passed into tube 3, until the yellow colour persists. Excess reagent is then removed in a stream of nitrogen."

When large amounts of diazomethane are needed, the procedure of De Boer and Backer[171] can be recommended provided suitable precautions are taken.

4. Special cases

(I) SHORT-CHAIN FATTY ACIDS

Short-chain acids are completely esterified by any of the procedures described above, but quantitative recovery of the esters from the reaction medium is very difficult because of their high volatility and partial solubility in water. As short-chain acids are major components of commercially important fats and oils such as milk fats or coconut oil, a great deal of attention has been given to the problem. Diazomethane can be used to esterify free fatty acids quantitatively in ethereal solution and the reaction medium injected directly into the gas chromatograph so that there are no losses, but if the free acids have to be obtained by hydrolysis of lipids it is not easy to ensure that there are no losses at this stage. The best methods are those in which there are no aqueous extraction, or solvent removal steps and in which the reagents are not heated. The alkaline transesterification procedure of Christopherson and Glass,[143] on which the following method is based, best meets these criteria.

"The oil (20 mg) is dissolved in hexane (2.5 ml) in a stoppered test-tube, and 0.5M sodium methoxide in methanol (0.1 ml) is added. The mixture is shaken gently for 5 min at room temperature then acetic acid (5 μl) is added followed by powdered anhydrous calcium chloride (about 1 g). The mixture is allowed to stand for 1 hour then is centrifuged at 2000–3000 rpm for 2–3 min to precipitate the drying agent. An aliquot of the supernatant liquid is taken for GLC analysis."

If the sample contains both O-acyl bound and unesterified fatty acids, the latter can be esterified with diazomethane first, before the former are transesterified. Alternatively, for safety reasons, the procedures of McCreary et al.[471] or Thenot et al.[766], described earlier, might be adapted for the purpose. Although the method described above is widely used, it has been argued recently that more reproducible gas chromatographic analyses are obtained by preparing butyl ester derivatives.[341]

(II) UNUSUAL FATTY ACIDS

The methods described above can be used to esterify all fatty acids of animal origin without causing any alteration to them. Many fatty acids from plant sources and certain of bacterial origin are more susceptible to chemical attack. For example, cyclopropene, cyclopropane and epoxyl groups in fatty acids are disrupted by acidic conditions, and lipid samples containing such acids are best transesterified with basic reagents; the free fatty acids can be methylated safely with diazomethane. Conjugated polyenoic fatty acids such as α-eleostearic acid (see Chapter 1) undergo cis-trans-isomerisation and double bond migration when esterified with methanolic hydrogen chloride,[399] and all acidic reagents can cause addition of methanol to conjugated double bond systems[405] but no side effects occur when basic transesterification is used. Similar reactions occur under acidic conditions with fatty acids containing a hydroxyl group immediately adjacent to a conjugated double bond system (e.g. dimorphecolic acid, 9-hydroxy, 10-trans, 12-trans-octadecadienoic acid), and dehydration and other unwanted side reactions may also take place.

(III) AMIDE-BOUND FATTY ACIDS

Sphingolipids, which contain fatty acids linked by N-acyl bonds, are not easily transesterified under acidic or basic conditions. If the fatty acids alone are required for analysis, the lipids may be refluxed with methanol containing concentrated hydrochloric acid (5:1 v/v) for 5 hr[756] or by maintaining the reagents at 50°C for 24 hr,[129] and the products worked up as described above for the anhydrous reagent. Traces of degradation products of the bases that might interfere with subsequent analyses can be removed by TLC. If the long-chain bases are also required for analysis, sphingomyelin should first be dephosphorylated with phospholipase C (see Chapter 9) before being hydrolysed under basic conditions as described above (Section A). Further details of such procedures are given in Chapter 7. With N-acyl phosphatidylserine and related lipids, the O-acyl bound fatty acids can be released by mild alkaline methanolysis and so distinguished from the N-acyl components which require much more vigorous hydrolytic conditions.

After lipids have been separated by thin-layer chromatography, the conventional procedure is to elute them from the adsorbent before transesterifying for GLC analysis. A number of methods have been described for transesterifying lipids on silica gel without prior elution but it has been the author's experience that with some batches of silica gel, poor recoveries of esters are sometimes obtained when basic transesterification reagents are used, probably because bound water causes some hydrolysis. Acid-catalysed procedures give better results but when the ratio of silica gel to lipid is very high ($>4 \times 10^3$), poor recoveries are again the rule.[104] In practice, such high ratios may not often be obtained and satisfactory methylation is achieved by direct transmethylation. The favoured technique is to scrape the band of adsorbent containing the lipid into a test-tube, then to add the reagent (e.g. 2% methanolic sulphuric acid) with efficient mixing and to carry out the reaction as if no adsorbent were present. On working up the aqueous mixture obtained when the reaction is stopped, it is necessary to centrifuge to precipitate all the silica gel and then to extract with a more polar solvent than hexane, for example diethyl ether, to ensure quantitative recovery of the methyl esters.[128] Unfortunately, cholesterol esters are not trans-esterified readily in the presence of silica gel, and it is still necessary to elute these from the adsorbent first. The elimination of the conventional elution step reduces the risk of loss of the more polar lipids and their fatty acid components, and of contamination of the sample by traces of impurity in large volumes of solvent.

(V) SIDE REACTIONS

Methyl esters are the derivatives of choice for gas chromatography but in choosing an appropriate reagent, it is necessary to consider its effect on lipid components other than fatty acids and on gas chromatography stationary phases since artefacts may be produced which interfere with subsequent analyses. Such factors are discussed in Chapter 5.

5. Preparation of esters other than methyl

Esters other than methyl may be required for a variety of reasons, for example to diminish the

volatility of short-chain fatty acids or to introduce aromatic groupings into the molecule so that the UV-detection systems can be used as in HPLC. Many of the methods described above can be adapted for the purpose simply by substituting the appropriate alcohol for methanol; for example, boron trifluoride-butanol has been recommended for the preparation of butyl esters,[341] or phenyldiazomethane for benzyl esters.[401] Where the alcohol has a high boiling-point or is comparatively expensive, the acid chloride can be prepared and reacted with a slight excess of the appropriate alcohol in the presence of pyridine.[129,159] Alternatively, the free acid can be reacted with an alkylating reagent in the presence of an amine or a catalyst, such as a crown ether or triphenyl-phosphine. Among the fatty acid derivatives prepared by this or related methods are benzyl[401], naphthacyl[151] and phenacyl[76,191] esters. Tri-methylsilyl esters of unesterified fatty acids have also been prepared for GC analysis.[423]

Amide derivatives of fatty acids can be prepared by similar methods. For example, N-acylpyrrolidines are prepared by reaction of methyl esters with pyrrolidine and acetic acid[32] for mass spectrometric analysis.

C. DERIVATIVES OF HYDROXYL GROUPS

The free hydroxyl groups of long-chain alcohols, hydroxy fatty acids and partial glycerides or glycerol ethers are frequently converted to non-polar derivatives for chromatographic analysis. Acyl migration of fatty acids in partial glycerides is prevented and more symmetrical peaks are obtained on gas chromatography than could be obtained with the original compounds. The choice of derivative will depend on the nature of the compound and the separation to be attempted, and on occasion it may be necessary to prepare several types of derivative to confirm identifications of components. The following are among the more useful.

1. Acetylation

Acetyl chloride and pyridine at room temperature, or after prolonged heating with acetic anhydride, can be used to acetylate lipids, but the mildest reagent is probably acetic anhydride in pyridine (5:1 v/v) which is used as follows.[619]

"The lipid (up to 50 mg) is dissolved in acetic anhydride in pyridine (2 ml, 5:1, v/v) and left at room temperature overnight. The reagents are then removed in a stream of nitrogen with gentle warming and the acetylated lipid is purified, if necessary, by preparative TLC on silica gel layers. A solvent system of hexane–diethyl ether (80:20, v/v) is suitable for most acetylated simple lipids."

Free amino groups are also acetylated with this reagent; a procedure for acetylating amino groups without acetylating hydroxyl groups at the same time is described in Chapter 7. Acetylmethane-sulphonate in microcolumns of celite, acetylates alcohols very rapidly and is suited to the routine analysis of large number of samples.[681]

Acetylated diacylglycerols can be prepared directly from phospholipids by heating them at 140°C with acetic anhydride and acetic acid in a sealed tube,[617] but considerable intramolecular acyl migration occurs under these conditions[553,618] and it is preferable to hydrolyse phospholipids to diacyl-glycerols enzymatically with phospholipase C before acetylating (see Chapter 9).

2. Trifluoroacetates

Trifluoroacetate derivatives of hydroxy acids, monoacylglycerols and glycerol ethers are comparatively volatile and are sufficiently temperature-stable to be subjected to gas chromatographic analysis. They are prepared by dissolving the hydroxy compound in excess of trifluoroacetic anhydride, leaving for 30 min and then removing the excess reagent on a rotary film evaporator.[846] Diacylglycerol trifluoroacetates are not sufficiently stable at high temperatures for gas chromatography, however.[411] Trifluoroacetates hydrolyse very rapidly, even in inert solvents such as hexane or benzene, and it is necessary to store them and to inject them onto the gas chromatographic column in trifluoroacetic anhydride solution. Column packings must be conditioned by repeatedly injecting trifluoroacetic anhydride into them before being used, but they are rendered acidic and may no longer be suitable for other analyses. Trifluoroacetic acid also appears to attack the methylene group between double bonds, causing losses of polyunsaturated components.[846] Trifluoroacetate derivatives (amides) of phos-phatidylserine and phosphatidylethanolamine are easily prepared and are much more stable chemically than the ester derivatives.[863]

3. Trimethylsilyl ether and related derivatives

Trimethylsilyl (often abbreviated to TMS) ether derivatives are a useful alternative to acetates for gas chromatographic analysis. They are much more volatile than acetates but are not as stable, particularly to acidic conditions, and will hydrolyse slowly on TLC adsorbents. The preparation and properties of TMS and related derivatives have been reviewed.[185,587,588] The most popular reagent for the preparation of TMS ethers consists of a mixture of hexamethyldisilazane, trimethylchlorosilane and pyridine (3:1:10 by volume) and is used as follows.

3:1:9 (Sigma Sd A)

"To the hydroxy-compound (up to 10 mg) is added pyridine (0.5 ml), hexamethyldisilazane (0.15 ml) and trimethylchlorosilane (0.05 ml). The mixture is shaken for 30 sec and allowed to stand for 5 min. An aliquot can then be injected directly into a gas chromatography column or the reaction mixture can be taken to dryness on a rotary evaporator, the products extracted with hexane (5 ml), the hexane layer washed with water (1 ml) and dried over anhydrous sodium sulphate, before the solvent is removed. The derivatives are stored in fresh hexane and are stable at –20°C for long periods."

A report that this reagent causes isomerisation of partial glycerides[816] is discounted by Myher.[528] A number of other reagents have been used for the purpose, but the only one to have been used widely in lipid analysis is bis(trimethylsilyl)acetamide.[764,810]

More recently, t-butyldimethylsilyl (t-BDMS) etherification has been applied to the preparation of lipid derivatives. They are approximately 10^4 times more stable than the corresponding TMS ethers and only hydrolyse at an appreciable rate under strongly acidic conditions. The following method of preparation is recommended.[153]

"The silylation reagent consists of *t*-butyldi-methylsilyl chloride (1 mmole) and imidazole (2 mmole) in NN-dimethylformamide (10 ml). The reagent (0.5 ml) is added to the lipid (up to 10 mg) and heated at 60° for 30 min. After rapid cooling, the mixture is mixed with hexane (5 ml) and is washed (3 times) with water (1 ml). The solvent is dried over anhydrous sodium sulphate, filtered or decanted and evaporated in a stream of nitrogen to yield the required derivative."

Unlike TMS derivatives, it is possible to purify *t*-BDMS derivatives by preparative TLC.

4. Isopropylidene compounds

Isopropylidene derivatives of vicinal diols, for example glycerol ethers, 1-monoacylglycerols or dihydroxy acids, are prepared by reacting the diol with acetone in the presence of a small amount of an acidic catalyst (see Fig. 4.1). Anhydrous copper sulphate is probably the mildest catalyst and is used as follows:

"The esters (10 mg) are dissolved in dry acetone (3 ml) and anhydrous copper sulphate (50 mg) is added. After 24 hr at room temperature (or 3 hr at 50°C), the solution is filtered, the copper salts are washed with dry ether and the combined solutions are evaporated *in vacuo*."

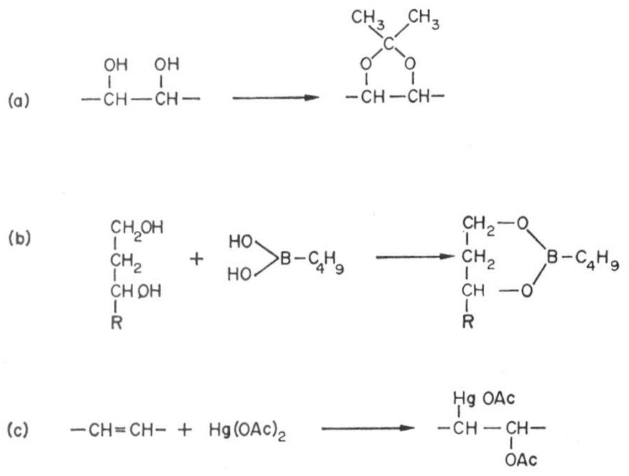

FIG. 4.1. Preparation of lipid derivatives. a. Isopropylidene derivatives. b. *n*-Butylboronate derivatives. c. Mercuric acetate adducts.

The use of perchloric acid as a catalyst permits a more rapid reaction,[838] but is potentially more hazardous. Isopropylidene compounds are stable under basic conditions, but are hydrolysed by aqueous acid and the original diol compound can be regenerated by shaking with a solution of concentrated hydrochloric acid in 90 per cent methanol (1 M).

5. n-Butylboronate derivatives

Alkyl-boronic acids, such as n-butylboronic acid, react with 1,2- or 1,3-diols of α- or ß-hydroxy acids to form 5- or 6-membered ring non-polar boronate derivatives as illustrated in Fig. 4.1b. They are prepared simply by adding n-butylboronic acid to a solution of the hydroxy-compound in dimethyl-formamide. The reaction is complete in 10–20 min at room temperature and the reaction mixture can be injected directly into a gas chromatographic column for analysis.[37,93]

D. DERIVATIVES OF FATTY ALDEHYDES

Fatty aldehydes are obtained on acidic hydrolysis of plasmalogens, and methods for their analysis are discussed in Chapter 6. In some circumstances, it may be necessary to prepare derivatives to adequately characterise them and the preparation of such compounds is now described.

1. Hydrazone derivatives

p-Nitrophenylhydrazones or 2,4-dinitro-phenyl-hydrazones are particularly useful derivatives of aldehydes as they can be determined spectrophoto-metrically (see Chapter 6). They are prepared from the aldehydes or directly from the natural plasmalogens by the following procedure.[613]

"To a solution of the phospholipids (1–5 mg) in 95 per cent ethanol (1.6 ml) is added freshly prepared 0.02 M *p*-nitrophenylhydrazine in 95 per cent ethanol (0.2 ml) followed by 0.5 M sulphuric acid (0.2 ml). The solution is heated at 70°C for 20 min, then cooled, water (1 ml) and hexane (2 ml) are added and the mixture is thoroughly shaken. The hexane layer is washed twice with water (2 ml) and dried over anhydrous sodium sulphate, before the solvent is removed *in vacuo* to give the required product."

Dimethylhydrazones are useful non-polar volatile derivatives of aldehydes as they can be readily separated by gas chromatography. They are prepared simply by dissolving the aldehyde in N,N-dimethylhydrazine and leaving at room temperature for 2 hr, before the excess reagent is removed in a stream of nitrogen.[362] Some hydrazines and their derivatives are potent carcinogens.

2. Acetals

Dimethyl acetals are the simplest acetal derivatives of aldehydes, and are prepared by heating the aldehydes under reflux with 5 per cent methanolic hydrogen chloride in the same manner as was described earlier for the preparation of methyl esters. They can also be prepared directly from plasmalogens, but then have to be separated from methyl esters, which are formed at the same time, by adsorption chromatography (see Chapter 6) or by saponification of the esters.

Cyclic acetals or 1,2-dioxolanes, prepared by condensing 1,3-propanediol with aldehydes, in the presence of an acidic catalyst, have greater thermal stability and are sometimes favoured for gas chromatographic analysis. They are prepared as follows.[743]

"The aldehydes (10 mg) with *p*-toluenesulphonic acid (0.5 mg) and 1,3-propanediol (50 μl) in 5 ml of chloroform are heated in a sealed tube at 80°C for 2 hr. On cooling, chloroform (3 ml), methanol (4 ml) and water (3 ml) are added and shaken. The lower layer, which contains the required derivatives, is taken to dryness."

Hydrazone derivatives of aldehydes can be converted directly to acetals by heating them with the required alcohol and an acid catalyst in the presence of acetone, which serves as an exchanger, by the following method.[478]

"The dinitrophenylhydrazone (50 mg) is mixed with 10 percent methanolic hydrogen chloride

(5 ml) and acetone (5 ml), and refluxed for 45 min. The mixture is cooled to 0°C, neutralised with 5 M sodium hydroxide in 90 per cent methanol (5 ml) and extracted twice with hexane (50 ml portions). The hexane layer is washed with water, dried over anhydrous sodium sulphate and decolourised with charcoal before the required derivatives are recovered."

Cyclic acetal derivatives are prepared similarly.[798] All acetal derivatives are stable to alkaline conditions, but are hydrolysed by aqueous acid.

E. DERIVATIVES OF DOUBLE BONDS

The double bonds of unsaturated fatty acids may be reacted to form various addition compounds as an aid to the isolation of individual fatty acids or as a method of establishing the configuration or location of the double bonds in the aliphatic chain.

1. Mercuric acetate derivatives

Mercuric acetate in methanol solution reacts with the double bonds of unsaturated fatty acids to form polar derivatives (see Fig. 4.1c), which are more easily separated by adsorption or partition chromatography, according to degree of unsaturation than the parent fatty acids. They can be prepared by the following procedure.[730]

"The lipid sample is refluxed with a 20 per cent excess of the theoretical amount of mercuric acetate in methanol (2 ml per g of mercuric acetate) for 60 min. After cooling to room temperature, a volume of diethyl ether 2.6-fold that of the methanol is added to precipitate mercury salts, and the solution filtered. The mercuric acetate adducts are obtained on removal of the solvent."

Although silver nitrate chromatography has supplanted the chromatography of mercuric acetate derivatives for most analytical purposes, such adducts may still be useful for the bulk preparation of pure fatty acids (see Chapter 5) and for certain synthetic applications. The original double bond is regenerated by reaction with aqueous acid with no double bond migration or *cis-trans* isomerisation if the following method is used.[730]

"10 g of the adduct in methanol (20 ml) is mixed with concentrated hydrochloric acid (50 ml) and a stream of hydrogen chloride gas is bubbled into the solution for a few minutes. The solution is extracted twice with hexane (50 ml portions), and the extracts are combined and washed with water. If a test of the organic layer with a solution of diphenylcarbazone in methanol reveals that mercury is still present, a single repeat of the process will remove it entirely."

Acetylenic groups react in a similar manner with 2 mols of mercuric acetate but cannot be regenerated as on acidic hydrolysis, a keto derivative is formed which may be used, particularly in conjunction with mass spectrometry, to locate the position of the original triple bond.[98, 397]

Mercuric acetate derivatives can be further reacted with inorganic halides yielding methoxy-halogenomercuri-derivatives, which are less polar than the original mercury compounds and are more easily separated by chromatography.[154,499,824] They can also be reacted with sodium borohydride and converted to methoxy compounds (oxymercuration-demercuration),[246,498] or reacted with halogens in methanol to form methoxyhalogen compounds.[497] Such derivatives are of value in locating double bonds by mass spectrometry.

2. Hydroxylation

Double bonds can be oxidised to vicinal diols by a variety of reagents of which the most useful are alkaline potassium permanganate and osmium tetroxide. With both reagents, *cis*-addition occurs to yield diols of the *erythro* configuration with *cis*-double bonds, and with *trans*-double bonds, diols of the *threo* configuration. The following procedure utilises alkaline permanganate and is particularly suited to reaction with monoenes and dienes.[548]

"The fatty acid (0.1–100 μ mole) is dissolved in 0.25 M sodium hydroxide solution (0.2 ml) and diluted with ice water (1 ml). 0.05 M potassium permanganate (0.2 ml) is added and, after 5 min, the solution is decolorised by bubbling sulphur

dioxide into it. The fatty acid derivatives are extracted with chloroform–methanol (7 ml, 2:1 v/v), and the lower layer is dried over anhydrous sodium sulphate before the solvent is removed."

Osmium tetroxide gives higher yields of multiple diols from polyunsaturated fatty acids.[548]

"The fatty acid (0.1–100 μ moles) is dissolved in dioxan–pyridine (1 ml , 8:1 v/v), and a 5 per cent solution of osmium tetroxide in dioxan (0.1 ml) is added. After 1 hr at room temperature, methanol (2.5 ml) and 16 per cent sodium sulphite in water (8.5 ml) are added, and the mixture allowed to stand for 1 hr more. After centrifuging to remove sodium sulphite, the supernatant solution is diluted with 4 volumes of methanol and filtered. The filtrate is evaporated to dryness and suspended in methanol (2 ml). Chloroform (4 ml) is added, the suspension is filtered and the solvent evaporated."

Such compounds in the form of less polar derivatives, for example isopropylidene[467] or methoxy compounds,[548] are used in conjunction with mass spectrometry to locate the position of the original double bond in the fatty acids (see Chapter 5). Because the isopropylidene derivatives of *threo* and *erythro* compounds are separable by gas chromatography, the configuration of the original double bond can also be established.[467,838]

3. Epoxidation

Epoxides are formed from olefins by the action of certain per-acids. *Cis*-addition occurs, and *cis*-epoxides are formed from *cis*-olefins and *trans*-epoxides from *trans*-olefins. The following procedure is based on that of Gunstone and Jacobsberg.[249]

"The monoenoic ester (20 mg) is reacted with *m*-chloroperbenzoic acid (16 mg) in chloroform (2 ml) at room temperature for 4 hr. Potassium bicarbonate solution (5 per cent, 4 ml) is added, and the product is extracted thoroughly with diethyl ether. After drying the organic layer over anhydrous sodium sulphate and removing the solvent, the required epoxy ester is obtained by preparative TLC on silica gel G layers with hexane– diethyl ether (4:1 v/v) as developing solvent."

Peracetic or perbenzoic acids are used in a similar manner for the purpose. Mass spectrometry of epoxy derivatives is sometimes used to locate the position of double bonds in fatty acids,[48,390] and since cis- and trans-isomers can be separated by GLC, as a means of estimating fatty acids with trans-double bonds.[201,202]

4. Hydrogenation

Hydrogenation of lipids is undertaken prior to confirming the chain-lengths of aliphatic moieties, or to protect lipids (and simultaneously simplify the chromatograms) during high temperature GLC analysis of intact lipids. Many of the published hydrogenation procedures are needlessly complex and the following method is adequate for most purposes.

"The unsaturated ester (1–2 mg) is dissolved in methanol (1 ml) in a test-tube and Adams' catalyst (platinum oxide; 1 mg) is added. The tube is connected via a two-way tap to a reservoir of hydrogen (e.g. in a football bladder) at or just above atmospheric pressure and to a water-pump. The tube is alternatively evacuated and flushed with hydrogen several times to remove any air, then it is shaken vigorously while an atmosphere of hydrogen at a slight positive pressure is maintained for 2 hr. At the end of this time, the hydrogen supply is disconnected, the tube is flushed with nitrogen and the solution is filtered to remove the catalyst. The solvent is removed under reduced pressure, and the required saturated ester is taken up in hexane or diethyl ether for GLC analysis."

Hexane may be used as the solvent for the hydrogenation reaction if the fatty acid is still esterified to glycerol as in a triacylglycerol or diacylglycerol acetate, but the hydrogenated compounds must later be recovered from the catalyst with a more polar solvent such as chloroform.

In the above method, the amount of hydrogen taken up is not measured but this can be done in order that the number of double bonds in the molecule be accurately determined if a manometer apparatus, constructed for the purpose, is used. Alternatively, the procedure of Brown et al.,[94] in which the platinum catalyst is generated in situ by treatment of platinum salts with sodium borohydride, can be adopted; hydrogen is also generated in situ from a standardised solution of sodium borohydride, which is introduced into a specially designed apparatus with a valve at the end of a burette so that reagent is added only as long as the sample is taking up hydrogen. The amount of hydrogen can be calculated accurately from the volume of standard sodium borohydride solution added. A modification of this procedure, that has been used for the determination of unsaturation in fats and oils, has been described.[501]

5. Deuterohydrazine reduction

Deuterium can be added across double bonds to assist in their location by mass spectrometry, by means of deuterohydrazine reduction.[176]

"The unsaturated ester (0.5 mmole) is dissolved in anhydrous dioxan (5 ml) at room temperature, and deuterohydrazine (5 mmole) in twice its volume of deuterium oxide is added. The mixture is stirred at 55–60°C in air but in the absence of atmospheric moisture for 8 hr when the reagents are evaporated in vacuo. The product is purified by preparative silver nitrate TLC" (see Chapter 5 for practical details).

CHAPTER 5

The Analysis of Fatty Acids

A. INTRODUCTION

THE ADVENT of gas–liquid chromatography revolutionised the analysis of the fatty acid components of lipids and it is undoubtedly the technique that would be chosen in most circumstances for the purpose. However, it should be recognised that individual fatty acids can only be tentatively identified by this means and that in the first analysis of any new sample, confirmation of fatty acid structures should be obtained by unequivocal chemical degradative and spectroscopic procedures. Ideally, individual pure fatty acids (usually in the form of the methyl ester derivatives, prepared as described in Chapter 4) should be isolated by a combination of complementary chromatographic methods, and examined first by non-destructive spectroscopic techniques before chemical degradative procedures are applied. For example, adsorption chromatography will separate normal fatty acids from those containing polar functional groups. Unsaturated fatty acids of animal origin may be provisionally identified by their gas chromatographic retention characteristics on two or more stationary phases and on the basis of possible biosynthetic relationships. Silver nitrate chromatography can be used to segregate fatty acids according to the number and geometrical configurations of their double bonds; a portion of each fraction should be hydrogenated so that the lengths of the carbon chains of the components can be confirmed. Finally some form of partition chromatography must be utilised to separate components of different chain-lengths, so that the position and configuration of the double bonds may be determined by spectroscopic (principally IR and NMR spectroscopy and mass spectrometry) and oxidative-degradation procedures. Appropriate spectroscopic and chemical techniques can also be used to detect and locate other functional groups in the fatty acyl chains.

Precautions should be taken at all times to prevent or minimise the effects of autoxidation (see Chapter 3).

B. ANALYTICAL GAS-LIQUID CHROMATOGRAPHY

1. The common fatty acids of plant and animal origin

(I) COLUMN PACKING MATERIALS

The liquid phase used in preparing gas chromatographic column packing materials is the principal factor determining the nature of the separations that can be achieved. Silicone liquid phases such as SE–30, OV–1, JXR or QF–1 permit the separation of fatty acid esters mainly on the basis of their molecular weights, although there can be separation of unsaturated from saturated fatty acids of the same chain–length when the amount of the stationary phase on the support is low (1–3 per cent). High molecular weight hydrocarbon liquid phases such as the Apiezon greases (of which the most popular is Apiezon L) also separate saturated and unsaturated components of the same chain–length, unsaturated esters eluting before the related saturated compounds, but in packed columns there is very little separation of esters of a given chain–length differing in the number of double bonds in the molecule. Such phases are now only used for the analysis of oxygenated or of high molecular weight fatty acids.

Polar polyester liquid phases are much more suited to fatty acid analysis as they allow clear separations of esters of the same chain–length, but with zero to six double bonds, unsaturated components eluting after the related saturated ones.

These phases can be subdivided into three classes: group a, highly polar phases, e.g. polymeric ethyleneglycol succinate (EGS), diethyleneglycol succinate (DEGS) and EGSS-X (a copolymer of EGS with a methyl silicone); group b, medium polarity phases, e.g. polyethyleneglycol adipate (PEGA), butanediol succinate (BDS) and EGSS-Y (a copolymer of EGS with a higher proportion of the methylsilicone than in EGSS-X); group c, low polarity phases, e.g. neopentylglycol succinate (NPGS) and EGSP-Z (a copolymer of EGS and a phenyl silicone). EGSS-X and EGSS-Y have now become widely accepted as the most useful representatives of the first two groups because of their relatively high thermal stability, particularly in packed columns, and it is to be hoped that more research workers will adopt these as standard phases for reporting gas chromatographic retention data. Low polarity phases are utilised principally in open-tubular and support-coated open-tubular (SCOT) columns as when they are used in packed columns, saturated and monoenoic components of the same chain-length are poorly separated.

More recently, a range of new stationary phases has been developed and marketed that are apparently alkylpolysiloxanes containing various polar substituents, especially nitrile groups. These are available with a wide range of polarities, but the most useful are somewhat more polar than EGSS-X; they are sold under various trade designations e.g. Silar 10C, Silar 9CP, SP2340 and OV275. In packed columns, they are stable at temperatures above those possible with EGSS-X, for example, and afford excellent separations of polyunsaturated fatty acids,[294, 533] complementing those obtained with the more common stationary phases, and they are of particular value in the separation of cis- and trans-isomers (see below).

It is occasionally necessary to subject fatty acids in the unesterified form to GLC analysis. In this instance, acidic liquid phases such as DEGS containing 3 per cent phosphoric acid,[493] or carbowax 20 M-terephthalic acid or the structurally-related phase SP-1000 (Supelco Inc., U.S.A.), are used. The problems involved in such analyses have been comprehensively reviewed.[418]

The quality and to some extent the nature of the separations achieved is influenced by the amount of liquid phase applied to the support. Low levels of polyesters (1–3 per cent by weight relative to that of the support) are occasionally suggested for use with highly inert supports, as methyl esters of long-chain polyunsaturated fatty acids will elute from them at comparatively low temperatures. On the other hand, greater amounts of polyester (10–15 per cent by weight) offer more protection to polyunsaturated esters at high temperatures so are generally preferred. The retention times of fatty acid esters relative to a chosen standard ester (usually 16:0 or 18:0) tend to decrease as the amount of liquid phase on the support is decreased and therefore, for reproducible work, it is advisable to determine the optimum amount of liquid phase necessary for a given separation and to standardise the chromatographic conditions accordingly. Unfortunately, columns age with use as the stationary phase polymerises further or bleeds from the column and some changes in the retention characteristics of esters inevitably occur.

The nature of the support material for the stationary phase can influence the quality of the separations (see Chapter 3), but acid-washed and silanised support materials are widely available; they are almost completely inert so that separations are controlled largely by the liquid phase. A uniform fine grade (100–120 mesh is considered to the optimum) is to be preferred for analytical columns.

It would aid interlaboratory comparisons, if more general use were made of a select number of phases; the author would recommend columns packed with either 15 per cent EGSS-Y, EGSS-X and one of the polysiloxanes (e.g. Silar 10C or OV275) on 100–120 mesh acid-washed silanised supports for the analysis of polyunsaturated fatty acids. Indeed columns packed with each of these phases should be available in the well-equipped lipid laboratory, as fatty acids which co-chromatograph on one of these columns will generally be separable on one of the others.

There have been a number of reports of losses of esters of polyunsaturated fatty acids on gas chromatographic columns. They can be attributed partly to the use of too active support materials and partly to transesterification of methyl esters with the polyester liquid phase. The latter effect is caused by residues of the catalyst required for the preparation of the polyester, so that those components remaining longest on the columns will suffer the greatest losses. Fortunately, polyester liquid phases made without catalyst are commercially available so that such losses need not be significant. If column packing materials are made with catalyst-free liquid phases coated on acid-washed silanised supports, quantitative recovery of most fatty acid esters is possible. Such materials are naturally more expensive, but it is a false economy to use inferior grades as columns will last for years if looked after properly. With capillary columns (WCOT and SCOT), some losses due to absorption on the walls of the column are inevitable and are less easily controlled. This problem is discussed below.

(II) PROVISIONAL IDENTIFICATION USING STANDARDS OR RETENTION TIME RELATIONSHIPS

Standard mixtures containing accurately known amounts of methyl esters of saturated, monoenoic and polyenoic fatty acids are commercially available from a number of reputable biochemical suppliers. These are invaluable for checking the quantification procedures used (see below), and also for the provisional identification of fatty acids by direct comparison of the retention times of their methyl esters with those of the unknown esters on the same columns under identical conditions. Comparisons should be made on at least two columns with different types of packing materials e.g. EGSS-X versus EGSS-Y or one of the polar polyalkyl-siloxanes.

The lipids in animal tissues usually contain a much wider spectrum of fatty acids than is available commercially. It is therefore helpful to obtain a secondary external reference standard consisting of a natural fatty acid mixture of known composition. This can be a common natural product that has been well characterised or a mixture of natural esters, the composition of which has been accurately established by the procedures described below. Ideally, it should be similar to the samples under investigation; for example, Ackman and Burgher[10] have used cod-liver oil in this way to identify the fatty acids of other marine animals, and Holman et al.[316, 317] have used the fatty acids of bovine and porcine testes for a similar purpose. Rat liver fatty acids are frequently used in the same way. The author uses a simple mixture made up of the fatty acids of pig liver lipids and cod-liver oil together with a little linseed oil, as this contains significant amounts of all the major fatty acid classes (saturated and mono-, di-, tri-,

tetra-, penta- and hexaenoic components of both the (n-3) and (n-6) families), including most of the fatty acids likely to be encountered in animal tissues.

Figure 5.1 illustrates separations of such a standard mixture, on 15 per cent EGSS-X and EGSS-Y in packed columns. Both columns give excellent separations of fatty acid esters of a given chain-length that differ in degree of unsaturation. Separation of esters which differ only in the positions or configurations of the double bonds where these are approximately central, is not easily achieved with monoenoic acids, but is possible with esters of polyunsaturated fatty acids. For example, on both of these columns, 18:3(n-3) and 18:3(n-6) are separated, as are three isomers of 20:3, two isomers of 20:4 and two isomers of 22:5 as is apparent in Fig. 5.1. With the methyl esters of the more common families of polyunsaturated fatty acids, the shorter the distance between the last double bond and the end of the molecule, the longer the retention time of the isomer.

The principal disadvantages of these columns is that there is some overlap of fatty acids of different chain lengths.[5] For example, on EGSS-X 18:3(n-3) and 20:1(n-9) coincide, as do 20:4(n-6) and 22:1(n-9). On EGSS-Y, these pairs can be separated, but 20:0 and 18:3(n-3), 18:4(n-3) and 20:1(n-9) or 22:1(n-9) and 20:4(n-3) are not separable. If Silar 10C had been selected as a stationary phase rather than EGSS-Y, a different range of separations again would be seen, complementing those on EGSS-X. By a judicious use of two columns differing in polarity, most fatty acids can be separated and estimated. However, there may still be some troublesome components for which it is necessary to use some form of open-tubular GLC or some other technique, such as silver nitrate or liquid–liquid partition chromatography (see below), in combination with gas chromatography to completely separate, identify and estimate all components. It is also possible to eliminate the chain-length problem by using low polarity polyester liquid phases such as those of group c above, but packed columns in which these are used are generally of low efficiency and the resolution of peaks is inferior to that obtained with more polar liquid phases. With group c liquid phases in WCOT or SCOT columns, much greater resolution can be achieved, but there are other difficulties with these columns (see Chapter 3 and below).

It should be noted, however, that the retention

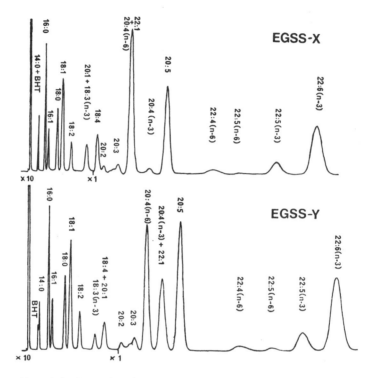

FIG. 5.1. Gas chromatographic analysis of a complex mixture of natural fatty acids (as the methyl esters) on packed columns with EGSS-X and EGSS-Y as stationary phases (see footnote to Table 5.1 for further chromatographic details).

times of esters and the separations achieved are all dependent on the precise column conditions used and may vary with such factors as the temperature or the age of a column and the amount of stationary phase on the support.

The absolute retention time of an ester on any gas chromatographic column has very little meaning as a measure of its elution characteristics, because slight changes in the operating conditions or in the character of the packing material (in its origin or on ageing) can affect this parameter drastically. On the other hand, the retention time of a fatty acid ester relative to that of a chosen standard commonly occurring component (usually 16:0 or 18:0) has a greater absolute significance and is a quantity more suited to inter-laboratory comparisons, i.e.

$$\text{relative retention time} \atop (r_{18:0})} = \frac{\text{retention time of ester}}{\text{retention time of 18:0}}$$

Theoretically, retention times should be measured from the time of injection of the sample into the gas

chromatograph column to the time when the peak is at its maximum, but as very large peaks may be skewed, it has been suggested that the distances should be measured to the point where the tangent drawn to the leading edge of the peak intercepts the base line.[3] This should rarely be necessary with modern packed columns, but can be useful with capillary columns. Also, as the retention times of esters are influenced to some extent by components eluting immediately adjacent to them, relative retention times should be measured on pure compounds or on simpler fractions isolated by silver nitrate chromatography (see below), whenever this is possible. The relative retention time $(r_{18:0})$ of the component esters of the external reference standard, separated on EGSS-X and EGSS-Y columns as illustrated in Fig. 5.1, are listed in Table 5.1, as are similar values for a Silar 10C column. An interesting feature of the Silar 10C column is that the retention time of 22:6(n-3), the last component to emerge in most analyses of animal lipids, relative to that of 18:0 is much lower than that obtained with the other

TABLE 5.1. EQUIVALENT CHAIN LENGTHS AND RELATIVE RETENTION TIMES OF SOME UNSATURATED ESTERS ON PACKED COLUMNS WITH EGGS-X, EGGS-Y AND SILAR 10C AS STATIONARY PHASES

Methyl ester	ECLs			Relative retention times*		
	EGSS-X	EGSS-Y	Silar 10C	EGSS-X	EGSS-Y	Silar 10C
16:0	16.00	16.00	16.00	0.58	0.57	0.67
16:1(n-9)	16.57	16.62	17.08	0.69	0.68	0.84
16:2(n-6)	17.65	17.45	—	0.90	0.85	—
18:0	18.00	18.00	18.00	1.00	1.00	1.00
18:1(n-9)	18.53	18.52	19.00	1.19	1.16	1.23
18.2(n-6)	19.42	19.20	20.07	1.52	1.41	1.56
18:3(n-6)	20.00	19.67	21.10	1.80	1.60	1.75
18:3(n-3)	20.40	20.02	21.28	2.02	1.78	2.01
18:4(n-3)	21.05	20.52	22.12	2.40	2.04	2.37
20:0	20.00	20.00	20.00	1.77	1.76	1.53
20:1(n-9)	20.50	20.45	20.90	2.08	2.01	1.84
20:2(n-6)	21.40	21.15	—	2.67	2.44	—
20:3(n-9)	21.63	21.33	22.22	2.87	2.57	2.22
20:3(n-6)	21.77	21.53	22.72	2.99	2.72	2.42
20:3(n-3)	21.95	21.60	23.10	3.13	2.78	2.64
20:4(n-6)	22.43	22.00	23.53	3.59	3.10	3.14
20:4(n-3)	23.00	22.47	24.13	4.18	3.54	3.55
20:5(n-3)	23.50	22.80	24.77	4.85	3.91	4.06
22.0	22.00	22.00	22.00	3.14	3.12	2.31
22:1(n-9)	22.43	22.35	22.80	3.59	3.44	2.71
22:4(n-6)	24.45	24.00	25.70	6.34	5.48	4.94
22:5(n-6)	24.57	23.85	25.96	6.55	5.27	5.14
22:5(n-3)	25.53	24.80	26.83	8.60	6.92	6.20
22:6(n-3)	26.18	25.20	27.40	9.10	7.74	6.89
24:0	24.00	24.00	24.00	5.59	5.50	3.50

* Relative to 18:0

EGSS-X and EGSS-Y: Data obtained with 2m × 4mm i.d. glass columns packed with 15% (w/w) stationary phase on Chromasorb W (100–120 mesh, acid washed and silanised). Carrier gas, Nitrogen at 50 ml/min; column temperatures, 178°C (EGSS-X) and 194° (EGSS-Y).

Silar 10C: Data obtained with 3m × 4mm i.d. glass column packed with 10% Silar 10C on Gaschrom Q (100–120 mesh). Carrier gas, nitrogen at 11.5 ml/min; column temperature, 210°C.

polyester columns. In practice, this can mean that shorter analysis times are possible while maintaining a good spread of peaks in the chromatogram. Relative retention times vary somewhat with the conditions of the column packing materials and with other operating parameters, such as temperature or flow rate, but these variations are comparatively small and are in the same direction for all components.

Kovats' retention indices are more generally accepted as a standard means of recording GLC retention data, but have been little used for the esters of fatty acids. Analogous parameters known as *equivalent chain lengths* (abbreviated to ECLs)[502] or carbon numbers[851] (the latter term is now little used because of its ambiguity) have considerable utility,

however. ECL values can be calculated from an equation similar to that for Kovats' indices,[349] but are usually found by reference to the straight line obtained by plotting the logarithms of the retention times of a homologous series of straight-chain saturated fatty acid methyl esters against the number of carbon atoms in the aliphatic chain of each acid. (Semilog. paper is particularly convenient for the purpose). The retention times of the unknown acids are measured under identical operating conditions and the ECL values are read directly from the graph. The ECL values of the esters of the component acids separated as illustrated in Fig. 5.1 are listed in Table 5.1, and were obtained in this way. Such values have more obvious physical meaning than relative retention times and are more easily remembered. It should be noted, however, that outwith the normal range of chain-lengths (C_{12}–C_{22}), a straight line relationship between log retention time and number of carbon atoms may no longer hold.[349]

The increment in ECL value of a given ester over that of the saturated ester of the same chain-length, sometimes known as the *fractional chain-length* (or FCL) value, is dependent on the structure of the compound and is influenced by the number of double bonds in the aliphatic chain and the distance of the double bonds from the carboxyl and terminal ends of the molecule. The retention characteristics of the methyl esters of the complete series of monoenoic C_{18} acids (i.e. Δ^2 to Δ^{17}),[248] all the methylene-interrupted *cis,cis*-dienoic acids (i.e. $\Delta^{2,5}$ to $\Delta^{14,17}$)[125] and for many acids with more than one methylene group between the double bonds[251] as well as of many natural polyunsaturated acids,[310] have been obtained on a number of liquid phases. In addition, the effect of double bond positions on ECL values has been well documented by Jamieson[348] in a comprehensive review. A single double bond in the centre of a long aliphatic chain gives an FCL value, when compared with that of the corresponding saturated compound, of about 0.5 on a DEGS column, but as the double bond nears the carboxyl end of the molecule, the FCL value tends to decrease (except for the Δ^3 isomer); as it nears the terminal end of the molecule it tends to increase to a value of nearly 1.0 (the FCL value of the Δ^{16} isomer is even higher).[248] Proximity of a functional group of any kind to the terminal end of the molecule appears to have a greater effect on ECL values than when it is a similar distance from the carboxyl end. FCL values for individual double bonds are approximately additive; for example, if the data for the Δ^9 and Δ^{12} monoenes are used to predict the ECL value of the $\Delta^{9,12}$ diene, the calculated value is just a little lower than that obtained experimentally, and the difference between the actual and calculated results is consistent over the entire methylene-interrupted series.[125] Some interaction between the double bonds increasing the polarity of the molecule may be responsible for the minor discrepancy. When double bonds are separated by more than one methylene group, the difference between the calculated and experimental values is negligible.[251]

With any homologous series of fatty acids that contain a substituent in the alkyl chain, the distance of the substituent from either the proximal or the terminal end of the chain must vary. The logarithms of the retention times of the esters of such series plotted against the numbers of carbon atoms in the chain do not lie in a straight line unless the series are short. The deviations from a straight line are greatest for short-chain esters, but with longer-chain compounds, the FCL values obtained for esters with similar groups of double bonds and terminal structures are approximately constant. Using the data in Table 5.1, on EGSS-X, 20:4(*n*-6) has an ECL value of 22.43 (i.e. 20 + 2.43) and that of 22:4(*n*-6) is 24.45 (i.e. 22 + 2.45); 20:4(*n*-3) has an ECL value of 23.00 (i.e. 20 + 3.00) so we can predict that 22:4(*n*-3), which does not occur in the standard mixture, will have an ECL value of close to 25.00 (i.e. 22 + 3.00). With a suitable secondary reference standard, the ECL values of the esters of most of the fatty acids likely to be encountered in animal tissues can be measured or predicted. Marine oils, which may contain more families of polyunsaturated fatty acids, present more complicated identification problems and there are difficulties in applying the factors to shorter chain-length esters. Ackman[3] has proposed the use of a series of systematic separation factors, which can be calculated from the relative retention times of esters, to assist in overcoming these problems.

Again it must be emphasised that ECL values may vary a little with column conditions (e.g. temperature and carrier gas flow rates), with the nature of the support, with the amount of liquid phase and with the age of columns. ECL values obtained in allegedly similar circumstances in other laboratories may be taken as a guide, but should be applied with caution. A method of overcoming some of the difficulties of interlaboratory comparisons has been developed in Jamieson's laboratory.[349, 353] Polyester liquid phases are considered as a single class of substances varying continuously in polarity, and the ECL value of linolenic acid (18:3(*n*-3)) on a given column is a function of this polarity. If the ECL value of this component is known, those of other fatty acids can be obtained from tabulated values, or from a simple equation and computer-derived constants.

The provisional identification of fatty acids can often be aided by ancillary techniques such as silver nitrate chromatography (see below).

The applications of WCOT and SCOT columns to the analysis of fatty acids have been reviewed in some detail,[3, 8, 11] and some remarkable GLC traces of natural lipid samples have been published, illustrating how much more complex these are than might be expected from more conventional analyses. Positional and geometrical isomers of unsaturated fatty acids are much more easily separated by this means than on packed columns, for example. Fortunately, ECL values obtained with packed columns are similar to those with open-tubular columns so that some comparisons between the two are possible. Overlap problems between compounds of different chain-length tend to be much less serious with open-tubular columns, when certain of the less polar stationary phases such as BDS[5] or Silar 5CP[13] are used. In consequence, an analysis of a complex sample can be completed in a single run whereas multiple analyses on different phases with packed columns might be required. In much of the work with open-tubular gas chromatography published to date, stainless steel columns have been used and there are problems in quantification because of loss of material by oxidation or adsorption at the walls.[3, 8, 11] Minor components can be lost entirely when the connection between the end of the column and the detector is made of metal, but not when it is lined with glass.[15]

Glass or better fused silica capillary columns are now much more widely available and seem to be comparatively free of problems associated with losses at the column walls[347, 442, 711] although there would appear to be a need for a definitive study with respect to oligounsaturated fatty acids, in particular. Detector response factors must be determined experimentally for different fatty acids at regular intervals and they can change by a factor of 10% over a period of about a year.[347] The useful life of a glass capillary column is much longer (a year or more) than that of a corresponding steel column (2–3 months only).

Open-tubular columns have proved especially useful for the analysis of geometrical isomers of unsaturated fatty acids, as is discussed below.

(IV) QUANTITATIVE ESTIMATION OF FATTY ACID COMPOSITION

As discussed earlier (Chapter 3), with reliable modern gas chromatographs equipped with flame ionisation detectors, the areas under the peaks on the GLC traces are, within limits, linearly proportional to the amount (by weight) of material eluting from the columns. Problems of measuring this area arise mainly when components are not completely separated, and there is no way of overcoming this difficulty entirely. When overlapping peaks have distinct maxima, the height multiplied by retention time method of area measurement is probably the most accurate manual technique in isothermal analyses, but computer analysis of peak shapes[81] may improve the accuracy of the estimation. Where one component is visible only as a minor shoulder or broadening of a major peak, no manual or computer method is likely to give very precise results for the individual components, although electronic integration can at least give an accurate measure of the total amount of material present in a multiple peak. Wherever feasible, column conditions should be altered in an attempt to improve resolution, or another liquid phase can be tried. The magnitude of the problem can be much less with capillary columns.

Other problems may arise through loss of components on the columns, but effects of this kind can be minimised with packed columns as discussed above, by using catalyst-free liquid phases and highly inert supports. It is necessary to check frequently whether losses are occurring by running standard mixtures of accurately known composition through the columns. These standards should be similar to the samples to be analysed; for example, saturated standards should not be used to calibrate columns for the analysis of polyunsaturated fatty acids, and the calibration should be checked regularly. Difficulties arise most often with polyunsaturated fatty acids, and it is possible to circumvent them by adding an appropriate internal standard. For example, it has been suggested that a 22:1 fatty acid be used in the determination of 22:6[147] and a 20:3 fatty acid for 20:4(n-6).[230]

If necessary, calibration factors may have to be calculated for each component to correct the areas of the relevant peaks in the mixtures analysed. When flame ionisation detectors are used, small correction factors are sometimes applied to compensate for the fact that the carboxyl carbon atom in each ester is not combusted to a component to which the detector responds.[4, 8, 14] With open-tubular columns of all types, losses cannot always be avoided, and frequent calibration and calculation of correction factors are essential for quantitative work.[17] Methyl ester mixtures made originally to the specifications of the National Institutes of Health in the U.S.A. are available from several commercial sources for the purpose.

Good sample injection technique can improve the recoveries of fatty acids and the resolutions attainable. The syringe needle should be inserted rapidly but steadily into the column until it reaches a predetermined depth when the plunger is pressed in firmly; the needle is left in place for about one second and is then removed rapidly.

It is not easy to give an objective assessment of the standard of accuracy that should be possible in routine analyses of fatty acids, but in a collaborative study of IUPAC methodology for fatty acid analysis, typical coefficients of variation (%) at various concentrations were 15 (2% level), 8.5 (5% level), 7 (10% level) and 3 (50% level).[210]

Results can be expressed directly as weight percentages of the fatty acids present, but it is often necessary to calculate the molar amounts of each acid as, for example, in most lipid structural studies. This is performed simply by multiplying the area of each peak by an arithmetic factor, obtained by dividing the weight of some standard ester (say 16:0) by the molecular weight of the component. It should be noted that if fatty acid compositions are calculated on a weight basis, it is not always necessary to positively identify each compound, but this cannot be avoided if molar proportions are required.

2. Some less common fatty acids

Cis and trans-isomers of unsaturated fatty acids are not in general separable on the conventional polyester stationary phases but some excellent separations have been recorded with the newer cyanoalkylpolysiloxane phases such as Silar 10C, SP2340 or OV-275. For example, near base-line separations of methyl oleate and methyl elaidate have been achieved on longer than usual packed columns (6–7 m long × 0.2 mm internal diameter) containing these phases,[150, 178, 564, 576] although even better separations can be obtained with capillary col-

umns.[11, 178, 295, 347] It should be noted, however, that many of these published separations have been accomplished with simple model mixtures, and such columns might be less efficient with lipids of potential commercial interest (e.g. butter fat or partially-hydrogenated vegetable oils), in which there is a range of positional isomers. The presence of conjugated double bond systems in the alkyl chain increases the retention time of an ester considerably over that of a similar compound with methylene-interrupted double bonds. Methyl 9-cis, 11-trans-octadecadienoate, for example, has ECL values of 20.48 on EGSS-X and 20.24 on EGSS-Y,[130] i.e. appreciably greater than the corresponding values for methyl linoleate. In addition, the configuration of the double bonds has a much more pronounced effect and some separation of geometrical isomers of conjugated esters may be possible on columns packed with conventional polyester phases as well as on the cyanoalkylpolysiloxanes. Conjugated trienes reportedly undergo cis-trans isomerisation and double bond migration on polyester columns,[494, 510] and dehydration of compounds with allylic hydroxyl groups may occur leading to spurious peaks.[510] The latter problem was apparently not alleviated if the acetate or methoxy derivatives were prepared, but it might be instructive to repeat this work with more modern catalyst-free liquid phases on inert supports.

Retention data, including ECL values, have been reported for the complete series of isomeric methyl-branched octadecanoates,[2, 6] and all are eluted before methyl nonadecanoate on polar and non-polar columns. The iso- and anteiso-isomers are those most often found in nature and they are easily separated from each other. The iso-compound is eluted first, and the C_{19} fatty ester with this structure has an ECL value of approximately 18.5 on 15 per cent EGSS-X and EGSS-Y, while the related anteiso-compound has an ECL value of 18.7 on these columns (W. W. Christie, unpublished data). FCL values taken from the literature have been used to estimate ECL values for multibranched acids with some success.[7] Diastereoisomers of branched-chain acids have been separated by open-tubular gas chromatography.[7, 12, 718] The problems involved in the analysis of complex mixtures of branched-chain fatty acids have been reviewed by Smith and Lough.[713]

The ECL values of the complete series of methyl

esters of C_{19} isomeric cyclopropane fatty acids have been recorded on polar and non-polar liquid phases;[132] they are all approximately one unit greater than those of the corresponding C_{18} monoenoic esters from which they are derived, synthetically or biosynthetically. The 9,10- and 11,12-methyleneoctadecanoates, which are occasionally found together in bacterial lipids, can be separated on open-tubular columns.

Tulloch[777] has reported ECL values for the complete series of methyl hydroxy-, acetoxy- and oxo-stearates on polar and non-polar liquid phases. As is usual with substituted compounds, components with central functional groups are not easily separated, but where the substituents are close to either end of the molecule, positional isomers can be resolved. A hydroxyl group increases the ECL value of a fatty acid greatly, especially on polar columns, and on EGS, methyl 12-hydroxystearate has an ECL value of 26.25 although its acetate derivative has an ECL value 1.5 units lower than this and the trimethylsilyl ether (TMS) derivative is eluted even more rapidly. Compounds with free hydroxyl groups present several problems to the gas chromatographer. Adsorption effects on the support or on the walls of the column and hydrogen-bonding effects may come into play so that unsymmetrical peaks are obtained and recoveries are incomplete. Losses can occur because of transesterification of the hydroxyl group with polyester liquid phases. Although such effects are less apparent with modern catalyst-free liquid phases and highly inert supports, it is still advisable to chromatograph the compounds in the form of a non-polar volatile derivative such as the acetate,[777] trimethylsilyl ether[844] or trifluoroacetate[839] derivatives, as sharper peaks, better recoveries and better resolutions of positional isomers are obtained. n-Butylboronate derivatives[37,93] are invaluable for characterising 2- and 3-hydroxy fatty acids (see Chapter 4 for details of preparations). Polyhydroxy esters have been subjected to GLC in the form of the trimethylsilyl ethers,[839] as trifluoroacetates[839] or as isopropylidene derivatives.[838] Erythro- and threo-forms of vicinal diols can be separated on packed columns when they are converted to either of the last two derivatives and, as these compounds can be prepared quantitatively to a high degree of stereochemical purity from cis or trans-olefins

respectively, this provides a basis for gas chromotographic separation and estimation of stereoisomers of unsaturated acids on packed columns.[838] Epoxide derivatives can be used in a similar manner.[201,202] It appears to be a general rule that derivatives with the trans- or threo-configuration are eluted before those of the cis- or erythro-configuration, especially on non-polar stationary phases.

Optical enantiomers of the methyl esters of mono-hyroxy fatty acids can be resolved by gas chromatography if they are first converted to the (—)-menthyl or D-phenylpropionyl derivatives.[272,718] Some limited success has also been achieved with polyhydroxy fatty acids.

An isolated triple bond has a similar effect on the retention characteristics of a fatty acid as three methylene-interrupted double bonds, and methyl stearolate (methyl octadec-9-ynoate) is eluted with or slightly after methyl linolenate on DEGS[259] or PEGA[511] columns. Separation factors have been calculated for such phases.[354] Distinctive separations are obtained on the cyanoalkylpolysiloxane phases.[350]

Methyl esters of cyclopropene fatty acids are less easily subjected to gas chromatography as they tend to decompose or rearrange on the columns to give spurious peaks, and it is more usual to modify them chemically by hydrogenation or by reaction with silver nitrate[360] or methanethiol[606] before analysis. If highly inert supports and silicone liquid phases are used, however, successful GLC of the native esters is possible;[615] the author has successfully chromatographed methyl sterculate on a well-conditioned EGSS-X column where it eluted just after methyl linoleate, although the same compound decomposed on a new EGSS-Y column. Other saturated and unsaturated cyclic esters are more stable to gas chromatography. Methyl 11-cyclohexylundecanoate is eluted with 18:2 on a PEGA column and with 18:0 on ApL[277]; the cyclopentenoic acids of seed oils of Hydnocarpus species have also been subjected to GLC analysis.[870]

When short and medium chain-length fatty acids are present in a sample in addition to the common range of longer chain-length components, it is possible to obtain an analysis by isothermal gas chromatography at two or more different temperatures, if a peak common to each chromatogram is used as a reference point for quantification purposes. It is generally much more convenient and accurate, however, to use temperature-programming in such circumstances, i.e. to start the analysis at a low column temperature and raise it at a fixed reproducible rate to an appropriate final temperature. The optimum temperature differentials and programming rates will depend on the nature of the sample and of the chromatographic column, and must be determined empirically in each instance. Ideally, members of homologous series should emerge at approximately equal time intervals, giving symmetrical peaks of approximately equal width on the chromatographic trace. Individual peaks can be quantified by triangulation or better by electronic integration. Substantial correction factors may be required for the shorter chain-length components and these must be obtained and checked at regular intervals by analysis of suitable standard mixtures. Some valuable advice on the problems associated with the temperature-programmed analysis of the fatty acid components of some common fats of commercial interest, such as coconut oil or butter fat, is contained in a review article.[340] Great care is, of course, necessary in the initial preparation of the methyl ester derivatives (see Chapter 4).

3. Spurious peaks on recorder traces

Spurious peaks can appear on recorder traces for a variety of reasons, but most often are a consequence of unwanted interaction between the esterifying reagent and a functional group in a fatty acid chain or other lipid component, or of contamination of a sample by impure reagents. Some extraneous substances, such as antioxidants, are introduced deliberately; BHT tends to elute near to, or coincidently with, methyl myristate on polyester stationary phases.

Some of the difficulties which can be encountered during methylations are described in Chapter 4, and these can be exacerbated if longer than necessary reaction times or old and deteriorating reagents are used. Particular care in the choice of reagents is necessary with fatty acids containing potentially-labile functional groups, such as cyclopropane or cyclopropene rings or epoxy groups. Methoxy-compounds are sometimes formed from unsaturated fatty acids when boron trifluoride–methanol[222, 453] or sulphuric acid–methanol[262, 278] are used as esterifying reagents, and polymeric by-products may be formed if diazomethane[519, 663] or dimethoxypropane and methanolic hydrogen chloride[519] are used. Difficulties of this kind need not occur if the procedures described in Chapter 4 are selected with care. BHT may be partially methylated by boron trifluoride-methanol reagent giving rise to an extraneous peak which chromatographs with methyl palmitoleate.[295] Traces of the transesterifying agent inadvertently injected onto a polyester column can also give rise to spurious peaks. Sodium methoxide, for example, interacts with cyanoalkylpolysiloxane phases to produce peaks which could be mistaken for short-chain esters.[773]

If cholesterol esters are transesterified directly and the free cholesterol is not removed prior to GLC analysis, it may dehydrate to form cholestadiene on columns made up of low percentages of EGSS-X or EGSS-Y and obscure the C_{22} polyenoic esters.[409] With higher amounts of liquid phase this does not occur and cholesterol is eluted so late that it does not significantly affect the base line. Unfortunately, cholestadiene and cholesterol methyl ether can also be generated when most acidic reagents are used for transesterification purposes[291, 520, 686] and similar by-products are formed from plant sterols[406], but this does not occur with base-catalysed transmethylation. Similarly, when plant lipid extracts are transesterified directly, spurious peaks can be found on chromatographic traces as a result of reaction with phytol and other non-saponifiable materials.[745]

Other potential contaminants can appear in any of the reagents used in lipid analysis as discussed in Chapter 3. Vigilance must be exercised continuously to detect and eliminate them.

Methyl esters are purified prior to analysis by preparative TLC on thin layers of silica gel G; hexane–diethyl ether (9:1, v/v) is a suitable solvent for development. The esters are recovered from the adsorbent by elution with diethyl ether. Alternatively, a short column of silicic acid in a disposable pasteur pipette can be used, and the required esters are eluted with hexane–diethyl ether (95:5, v/v); in this instance, it may be less easy to eliminate BHT, cholestadiene or other hydrocarbons.

C. ISOLATION OF INDIVIDUAL FATTY ACIDS FOR STRUCTURAL ANALYSIS

It is often necessary to isolate fatty acids in a pure state before they can be identified unequivocally. Adsorption chromatography is invaluable for the isolation of fatty acids with polar functional groups, but silver nitrate chromatography, partition chromatography or combinations of both may be required to separate individual unsaturated fatty acids. The use of urea to obtain concentrates of polyunsaturated or branched-chain acids is discussed in section G.2 below.

1. Adsorption chromatography

In general, fatty acids that differ only in chain length or degree of unsaturation cannot be separated by adsorption chromatography although short-chain and polyunsaturated fatty acids migrate more slowly on silica gel and other adsorbents than C_{16} to C_{20} saturated or monoenoic components, and fractions enriched in these compounds can sometimes be obtained. If polar functional groups occur in the alkyl chain, however, some useful separations are possible. It is again more usual to separate fatty acids as the methyl ester derivatives, but unesterified fatty acids can also be chromatographed if one per cent of acetic or formic acids are incorporated into the solvent systems when silicic acid is the adsorbent. Sufficient material for many structural analyses can be obtained by preparative TLC and the elution characteristics of polar fatty acid esters on thin layers of silica gel have been thoroughly examined.

In particular, Morris et al.[515] have studied the chromatographic behaviour of the complete series of methyl hydroxy-, acetoxy- and keto-stearates. When each series is arranged in order as a line of spots on a single TLC plate, which is then developed in various hexane–diethyl ether solvent mixtures, the compounds emerge as a sinusoidal curve with a minimum at the 5-substituted isomer and a maximum at the 13- or 14- substituted isomer. Isomers with functional groups in positions 2 to 7 can often be separated from each other and from the remaining compounds. This sinusoidal effect is exhibited by other isomeric series of polar aliphatic compounds. Methyl 6-hydroxy-,

14-hydroxy- and 2-hydroxy-stearates have been separated by TLC on silica gel layers, with hexane–diethyl ether (85:15, v/v) as solvent system.[802] Figure 5.2 illustrates the elution characteristics of some hydroxy esters with this solvent system. Positional isomers of dihydroxy esters, i.e. methyl 6,7-dihydroxy- and 9,10-dihydroxy-stearate have been separated by TLC, with diethyl ether–hexane (4:1, v/v) as developing solvent.[513] Up to 20 mg of esters can be separated on a 20×20 cm plate coated with a layer of silica gel 0.5 mm thick. Bands are detected by means of the 2′, 7′-dichlorofluorescein spray and recovered by eluting the adsorbent in a column with diethyl ether or chloroform–methanol mixtures (about 9:1, v/v).

Column chromatography on silicic acid has also been used to isolate polar fatty acid esters (Radin[603] has reviewed some applications), and methyl 9-hydroxy- and 13-hydroxy-stearate[182] and the acetate derivatives of methyl 9,10-dihydroxy-myristate and 9,10-dihydroxy-palmitate[134] have been resolved by this technique. A small-scale column procedure, with Florisil as adsorbent, has been described for separating normal from 2-hydroxy fatty acid esters i.e. components of sphingolipids.[80]

Threo- and *erythro-* isomers of vicinal dihydroxy-esters can be separated on thin layers of silica gel impregnated with boric acid[507] (see chapter 3 also), with hexane–diethyl ether (60:40, v/v) as developing solvent (the *threo-*isomer migrates more rapidly),

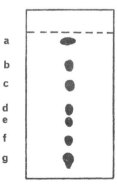

FIG. 5.2. Thin-layer chromatography of methyl esters of oxygenated fatty acids on silica gel G. Solvent system, hexane–diethyl ether (85:15, v/v). a, methyl palmitate; b, methyl 9,10-epoxystearate; c, methyl 12-keto stearate; d, methyl 2-hydroxy stearate; e, methyl 14-hydroxy stearate; f, methyl 6-hydroxy stearate; g, methyl 9,10-dihydroxy stearate.

although the compounds cannot be separated on silica gel alone. Sodium arsenite-impregnated layers give even more remarkable separations of diastereo-isomers.

Where the fatty acids contain double bonds in addition to more polar groups, silver nitrate chromatography may be of additional assistance in achieving separations (see below).

2. Silver nitrate chromatography

Since its introduction by Morris in 1962, TLC on silica gel impregnated with silver nitrate (see Chapter 3 for practical details) has been of enormous value to the lipid analyst. Fatty acids can be separated according to both the number and the configuration of their double bonds and often with care, according to the position of the double bonds in the alkyl chain. It is reportedly possible to separate fatty acid methyl esters with zero to six double bonds on a single plate with a double development in hexane-diethyl ether-acetic acid (94:4:2) provided that the atmospheric relative humidity is below 50%.[187] It is more usual, however, to attempt to separate methyl esters of fatty acids with zero to two double bonds on one plate, and those with three to six double bonds on

another as illustrated in Fig. 5.3, although complete separation of components with four or more double bonds is never easy. Hexane–diethyl ether (90:10, v/v) will separate components with up to two double bonds, and the same solvents in the ratio 40:60 will separate polyunsaturated esters. Bands are visualised under UV light after spraying with 2',7'-dichlorofluorescein solution, and components with zero to two double bonds are eluted from the adsorbent with diethyl ether or chloroform; chloroform–methanol (9:1, v/v) may be necessary for complete recovery of polyunsaturated compounds. Methods of eliminating unwanted silver ions from the extracts are discussed in Chapter 3. Up to 10 mg of esters can be separated on a 20 × 20 cm plate coated with a layer 0.5 mm thick of silica gel containing 10 per cent (w/w) silver nitrate. The GLC recorder traces (EGSS-X column) of fractions obtained in this way from the GLC reference mixture (Fig. 5.1) are also illustrated in Fig. 5.3. It is easy to see that considerable simplification of the sample has been achieved, and that separations are consistent with the tentative identifications made for each component.

The elution characteristics of a wide variety of unsaturated esters have been studied. For example, the complete series of methyl *cis*- and *trans*-

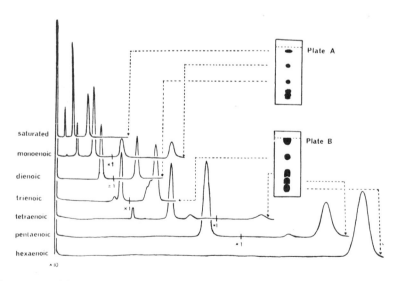

Fig. 5.3. GLC recorder traces (the EGSS-X column of Table 5.1 and Fig. 5.1) of esters separated into saturated, mono-, di-, tri-, penta-and hexaenoic fractions by TLC on silica gel G impregnated with 10 per cent (w/w) silver nitrate. A, solvent system, hexane–diethyl ether (90:10, v/v); B, solvent system, hexane–diethyl ether (40:60, v/v).

octadecenoates[248] and methylene-interrupted *cis,cis*-octadecadieonates[124] have been subjected to silver nitrate TLC. When run in order on a single TLC plate, each series migrates in the form of a sinusoidal curve similar to that observed with polar esters on silica gel alone, with a minimum at the Δ^6 isomer ($\Delta^{6,9}$ in the case of the dienes) and a maximum at the Δ^{13} isomer. *Trans*-isomers migrate consistently ahead of the corresponding *cis*-isomers, with the exception of the *cis*-2 component which not only migrates ahead of its *trans*-analogue but also ahead of methyl stearate. Natural monoenoic fatty acids containing *trans*-double bonds can be estimated by separating them from the *cis*-compounds by means of silver nitrate TLC with hexane–diethyl ether (9:1 v/v) as developing solvent and eluting them, together with the band containing the saturated components, from the adsorbent. If the samples are analysed by GLC before and after the separation, the amount of the *trans*- acids in the mixture can be determined.[140] Alternatively, the *cis*- and *trans*-components can be eluted separately and quantified by GLC with an internal standard.

With care, it is possible to separate positional isomers of unsaturated fatty acids by silver nitrate TLC. The most consistent and successful separations of this kind have been achieved by Morris *et al.*,[514] who utilise silica gel impregnated with very high concentrations of silver nitrate (up to 30 per cent) and develop the plates several times, in the same direction if necessary, with toluene as developing solvent at low temperatures (–5° to –25°C). With this system, the methyl 6-, 9- and 11- *cis*-octadecenoates can be well separated. Such separations are particularly useful in that the first two of these fatty acids occur together in many seed oils of the family Umbelliferae. Layers containing such high proportions of silver nitrate are very friable, but are not required for more routine separations.

An acetylenic group is less polar than a *cis*-double bond on silver nitrate TLC so methyl stearolate migrates just ahead of methyl oleate[511] and methyl crepenynate ahead of methyl linoleate.[250] The allenic ester, methyl labellenate, also migrates ahead of methyl oleate.[251] Conjugated double bonds are less strongly retained than similar compounds with isolated double bonds; for example, methyl 9-*cis*, 11-*trans*-octadecadienoate migrates with methyl oleate when hexane–diethyl ether (9:1 v/v) is the developing solvent, and just ahead of it when toluene is the developing solvent.[130] It is worth noting that compounds which are not resolved with hexane–diethyl ether solvent systems are frequently separable with aromatic solvents such as toluene and vice versa. Cyclopropene fatty acids are destroyed by silver nitrate chromatography and cannot be purified by this means, although the reaction has been used in a method of estimating them.[360] Cyclopropane and saturated branched-chain esters co-chromatograph with normal saturated straight-chain compounds.

Larger amounts of esters can be separated by column chromatography on silver nitrate-impregnated adsorbents, of which the most useful is acid-washed Florisil containing 16 per cent (w/w) silver nitrate (see Chapter 3 for further practical details). Up to a gram of esters can be separated on 40 g of adsorbent; saturated esters elute with hexane–diethyl ether (199:1, v/v), *cis*-monoenes with the same solvents in the ratio 99:1 and dienes with the solvents in the ratio 94:6.[830] Silver nitrate bound to ion-exchange resin has been used in columns for the semi-preparative scale separation of *cis*- and *trans*-isomers of unsaturated fatty acids, with monitoring of the eluate by differential refractive index detection.[204, 678] Similar analytical separations have been achieved by HPLC on columns of silicic acid impregnated with silver nitrate.[293] p-Bromophenacylesters (for UV detection) of oleic and elaidic acids were separated particularly effectively in a novel application of HPLC in the reversed-phase mode but with silver nitrate in the moving phase.[116] Such methods have some potential for routine measurements of the *trans*-contents of natural fats. (Conacher[148] has reviewed other chromatographic procedures.)

Before silver nitrate chromatography was developed, some similar separations of unsaturated compounds were achieved by thin-layer chromatography of mercuric acetate derivatives. The procedure is rather tedious as the derivatives must be prepared prior to the analysis (see Chapter 4), then decomposed when the fractionation is over before components can be analysed further by other procedures. Also, the resolutions that can be obtained are not as good as those accomplished with silver nitrate chromatography (the topic has been reviewed by Mangold[483]).

3. Preparative gas chromatography

Gas chromatography was developed initially as an analytical tool, but many commercial instruments can be adapted to permit analytical columns to be used preparatively. Specially designed equipment is available that can be used to purify gram quantities of methyl esters, but sufficient material can usually be obtained from the simpler analytical instruments for fatty acid structures to be determined by microchemical methods. Such procedures have been reviewed elsewhere.[300, 301]

If an analytical column is to be used preparatively, the gas chromatograph must have certain essential features. These include an efficient flash heater to ensure that the larger amounts of material to be separated are instantly volatilised and enter the column as a narrow band, a stream splitter at the exit end of the column (if the instrument has a flame ionisation detector) so that only 1 per cent or so of the column effluent passes into the detection system (although a continuous recorder trace of the components is obtained as they elute), and finally, the outlet from the stream splitter to a collection system should be heated so that components do not condense prematurely and contaminate the outlet. Conventional 4 mm i.d. analytical columns may be used to separate 1–2 mg quantities of esters preparatively, but an appreciable loss of resolution will occur. Better results are obtained if very high amounts of liquid phase (30 per cent or so) are used on a slightly coarser support (60–80 mesh) than is usual in analytical columns. If the oven dimensions are suitable, 6 mm i.d. columns, which have a much greater capacity than those of 4 mm, may be used to purify larger amounts of esters.

The precise amounts that can be separated conveniently on a preparative scale depend partly on the nature of the column and partly on the nature of the sample. If the latter is a complex natural mixture containing a number of saturated and unsaturated components, it will be necessary to use a polyester liquid phase and inject only 2–3 mg at a time to ensure adequate resolution. The process may have to be repeated several times, with pooling of corresponding fractions, to obtain sufficient material for structural analyses. Several difficulties may become apparent. With larger amounts of polyester liquid phase in wider columns, it is generally necessary to

heat the columns to a temperature close to the natural thermal limit of the polyester in order to elute components in a reasonable time and, as a result, fractions may be contaminated by liquid phase bleeding from the columns. Methyl esters can be purified by preparative TLC after separation if necessary (see Section B.3 above), but this introduces an extra step and may compound any other losses. Another approach is to simplify the mixture by silver nitrate or liquid–liquid partition chromatography, and to separately subject the fractions obtained to preparative gas chromatography. By this means, the resolution problem is greatly lessened, low-bleed temperature-stable silicone liquid phases can be used in the column packings and much larger amounts of material (5 mg or so of each ester on 6 mm i.d. columns) can be injected. Indeed, as a precise quantitative record may not be required, it is often possible to separate 20 mg or more of each component for, although highly unsymmetrical peaks will be obtained on the recorder tracing, the resolving powers of the columns will probably be sufficient to cleanly separate compounds differing in chain-length by two carbon atoms. Polyester liquid phases must be used if separation of positional isomers of polyunsaturated esters is required.

The collection of methyl esters, issuing from the columns, also presents certain problems as on rapid cooling they tend to form aerosols which are not easily condensed so losses can be high. Some manufacturers of GLC equipment supply specially designed traps which lead the effluent through a tortuous path so that the vapour has a greater opportunity to come in contact with the sides of the trap and condense, but very long straight tubes are often as effective as these. Other efficient but more complex devices have been described that utilise alternative heating and cooling[664], electrostatic precipitation[77] or millipore filters.[264] Good reproducible recoveries can be obtained if the carrier gas flow is interrupted by passing it through a short distance of column packing material on which the sample will condense, and from which it can later be washed with a suitable solvent;[381] glass beads moistened with toluene can be used in a similar manner.[131] If a small amount of hexane is injected into the column immediately before a peak is collected, any material that has condensed prematurely from earlier fractions will be removed and will not cause any contamination of the next peak.[607] It is frequently preferable, particularly with isotopically-labelled esters, that recoveries of individual components of a mixture be reproducible rather than complete.

There remains a slight possibility that methyl esters of polyunsaturated esters may be altered during chromatography at high temperatures and so will not be suitable for further structural analyses. The evidence is somewhat contradictory in that there are reports that no changes occur and others, equally authoritative, that esters such as 22:5 and 22:6 undergo isomerisation similar to that which occurs during distillation of these compounds. It is possible that the design of the gas chromatographs used for the purpose, particularly at the outlet end, may have some significance with regard to the problem but, whatever the reason, esters with fewer than three methylene-interrupted double bonds are unlikely to be affected.

4. Liquid–liquid partition chromatography (including HPLC and TLC)

Before gas chromatography was developed, liquid–liquid partition chromatography was the most useful technique for separating individual (or critical pairs of) fatty acids from natural mixtures. In the earlier stages of development the technique was rather cumbersome, and it was necessary to collect innumerable fractions and monitor these by tedious manual methods. More recently, the instrumentation developed for HPLC has been applied to effect excellent separations of fatty acid derivatives on an analytical or semimicro-preparative scale (reviewed by Aitzetmuller[19]).

Of the earlier reverse-phase column chromatographic procedures, that developed by Privett and Nickell,[598] in which heptane is supported as the stationary phase on silanised celite and acetonitrile–methanol (85:15, v/v) is the mobile phase, is the most stable and reproducible and has been most widely used, and so can be recommended for medium-scale preparative purposes. Both phases are volatile and can be used with numerous liquid detection systems, while the column is readily regenerated and can be used repeatedly. The column is packed with 100-mesh celite made hydrophobic by treatment with dimethyldichlorosilane, then heptane saturated with acetonitrile–methanol (85:15, v/v) is passed through the column; when this reaches the bottom, the acetonitrile–methanol phase saturated with heptane is allowed to flow through the column. When the latter solvent emerges at the bottom, the sample is placed on top of the column and allowed to percolate in a few cm before the solvent flow is restarted. Methyl laurate emerges in 8 column values, methyl myristate in 11 and methyl palmitate in 16. Up to 2 g of methyl esters can be applied to a 120 × 2.5 cm column. As they elute, compounds can be detected by removing the solvents and weighing them or by oxidising them with chromic acid (see Chapter 3), or continuously by means of a differential refractometer or by a moving wire detection system. It seems not unlikely that this procedure will be supplanted by others based on the use of polymers, such as alkylated sephadex,[60] as the stationary phase.

A number of applications of HPLC to the analysis of fatty acid esters have now been described, generally on columns containing C_{18}-bonded stationary phases. For example, methyl 9-hydroperoxy-10,12-octadecadienoates and methyl 13-hydroperoxy-9,11-octadecadienoates were separated by this technique, according to the position of the hydroperoxy group and the configuration of the double bonds;[115] an ethanol–water mixture was the mobile phase, while the high extinction coefficient of the conjugated double bond system was made use of in monitoring the column eluate with a UV-detector. HPLC would appear to be ideally suited to this analysis because such compounds decompose at temperatures appropriate for GLC.

Methyl ester derivatives of the more common range of fatty acids are separable by reverse-phase HPLC on C_{18}-bonded phases, with monitoring of the column effluent by means of a differential refractive index detector,[676, 677] but it is necessary to use a solvent mixture of constant composition as the mobile phase and this limits the resolution. Certain critical pairs, e.g. methyl palmitate and methyl oleate, which are not readily separated by conventional reverse-phase chromatography have been resolved by this technique. Such systems appear to have little practical value for analytical purposes, but they can be used as an alternative to preparative gas chromatography (at high temperatures, double bond migration may be a problem) for the isolation of particular fatty acids for further analysis by other chromatographic procedures, or for characterization or structure determination by chemical means.

Alternative procedures for reverse-phase HPLC of fatty acids have been developed, in which fatty acid esters containing strongly UV-absorbing substituents in the alcohol moiety are prepared, so that components emerging from the columns can be measured by means of a UV-detector. For example, Borch[76] has obtained some remarkable separations of fatty acids, in the form of the phenacyl esters, by means of HPLC with a C_{18}-bonded stationary phase and acetonitrile–water as moving phase, in conjunction with UV-detection (see Fig. 5.4). Solvent programming is used to improve the resolution and shorten the chromatography time. Critical pairs are resolved in this system and a wide range of components are separable, including polyunsaturated fatty acids and isomers in which the position or configuration of the double bond varies.

Indeed, oleic and petroselinic acid are separated in this system, a feat that is not readily achieved even by open-tubular gas chromatography or high-resolution silver nitrate thin-layer chromatography. Others have reported comparable separations of fatty acids in the form of derivatives such as the 2-naphthacyl[151], p-bromophenacyl[115, 574] (separation of C_{30}–C_{56} fatty acids have been reported, for example[602],) or methoxyphenacyl[102, 495] esters, generally with methanol–water mixtures as the mobile phase. One advantage of the method is that any impurities not converted to UV-absorbing derivatives are not detected, so cannot obscure other separations. On the other hand, it is likely again that the principal use of such methods will be for small-scale preparative separations of single components for structural analysis or for liquid scintillation counting. Gas chromatography is likely to remain the method of choice for analytical separations of most natural fatty acids in the foreseeable future, with the possible exception of certain specific separations of configurational or positional isomers of unsaturated fatty acids or of fatty acids containing thermally-unstable functional groups.

Thin-layer liquid–liquid partition systems, operating on the reverse-phase principle, are useful for analytical or small-scale preparative purposes, and free fatty acids or their methyl esters are readily separated into critical pairs. Silicone oil (Dow Corning 200 fluid) has been the most popular stationary phase, though undecane and tetradecane are also used, with acetonitrile–acetic acid–water mixtures as the mobile phase (for example methyl esters can be separated with these solvents in the proportions, 70:10:25 by vol.).[482] Further details of alternative solvent systems are available elsewhere.[481, 483] Normally, samples are applied as discrete spots or as streaks in a volatile solvent, up to 10 mg of esters preparatively to a 20 × 20 cm plate, and the separation is complete in 2–3 hr. The principal disadvantage of the technique is that it is somewhat messy and that components are difficult to detect. The procedure of Ord and Bamford[561] overcomes the first of these; an ordinary silica gel G TLC plate is silanised by standing it in a tank of dimethyldichlorosilane vapour for several hours. (Plates prepared in this way are now available commercially). The methyl groups, which are then chemically bonded to the silica gel, serve as the non-

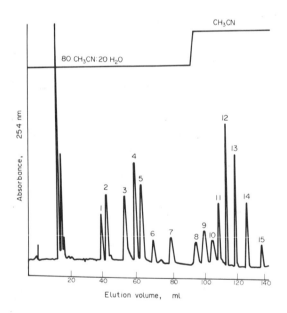

FIG. 5.4. HPLC of phenacyl esters of fatty acids. Column 90 cm × 0.64 cm μ-Bondapack C-18; eluent, acetonitrile–water; flow rate 2.0 ml/min. Peak 1, lauric (12:0); 2, myristoleic, (14:1); 3, linolenic (18:3); 4, myristic (14:0); 5, arachidonic (20:4); 6, linoleic (18:2); 7, eicosatrienoic (20:3); 8, palmitic (16:0); 9, oleic (18:1(n-9)); 10, petroselinic, (18:1(n-12)); 11, eicosadienoic (20:2); 12, stearic (18:0); 13, arachidic (20:0); 14, behenic (22:0); 15, lignoceric, (24:0). (Reproduced by kind permission of R. F. Borch and *Analytical Chemistry*.[76]

polar stationary phase on which the fatty acids or esters are separated with acetonitrils–methanol–water (6:3:1 by vol.) as mobile phase. Figure 5.5 illustrates the kind of separation possible. For analytical purposes, 20 μg spots are applied to the plate and detected with a phosphomolybdic acid spray. Preparatively, 1–2 mg of esters can be separated on a 20 × 20 cm plate, but the edges of the plate must be exposed to iodine to detect the bands and, as they may not always be straight and parallel, some loss may occur.

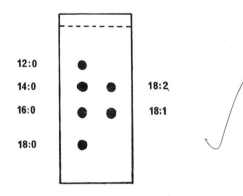

FIG. 5.5. Reverse-phase separation of methyl esters on thin layers of silanized silica gel G with acetonitrile–methanol–water (6:3:1, by vol.) as mobile phase.[561]

Liquid–liquid chromatography (other than HLPC) does not in general give pure single components but critical pairs from natural mixtures (see Chapter 3). As such components differ both in degree of unsaturation and chain-length, the fractions can be subjected to silver nitrate chromatography or to preparative gas chromatography so that pure compounds are ultimately obtained.

D. SPECTROSCOPY OF FATTY ACIDS

1. Infrared absorption and Raman spectroscopy

Fatty acids may be subjected to IR spectroscopy in the unesterified state if bound to glycerol or as the methyl ester derivatives, although an esterified form is to be preferred as a band due to the free carboxyl group between 10 and 11 μm may obscure a number of other important features in the spectra. Most information on the chemical nature of fatty acid derivatives can be obtained when they are in solution (see Chapter 3), and Fig. 5.6 illustrates the IR spectrum of soyabean oil in carbon tetrachloride solution. The sharp band at 5.75 μm is due to the esterified carbonyl function, which is also responsible for a band at 8.6 μm. With free fatty acids, the first of these bands is displaced to 5.9 μm and there are also broad bands at 3.5 μm and 10.7 μm. cis-Double bonds give rise to small bands at 3.3 μm and 6.1 μm but, although they are useful as diagnostic aids, they are not sufficiently distinct to be of use in quantitative estimations. Most of the remaining bands are absorption frequencies of the hydrocarbon chain.

Many other functional groups give rise to characteristic bands that can be used to identify or to estimate the amount of a given acid. The most important of these is a sharp peak given by trans-double bonds at 10.3 μm, and IR spectroscopy has long been the principal method for estimating compounds with this structural feature in the fatty acid chain. In the recommended procedure,[765] a standard is run at the same time as the unknown, and the amount of trans-isomer in the latter is calculated from the ratio of the absorptivities at 10.3 μm. In an alternative rapid procedure, the ratio of the absorbances at 8.6 μm (due to the carbonyl function) and at 10.3 μm in the unknown is measured, and the amount of trans-isomer present is obtained by reference to a standard curve.[24] For accuracy with this last procedure, however, the standard curve must be prepared with material very similar to that to be analysed.[333] If a trans-double bond is part of a conjugated system, the band maximum may be shifted to about 10.1 μm.

Free hydroxyl groups give rise to a sharp band at 10.9 μm, epoxy groups produce a double band with maxima at 11.8 and 12.1 μm, allenes give a small band at 5.1 μm, a cyclopropene ring produces a small band at 9.9 μm and a cyclopropane ring produces two small bands at 3.25 and 9.8 μm. Some of the absorption frequencies caused by the more unusual functional groups have been discussed by Wolff and Miwa;[836] more detailed general information can be found in a number of reviews.[117, 118, 216] Characteristic frequencies are not altered markedly by the position of a group in the aliphatic chain, unless this is at

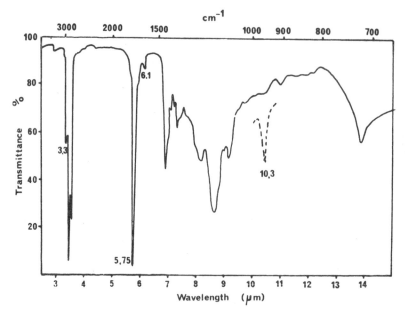

FIG. 5.6. Infrared spectrum of soyabean oil in carbon tetrachloride solution. The insert at 10.3 μm illustrates the absorption band due to a *trans*-double bond.

either extremity of the molecule or immediately adjacent to another functional group.

IR spectroscopy is therefore a valuable method for detecting the presence and, on occasion, for estimating the amount of certain functional groups in fatty acids, especially when they are in esterified form in a natural lipid mixture, but other spectroscopic or chemical procedures must be used to fix the exact position of the group in the alkyl chain.

Structural features in unsaturated fatty acid methyl esters (in carbon tetrachloride solution) can give rise to distinctive bands in Raman spectroscopy.[162-164] For example, characteristic Raman bands are found for *cis*-double bonds (1656 cm^{-1}), *trans*-double bonds (1670 cm^{-1}) and triple bonds (2232 and 2291 cm^{-1}); a terminal triple bond gives a single band at 2120 cm^{-1} (this group does not give a characteristic band in IR spectroscopy). The position of a double bond does not affect the spectrum unless it is at either extremity of the molecule.

2. Ultraviolet spectroscopy

UV spectroscopy is nowadays used principally to detect or to confirm the presence of fatty acids containing conjugated double bond systems in natural oils, or to detect chemical or enzymatic isomerisation of fatty acid double bonds in which conjugated systems are formed. With such acids, series of broad bands of increasing intensity the greater the number of double bonds in the conjugated system, are found in the UV region at successively higher wavelengths. For example, with conjugated dienes, λ_{max} is 232 nm ($\epsilon = 33,000$), with conjugated trienes (e.g. α-eleostearic acid), λ_{max} is at 270 nm ($\epsilon = 49,000$) and with conjugated trienes (e.g. α-parinaric acid), λ_{max} is at 302 nm ($\epsilon = 77,000$). Different geometrical isomers have slightly different spectra; the greater the number of *trans*-double bonds, the higher the extinction coefficient but the shorter the wavelengths of the band maxima. Isolated *cis*-double bonds exhibit a specific absorbance at 206 nm, but with a relatively low extinction coefficient so this feature is of little use as a diagnostic aid. Conjugated triple bonds also affect the spectra. The absorbance at 234 nm is used to detect the formation of the *cis,trans*-conjugated double bond system produced by the action of lipoxygenase, a property that is utilised in the estimation of polyunsaturated fatty acids with *cis,cis*-methylene-interrupted double bonds in lipids.[691] More detailed reviews of the applications of

UV spectroscopy in lipid analysis have been published.[117, 579]

3. Nuclear magnetic resonance spectroscopy

In recent years, NMR spectroscopy (especially PMR but also, to some extent, ^{13}C NMR) has been increasingly applied to the identification of lipid structures, and in particular to the detection, and often the location, of the double bond systems in fatty acid chains. Methyl esters are generally prepared for analysis, and Fig. 5.7 illustrates the NMR spectrum of methyl linoleate. There are six main features; a multiplet at 4.72τ for the olefinic protons, a sharp singlet at 6.40τ for the protons on the methoxyl group, a triplet at 7.25τ for the methylene group between the double bonds, a complex multiplet at 7.88τ for the protons on the carbon atoms immediately adjacent to the double bonds and carboxyl group, a broad band at 8.67τ for the chain methylene protons, and finally, a partially resolved triplet at 9.10τ for the terminal methyl protons.[795]

A wide range of unsaturated fatty acids have been subjected to NMR spectroscopy on 60 to 100 MHz instruments. They include the complete series of methyl *cis*-octadecenoates (Δ^2-Δ^{17}),[247] all the methylene-interrupted *cis,cis*-octadecenoates,[135] several *cis,cis*-octadecadienoates with more than one methylene group between the double bonds[252] and very many natural polyunsaturated fatty acids.[245, 322]

From studies with such compounds, it is now known how variations in the positions of double bonds affect the NMR spectrum of a long-chain fatty acid. For example, the Δ^2 to Δ^5 and Δ^{14} to Δ^{17} monoenoic isomers can be distinguished by this technique, largely because of small changes in the signal associated with the olefinic protons, but *cis*-isomers cannot be distinguished from the corresponding *trans*-isomers.[247] The terminal methyl group of polyunsaturated fatty acids of the (*n*-3) series produces a well-separated triplet at a slightly lower field (9.02τ) than usual, and this feature has been utilised in the estimation of such compounds in natural mixtures.[234] The intensity of the band due to the methylene group between the double bonds can be used to determine the degree of unsaturation of a given fatty acid or a natural oil. NMR spectra of some natural conjugated trienoic acids (α- and ß-eleostearic acid) have been determined; the olefinic protons give rise to a complex multiplet at approximately 4.0τ.[322]

Even more informative spectra can be obtained with more powerful instruments (220 MHz) and data is available for a wide range of unsaturated fatty acids, including *cis*- and *trans*-monoenes, polyenes and acetylenic compounds.[219, 220] Only the Δ^{10} and Δ^{11} isomers of the methyl octadecenoates cannot be distinguished, for example, while all the corresponding acetylenic compounds can be identified. Some evidence as to the configuration of the double bond can be adduced, especially in the presence of lanthanide shift reagents,[221] which also permit the location of centrally-placed double bonds.

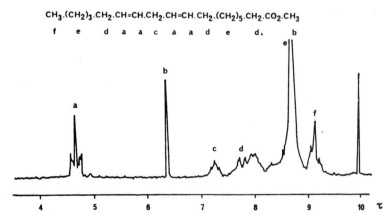

$$CH_3.(CH_2)_3.CH_2.CH=CH.CH_2.CH=CH.CH_2.(CH_2)_5.CH_2.CO_2.CH_3$$

FIG. 5.7. NMR spectrum of methyl linoleate in carbon tetrachloride solution (15% w/v) at 60 MHz (tetramethylsilane as internal standard). Features in the spectrum and corresponding protons in the fatty acid are labelled a–f.

Free hydroxl groups in fatty acid chains give rise to two separate signals; that due to the –OH proton is indistinct, and its intensity and position may vary because of hydrogen bonding effects, but that due to the -CHO- proton at 6.4τ is quite characteristic. All the isomeric hydroxy stearates have been examined by NMR spectroscopy, and all can be distinguished from each other by this technique when quinoline is used as the solvent.[778] Much valuable information on the structures of hydroxy acids can be obtained if lanthanide shift reagents are added.[831, 832] When the hydroxy esters are acetylated, the acetoxy protons give rise to a sharp signal at 7.9τ. Keto groups influence the α-methylene protons, which produce a signal similar to that for protons adjacent to a carboxyl group. Many other functional groups give rise to distinctive signals; epoxide ring protons at 7.2τ[324] cyclopropene ring protons at 9.2τ,[324] cyclopropane ring protons at 9.4 and 10.3τ[452] and olefinic protons in a cyclopentene ring at 4.3τ.[323]

Methyl branches on aliphatic chains do not give signals that are helpful in locating their positions, unless the branch is immediately adjacent to either end of the molecule; *iso*-compounds can, therefore, be recognised, but *anteiso*-compounds cannot be distinguished from fatty acids having the methyl group in a central position.[324]

Methyl esters are generally dissolved in carbon tetrachloride for analysis by NMR spectroscopy, but chloroform, although it produces a single peak at 2.75τ, or deuterated chloroform may also be used (but not with lanthanide shift reagents). The application of PMR spectroscopy to fatty acid analysis has been reviewed comprehensively elsewhere.[245, 322, 323, 400] As more powerful instruments become available, the technique will undoubtedly become even more valuable but it is used to greatest advantage on individual pure compounds, isolated by the procedures described above, rather than on natural mixtures.

There has recently been great interest in ^{13}C NMR spectroscopy of fatty acids as features in the spectra can be assigned to vitually every carbon atom.[57, 100, 101, 257, 258, 608, 781] *Cis*- and *trans*-isomers of unsaturated fatty acids are readily distinguished, for example. Unfortunately, the technique is relatively insensitive, requiring approximately 0.5 g of fatty acid for a usable spectrum, so it is not likely to gain acceptance for analytical purposes until further

technical advances are made. The applications of the technique to lipids have been reviewed by Klein and Kemp.[400]

4. Mass spectrometry

A wide variety of natural and synthetic fatty acids have been examined by mass spectrometry, which has proved invaluable in the study of such compounds. A high proportion of the available data has been obtained for the methyl ester derivatives of fatty acids, as these are easily prepared and are widely used in chromatographic analysis. On the other hand, it has become apparent that amide derivatives of fatty acids, especially pyrrolidides, give more distinctive spectra and more frequently permit definitive assignment of structure, while they are also suitable for chromatography. Mass spectrometry of lipids in general has been the subject of several reviews (see Chapter 3). Andersson has reviewed mass spectrometry of pyrrolidide derivatives.[30]

Long-chain saturated methyl esters are easily identified,[644] and are characterized by a prominent molecular ion (M_+) and other significant ions at $m/e = $ M-31 (loss of methanol) and M-43 (loss of C_2, C_3 and C_4 as a result of a complex rearrangement), together with a series of ions of general formula $-CH_3O_2C.(CH_2)_n^+$. The base peak at $m/e = 74$ is known as the 'McLafferty rearrangement ion', and is formed in a rearrangement reaction after cleavage of the parent molecule beta to the carboxyl group.[474] Pyrrolidide derivatives of saturated fatty acids also have prominent molecular ions and a base peak at $m/e = 113$, equivalent to the McLafferty rearrangement ion.[33]

Mass spectra of methyl esters of unsaturated fatty acids differ greatly from those of the corresponding saturated compounds, but are less useful for characterisation purposes. The molecular ion is often prominent, and there are generally major fragments at $m/e = $ M-31 or M-32 (loss of methanol) and, with monoenes and dienes, at $m/e = $ M-74 (loss of the McLafferty rearrangement ion).[267] There are no ions that can be used to locate the position of double bonds, which probably migrate when the molecule is ionised, although there are indications that some positional isomers of polyunsaturated fatty acids may have distinctive but not easily interpreted spectra.[317] Double bond positions can be determined if the unsaturated esters are first converted to a suitable derivative, preferably one which is sufficiently volatile to be subjected to gas chromatography. A wide variety of such derivatives have been tried; usually the double bond is oxidised to an epoxide or to a vicinal diol, which is then reacted further to form more stable or more volatile compounds. The most useful have proved to be vicinal diols in the form of the isopropylidene[467] or better trimethylsilyl ether[40, 103] derivatives (see Chapter 4 for methods of preparation). The former permits the configuration of the double bond to be established because of the different GLC retention characteristics of the stereoisomers; the latter gives more intense fragments that can be related to the position of the original double bond. Figure 5.8 illustrates the mass spectrum of the trimethylsilyl ether of the vicinal diol prepared from methyl oleate. Ions in the high mass range are of very low intensity and the molecular ion does not appear, but two ions at $m/e = 215$ and $m/e = 259$, which represent cleavage between the two carbon atoms that originally constituted the double bond, are particularly prominent. Indeed, these ions are sufficiently intense to be of use in estimating mixtures of positional isomers of unsaturated fatty acids. Trimethylsilyl ether derivatives have also been prepared from polyunsaturated acids, and appear to be well suited to the characterization of such compounds by means of mass spectrometry.[183, 355, 674] Other derivatives of double bonds suitable for the purpose have been prepared by various reactions with mercuric acetate adducts.[497, 498, 580]

Pyrrolidine, as opposed to methyl ester, derivatives of unsaturated fatty acids have distinctive mass spectra with prominent ions that can be used to fix the position of the double bonds. The mass spectrum of N-octadec-9-enoylpyrrolidine is illustrated in Fig. 5.9.[30, 32] In addition to the base peak at $m/e = 113$ and the molecular ion, there is a series of peaks spaced 14 a.m.u. apart in general, except in the vicinity of the double bond where the interval is 12 a.m.u., i.e. occurring between $m/e = 196$ and $m/e = 208$, and corresponding to fragmentation at carbons 8 and 9 of the carbon chain. For spectra of the Δ^5 to Δ^{15} isomers, the following rule was formulated. "If an interval of 12 a.m.u., instead of the regular 14 is observed between the most intense peaks of clusters of fragments containing n and $n-1$ carbon atoms in the acid moiety, a double bond occurred between carbons n and $n+1$ in the molecule". The remaining isomers also have distinctive spectra but they do not fit the rule. Methylene-interrupted polyunsaturated fatty acids, as the pyrrolidides, have characteristic mass spectra that can be interpreted in terms of the position of the double bonds with some modification to the above rule.[30, 31]

Triple bonds in fatty acid methyl esters can only be located by mass spectrometry after conversion to a suitable derivative, and oxymercuration reactions are generally favoured for the purpose (see Chapter 4).[98, 397] Pyrrolidides containing triple bonds can be characterized without further derivatisation.[787]

Most oxygenated fatty acids can be recognised and identified by mass spectrometry in the form of the methyl esters, as there are usually prominent ions formed by cleavage beta to the oxygen function,[647]

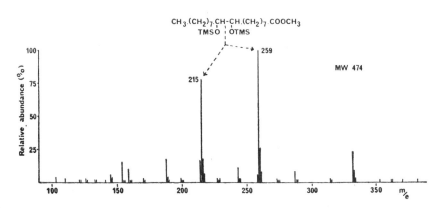

FIG. 5.8. Mass spectrum of the trimethylsilyl ether derivative of the vicinal diol prepared from methyl oleate.

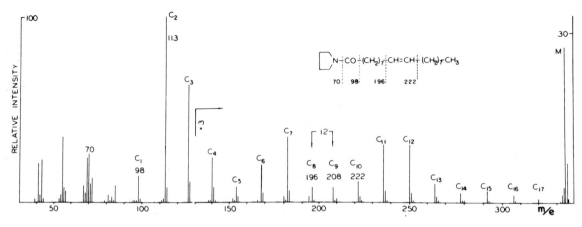

FIG. 5.9. Mass spectrum of N-octadec-9-enoylpyrrolidine.[30] (Reproduced by kind permission of B. A. Andersson and Pergamon Press.)

but characterisation of hydroxy esters is further simplified by conversion to the trimethylsilyl ethers.[398] Pyrrolidine derivatives of oxygenated fatty acids also have distinctive spectra.[30]

Methyl esters of branched-chain fatty acids have mass spectra superficially similar to those of the corresponding straight-chain compounds with the same number of carbon atoms, but changes in the intensities of some ions can be used to indicate the presence and fix the position of the branch.[38, 645, 646] Spectra, which are more readily interpreted in terms of the positions of the methyl branches, are obtained if the esters are reduced to the alcohols, by means of lithium aluminium hydride, and converted to the methyl ether derivatives,[379] or by using the pyrrolidine derivatives of the fatty acids.[33] Cyclopropane esters are not readily distinguished from mono-unsaturated esters with similar numbers of carbon atoms by mass spectrometry, apparently because on ionisation the cyclopropane ring opens to form such a monoenoic compound. Methyl esters can be used, if the ring is first opened by vigorous catalytic hydrogenation (forming methyl-branched compounds)[466] or by forming methoxy-derivatives by reaction with boron trifluoride–methanol.[496] Pyrrolidides of cyclopropane fatty acids give distinctive spectra without further derivatisation.[30] Cyclopropene esters must be converted to suitable derivatives,[196, 321, 607] but other larger-ring cyclic fatty acids, such as those with cyclopentene[142, 694] cyclohexane[675] or furanoid rings,[512] can be identified either in the form of the methyl esters or of the pyrrolidides.

E. IDENTIFICATION OF FATTY ACIDS BY CHEMICAL DEGRADATIVE PROCEDURES

1. Chain-length determination

One of the first steps in the determination of the structure of an unsaturated fatty acid is to establish its chain-length, and this can be ascertained simply by means of catalytic hydrogenation to form the saturated compound, which is then identified positively by gas chromatography. A procedure suitable for the purpose is described in Chapter 4. Ideally, the reaction should be carried out on the single fatty acid of interest but valid results can be obtained in some circumstances with natural mixtures or with fractions isolated from silver nitrate plates. A procedure suitable for oxygenated fatty acids is described in Section E.3. below.

2. Location of double bonds in fatty acid chains

The positions of double bonds in alkyl chains are generally determined by oxidative fission across the double bond, followed by gas chromatographic identification of the products. Lipid analysts have largely accepted two procedures as suitable for the purpose; oxidation with permanganate–periodate reagent (frequently termed von Rudiloff oxidation), and ozonolysis followed by oxidative or reductive cleavage of the ozonide. The former method yields mono- and dibasic acids as the products, while the latter can give either these or aldehydes and aldehyde esters. Both procedures have been reviewed in some detail by Privett.[594] Ozonolysis techniques have the advantages that over-oxidation and spurious by-product formation are negligible, and recovery of short-chain fragments is less of a problem, but it is necessary to have equipment to generate ozone (although this is probably a worthwhile investment if many samples must be analysed). Ozonolysis can also be used when other functional groups (e.g. hydroxy or epoxy) are present. The permanganate-periodate procedure, on the other hand, uses readily available inexpensive chemicals and over-oxidation is minimal. It is especially valuable when only an occasional sample must be analysed.

(I) Permanganate-Periodate Oxidation

In this procedure, the methyl ester of the unsaturated fatty acid in *tert*-butanol solution is oxidised by a solution containing a small amount of potassium permanganate together with a larger amount of sodium metaperiodate, which continuously regenerates the permanganate as it is reduced, while the whole is buffered by a solution of potassium carbonate. When the reaction is complete, the solution is acidified, excess oxidant is destroyed by addition of sodium bisulphite or preferably by passing sulphur dioxide into the solution, and the products are extracted thoroughly with diethyl ether before being methylated for GLC analysis. Very little overoxidation occurs with this reagent, but it is not easy to achieve quantitative isolation of short-chain mono- and dibasic acids or half-esters of these for GLC analysis. The problem has been partially resolved by injecting the free short-chain fatty acids directly on to a GLC column of Porapak Q,[270] although carbowax 20 M-terephthalic acid might be a better choice of liquid phase, or by pyrolysing the tetramethylammonium salts of the acids, thus converting them to methyl esters, in the heated injection port of a gas chromatograph.[184] Where the compound to be oxidised consists of more than one positional isomer, only the longer-chain fragments are obtained in reproducible yields, and wherever possible, these alone should be considered when determining the amounts of individual positional isomers.[290] The recommended procedure is as follows (only the highest purity reagents should be used).[808, 809]

"A stock oxidant solution is made up of sodium metaperiodate (20.86 g) and potassium permanganate (0.395 g) in water (1 litre). This solution (1 ml) together with potassium carbonate solution (1 ml; 2.5 g/l) is added to the monoenoic ester (1 mg) in *tert*-butanol (1 ml) in a test-tube, and the mixture is shaken thoroughly at room temperature for 1 hr. At the end of this time, the solution is acidified with one drop of concentrated sulphuric acid, and excess oxidant is destroyed by passing sulphur dioxide into the solution which is then extracted thoroughly with diethyl ether (3×4 ml). The organic layer is dried over sodium sulphate, before the solvent is removed on a rotary evaporator or in a stream of nitrogen at room temperature. The products are methylated for GLC analysis, preferably by reaction with diazomethane, freshly-prepared by the procedure of Schlenk and Gellerman[663] (see Chapter 4)."

The procedure can be scaled up for polyunsaturated fatty acids, but the proportion of water to *tert*-butanol should be kept as close to 2:1 (v/v) as possible. Malonic acid, formed by oxidation of methylene-interrupted double bonds, is oxidised further and is not detected. Ambiguity may result when polyunsaturated fatty acids with more than one methylene group between the double bonds are oxidised, as it is then not possible to state which dibasic fragment contained the original carboxyl group. However, this difficulty can be resolved by reducing the carboxyl group to an alcohol prior to the analysis.[254]

(II) Ozonolysis and Reductive or Oxidative Cleavage

Ozone attacks olefins rapidly and quantitatively to form ozonides with no over-oxidation and very few side reactions, if the reaction is carried out at low temperature in an inert solvent such as pentane. The ozonide can be cleaved reductively by catalytic reduction with Lindlar's catalyst and hydrogen, or chemically by a number of reagents of which the most convenient is probably dimethyl sulphide in methanol,[611] although triphenyl phosphine[731] has been more widely used. Reduction of the ozonides with sodium borohydride to yield products which are alcohols or alcohol esters, is reported to be less subject to interference in the gas chromatographic analysis.[363] The following procedure is generally satisfactory.

"A solution of ozone in pentane is prepared by bubbling oxygen containing ozone through purified pentane at –70°C until a blue colour is obtained. The unsaturated ester (1 mg) is dissolved in pentane (1 ml) and cooled to 0°C, when the ozone solution is added dropwise until the blue colour persists. After 1 min, the reagents are removed in a stream of nitrogen. Methanol (0.3 ml) then dimethyl sulphide (0.5 ml), both precooled to –70°C, are added and the mixture is maintained at this temperature for 20 min before it is allowed to warm up to room temperature. The excess reagents can be removed in a gentle stream of nitrogen, provided that the shortest-chain aldehydes are absent, and the products are dissolved in pentane or hexane for injection into the gas chromatograph for analysis."

With larger amounts of ester, it may be necessary to add more methanol to the reaction mixture to assist the reaction. There may be difficulties in recovering quantitatively, short-chain aldehydes or aldehyde esters for analysis. Dimethyl sulphoxide is the other product of the reaction, but does not interfere with the GLC analysis.

The products of the two types of reaction can be identified on a number of gas chromatographic columns; difunctional compounds are more rapidly eluted on silicone liquid phases, but are less likely to be confused with monofunctional compounds on polyester phases. With both types of liquid phase, temperature-programmed analysis is generally necessary if all the fragments must be determined. Dibasic standards are readily available for comparison, but aldehyde esters are not and it may be necessary to make up a suitable standard by ozonolysis of monoenoic esters of known structure that are commercially available, e.g. methyl petroselinate, methyl oleate and methyl vaccenate. A GLC trace of the ozonolysis products of such a mixture on an EGS column is illustrated in Fig. 5.10.

Oxidative fission of ozonides is used less frequently, but it has been shown that heating with boron trifluoride–methanol reagent brings about complete oxidation with conversion of the fragments to methyl ester derivatives and without secondary degradation. In this instance, standards are freely available for identification purposes. The procedure has been applied successfully to monoenoic fatty acids[9] and, with some modification, to polyunsaturated fatty acids with other than methylene-interrupted double bonds.[614]

(III) SPECIAL PROCEDURES FOR CONFIGURATIONAL ISOMERS OR FOR POLYUNSATURATED FATTY ACIDS

The procedures described above give optimum results with mono- and dienoic fatty acids. When the structures of polyunsaturated fatty acids must be

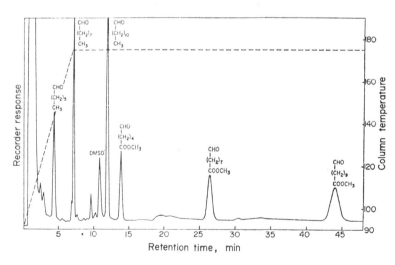

FIG. 5.10. GLC recorder trace of the ozonolysis products of a mixture of methyl petroselinate, methyl oleate and methyl vaccenate (EGS was the stationary phase)[611] (Reproduced by kind permission of D. G. Cornwell and the *Journal of Lipid Research*.)

determined, especially when they contain double bonds of both *cis-* and *trans-*configuration or are part of conjugated double bond systems, more positive identification can be obtained by partially reducing the compounds prior to oxidation.[599] Hydrazine, a reagent that does not cause any double bond migration or stereomutation, is used for the purpose, under conditions such that a high proportion of monoenoic compounds are formed; isomers are then found with double bonds in each of the positions in which they were present in the original polyunsaturated fatty acid. *cis-* and *trans-*Monoenes are separated by silver nitrate TLC (see above), and their structures separately determined by one of the above methods so that the original compound is fully identified. Hydrazine reduction, which is better performed on the free acid than on the ester, is carried out as follows.

"The free fatty acid is heated in air at 35°C with 100 volumes of 10 per cent hydrazine hydrate in methanol for a predetermined time (1.5–2 hr), found by trial and error with a standard polyunsaturated acid, such that there is an approximately 50 per cent yield of monoenes. Excess methanolic hydrogen chloride (6 per cent, w/w) is than added to stop the reaction, and the mixture is refluxed for 2 hr to convert the acids to the methyl esters which are recovered for further study as described earlier."

A number of variations on the procedure exist and have been reviewed by Privett.[594] For example, the positions of double bonds in monoenes containing both *cis-* and *trans-*isomers, can be determined by conversion to epoxides, of which configurational isomers are readily separable, and cleavage of these with periodic acid.[203] Partial oxymercuration of the double bonds followed by mass spectral identification of the products has been applied to the analysis of polyunsaturated fatty acids.[580]

3. Location of other functional groups in fatty acids

Spectroscopic aids to the recognition and location of functional groups in fatty acid chains are discussed above. Chemical methods for characterization are also of value.

Isolated triple bonds in fatty acids are not easy to recognise, as they do not exhibit particularly distinctive features when examined by any of the spectroscopic techniques other than Raman spectroscopy, but a specific TLC spray consisting of 4-(4'-nitrobenzyl)-pyridine (5 per cent) in acetone, which gives a violet colour with acetylenes, has been described.[680] The position of the triple bond can be located by mass spectrometry of the keto-compound prepared by acid hydrolysis of the mercuric acetate adduct (see Section D4), and triple bonds are readily cleaved by the permanganate–periodate reagent. Ozone reacts with triple bonds, although more slowly than with double bonds, and the products of reductive cleavage of the ozonide are mono- and dibasic acids rather than aldehydes and aldehyde esters, so that double and triple bonds in a single fatty acid can be differentiated by this technique.[592] In addition, double bonds are hydroxylated by peracids while triple bonds remain unchanged.[593]

Allenic groups have distinctive IR and NMR spectra, and all the natural fatty acids containing this functional group exhibit a marked optical activity. The position of the group in the fatty acid chain can be determined by partial reduction and oxidation of the monoene fragments as described above.[52]

A number of spectroscopic procedures are of value for the detection and location of oxygenated functional groups, such as keto, hydroxyl or epoxyl, in fatty acid chains (see Section D). In addition, several chemical techniques are available;[603] for example, the presence of hydroxy groups can be confirmed by GLC analysis before and after the preparation of volatile derivatives, as discussed above. The first step in identifying an acid of this type is to determine its chain-length. The acid is hydrogenated to eliminate any multiple bonds before free hydroxy or epoxy groups are converted to iodides by the action of iodine and red phosphorus. The iodine atom in the aliphatic chain is then removed by hydrogenolysis with zinc and hydrochloric acid in methanol, when the resulting saturated straight-chain compound is identified by gas chromatography[491] (see Fig. 5.11a). Keto groups should be reduced to hydroxy groups by the action of sodium borohydride,[714] prior to analysis by the above procedure. The hydroxy group can be located in the fatty acid chain unequivocally by the reaction illustrated in Fig. 5.11b. It is first oxidised with

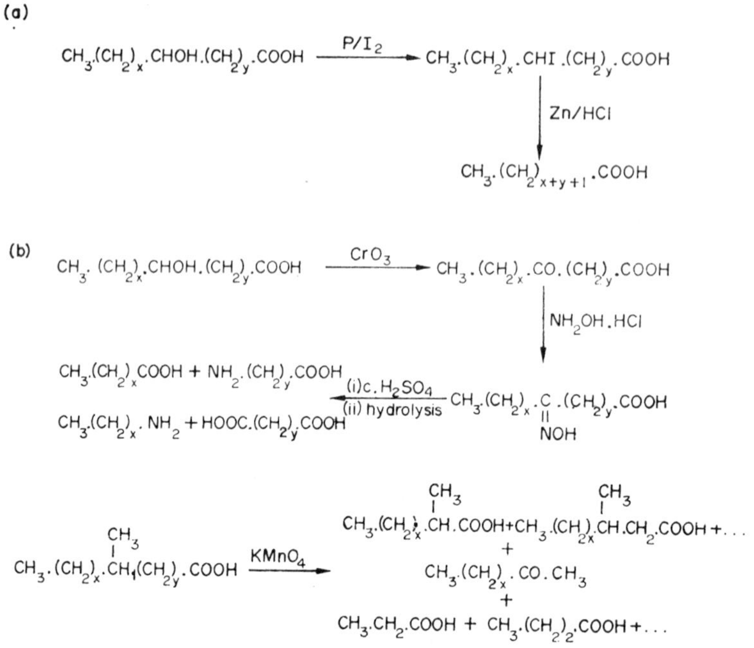

(a)

$$CH_3.(CH_2)_x.CHOH.(CH_2)_y.COOH \xrightarrow{P/I_2} CH_3.(CH_2)_x.CHI.(CH_2)_y.COOH$$

$$\downarrow Zn/HCl$$

$$CH_3.(CH_2)_{x+y+1}.COOH$$

(b)

$$CH_3.(CH_2)_x.CHOH.(CH_2)_y.COOH \xrightarrow{CrO_3} CH_3.(CH_2)_x.CO.(CH_2)_y.COOH$$

$$\downarrow NH_2OH.HCl$$

$$CH_3.(CH_2)_x COOH + NH_2.(CH_2)_y.COOH$$
$$CH_3.(CH_2)_x.NH_2 + HOOC.(CH_2)_y COOH \xleftarrow[\text{(ii) hydrolysis}]{\text{(i)c. } H_2SO_4} CH_3.(CH_2)_x.\underset{\underset{NOH}{\|}}{C}.(CH_2)_y.COOH$$

(c)

$$CH_3.(CH_2)_x.\overset{\overset{CH_3}{|}}{CH}(CH_2)_y.COOH \xrightarrow{KMnO_4}$$

$$CH_3.(CH_2)_x.\overset{\overset{CH_3}{|}}{CH}.COOH + CH_3.(CH_2)_x\overset{\overset{CH_3}{|}}{CH}.CH_2.COOH + \ldots$$
$$+$$
$$CH_3.(CH_2)_x.CO.CH_3$$
$$+$$
$$CH_3.CH_2.COOH + CH_3.(CH_2)_2.COOH + \ldots$$

FIG. 5.11. Location of functional groups in fatty acids by chemical means. a. Determination of chain length in a hydroxy fatty acid. b. Location of hydroxyl- of keto-groups in fatty acids. c. Location of methyl branches in fatty acids.

chromic oxide in acetic acid to a keto group from which the oxime derivative is prepared. This is heated with concentrated sulphuric acid so that it undergoes a Beckman rearrangement reaction with formation of isomeric amides, which are hydrolysed and the acidic products isolated, methylated and identified by gas chromatography[133] (it is only necessary to identify one of the products to be able to completely identify a monohydroxy acid). Simple fragmentation of an aliphatic chain at a keto-group can also be obtained by Bayer-Villiger oxidation with peroxytrifluoroacetic acid.[500] Epoxy-groups are cleaved directly with periodic acid in halogenated solvents[66] or in diethyl ether,[428] and the position of the ring is established by analysis of the products.

The chemistry of cyclopropane and cyclopropene ring-containing fatty acids has been reviewed comprehensively elsewhere.[127] The presence of such functional groups can be detected by a number of spectroscopic procedures and also by various chemical techniques. For example, cyclopropene rings give a pink coloration with carbon disulphide (the Halphen test) and a brown coloration with silver nitrate. The ring is disrupted by ozonolysis or permanganate–periodate oxidation, and a ß-diketo

compound is formed which can be identified by mass spectrometry,[321] or this compound can be subjected to more vigorous oxidation so that the molecule is cleaved between the keto groups, and the resulting acidic fragments can be recovered and identified. Cyclopropene rings can be hydrogenated and their positions determined by the methods used for natural cyclopropane fatty acids. The latter behave as normal saturated fatty acids on silver nitrate plates, but react with bromine and so can be distinguished by this means.[85] They form methoxy-derivatives with boron trifluoride–methanol reagent,[496] and they are converted to methyl-branched fatty acids by vigorous catalytic hydrogenation.[466] Both types of derivative can be characterized by mass spectrometry, and the latter can also be characterized by the chemical technique described below.

Methyl branches in fatty acids are not easily located by chemical means as such compounds are comparatively inert to most chemical reagents.[2] Their positions can be determined chemically by oxidising the fatty acids vigorously with acidic potassium permanganate. Homologous series of normal and branched-chain acids are formed together with a neutral keto compound that identifies the point of branching. Both the acidic and keto products can be isolated and identified by gas chromatography[526, 546] (see Fig. 5.11c). These and related procedures have been reviewed by Polgar.[583]

F. PHYSICAL CHARACTERIZATION OF FATTY ACIDS

Melting points of long-chain saturated or oxygenated fatty acids are often sufficiently distinct to be useful in their characterisation; they are commonly determined on the free acids rather than on the methyl esters, as the melting points of the former tend to be appreciably higher. Unsaturated acids are in general low melting, and while a melting point apparatus equipped with a cold stage may be used to characterize monoenes (*trans*-isomers have higher melting points than the corresponding *cis*-compounds) and dienes, the melting points of polyunsaturated fatty acids are in general too low to be measured with confidence, as traces of impurities such as autoxidised materials can greatly affect this property. An apparatus for semi-micro crystallisations of fatty acids at low temperatures for subsequent melting point determination has been described by Privett.[595] *p*-Phenylazophenacyl esters of polyunsaturated fatty acids have been suggested as suitable derivatives for characterization purposes,[801] as they have comparatively high melting points, but they have not been widely adopted.

Measurement of *critical solution temperatures* may be an acceptable alternative to melting points, when the latter cannot conveniently be determined. The method utilises the property that two liquids, which are almost immiscible at low temperatures, become more soluble in each other as the temperature is raised. The temperature at which they become soluble in all proportions is known as the upper critical solution temperature. Nitromethane or acetonitrile have been used as solvents to characterize a wide range of polyunsaturated fatty acids and related compounds in the systematic studies of Schmid *et al.*[669] Other physical methods of characterizing low-melting fatty acids have been reviewed by Holman and Rahm.[317] Chromatographic data (e.g. equivalent chain length values) are now frequently accepted as equivalent to physical constants.

G. PREPARATION OF LARGE QUANTITIES OF PURE FATTY ACIDS

Although a number of methods are available for the preparation of pure fatty acids in large quantities (50 g or more) as fatty acid standards or for biochemical or nutritional studies, many procedures require special equipment only available in a few laboratories. For example, highly efficient fractional distillation apparatus is used by most commercial suppliers in the preparation of fatty acids, but is not generally accessible. Several less sophisticated methods have been developed, however, that can be used to advantage by lipid analysts. The key to success frequently lies in the choice of suitable starting materials.

1. Saturated fatty acids

Coconut oil is commonly used as a source of C_8 to C_{14} fatty acids, and beef tallow or hydrogenated fats for longer chain esters. Certain seed oils are useful sources of specific acids; for example, bayberry tallow may contain up to 60 per cent myristic acid.,

Concentrates of specific shorter chain acids are obtained by high vacuum distillation of coconut oil methyl esters, with comparatively simple fractional distillation columns, but more expensive equipment is necessary to obtain highly pure esters in quantity by this technique.

The liquid–liquid chromatography system of Privett and Nickell,[594, 598] described above (section C4), is a useful alternative. A column (5.5 × 125 cm) containing 1000 g of solid support has been used to purify up to 15 g of esters, and an even larger column to purify up to 50 g of esters in a single run. The columns can be used repeatedly. However, most of the common saturated fatty acids are now obtainable in a pure form comparatively cheaply from commercial suppliers, so that simple laboratory methods of preparing them are becoming less important. Odd-numbered fatty acids can be prepared synthetically, if necessary, by chain-elongation of the more common even-numbered acids.

2. Polyunsaturated fatty acids

(I) SOURCES

Sources of polyunsaturated fatty acids and methods of preparing them have been described in an excellent review by Privett.[594] Again, it is advisable to start with some natural material rich in the acid required. For example, olive oil is an excellent source of oleic acid, safflower oil of linoleic acid, linseed oil of linolenic acid, pig liver lipids of arachidonic acid and codliver oil of penta- and hexaenoic acids. Usually it is necessary to obtain a fraction rich in the particular component required before proceeding to more exacting fractionation procedures.

(II) LOW TEMPERATURE CRYSTALLISATION

Fractions enriched in polyunsaturated fatty acids are readily obtained by low temperature crystallisation, a technique that demands little in the way of expensive equipment. Most methods employ acetone for cystallisation of samples in the form of methyl esters or acetone, diethyl ether or hexane for the free acids, which are generally preferred for the purpose, at temperatures down to −70°C, a temperature that can be attained with solid carbon dioxide as refrigerant. Brown and Kolb[95] have described a number of useful procedures.

Three fractions enriched in saturated, monoenoic and polyenoic fatty acids respectively are easily obtained. The free acids are taken down to the lowest working temperature (about −50° generally), at a concentration of 1 g per 10 ml of acetone, and are kept at this temperature for up to 5 hr with gentle swirling until equilibrium is reached. The solution is then filtered through a Buchner funnel, cooled to just below the solution temperature and the crystals washed with a small amount of cold solvent. The material in solution consists mainly of poly-unsaturated fatty acids, although it is always contaminated by small amounts of saturated and monoenoic components held in solution by the polyunsaturated compounds. If the volume of solvent is reduced until a 10 per cent solution is again attained, the process can be repeated and improved separations obtained. The crystals may be re-dissolved in fresh solvent and further fractionated at slightly higher temperatures into components enriched in monoenoic and then in saturated fatty

acids. With care some unexpected separations can be achieved; for example, petroselinic acid has been separated from oleic acid by this technique.[597] Gunstone et al.[253] have shown how the technique can be used to obtain concentrates (93–99%) of oleic or linoleic acids.

The main disadvantages of the method are that the fractions are not entirely pure, and that the solutions must be held at the low temperatures for very long periods before equilibrium is attained. In the method's favour, large quantities of fatty acids can be processed in a single operation, and very little harm can come to polyunsaturated acids at the low temperatures employed.

(III) UREA ADDUCT FORMATION

Urea, when permitted to crystallise in the presence of certain long-chain aliphatic compounds, forms hexagonal crystals with a channel, into which the aliphatic compounds may fit and be removed from solution, provided they do not contain functional groups that increase their bulk. Such crystals are known as urea inclusion complexes. Saturated straight-chain acids (as the methyl ester derivatives) form complexes readily. The double bonds of unsaturated fatty acids increase their bulk so that monoenoic fatty acids do not form complexes easily, but tend to form them more readily than dienes, which in turn form them more readily than compounds with three or more double bonds. Fatty acids with trans-double bonds form complexes before the analogous compounds with cis-double bonds. Unfortunately, the separations are complicated by the fact that shorter chain-length compounds do not complex as readily as those of longer chain-length and methyl oleate, for example, is adducted with approximately the same facility as methyl laurate.[342] For this reason, the procedure has never been developed as an analytical technique. The following method can be applied to obtain a concentrate of polyunsaturated fatty acids in the form of methyl esters from a natural mixture.

"The esters (100 g) are dissolved in methanol (1 litre) to which urea (200 g) is added. The mixture is warmed until all the urea has dissolved, when the solution is allowed to cool to room temperature with occasional swirling. After a minimum of 4 hr, the material is filtered through a Buchner funnel to remove the urea complexes, which are washed twice with 25 ml portions of methanol saturated with urea. The solution, which is greatly enriched in polyunsaturated esters, is then poured into 1 per cent hydrochloric acid (600 ml) and extracted alternatively with hexane (500 ml) and diethyl ether (500 ml). The combined organic layers are washed twice with water (50 ml portions) and dried over anhydrous sodium sulphate before the solvent is removed under reduced pressure."

The procedure can be scaled up or down considerably. As an example, a GLC recorder trace of material obtained in this way from 0.25 g of the standard mixture described above (see Fig. 5.1) is illustrated in Fig. 5.12. The fraction obtained, which was 20 per cent of the original esters is enriched in the polyunsaturated components but also contains some branched-chain esters, not readily apparent earlier, and some shorter chain-length esters.

The adducted esters can be recovered, if necessary, by breaking up the complexes with water and extracting the esters into hexane or diethyl ether.

Fractions enriched in particular fatty acids such as linoleic or linolenic acids can be obtained from suitable starting materials, although it may be necessary to vary the amount of urea added. Gunstone et al.[253] have described some specific applications which illustrate the potential of the method.

Urea fractionation procedures have been used to separate ω-hydroxy fatty acids (after acetylation to increase their bulk) from related compounds with the substituent elsewhere in the chain.[134] Another particularly valuable application consists in the isolation of concentrates of branched-chain and cyclic esters from natural mixtures, and some useful separations of the former have been described by Smith and Lough,[713] for example. Iso-branched esters complex more readily than the corresponding anteiso-compounds so some change in the ratio of these may occur.

Attempts to use urea in thin-layer adsorbents or in columns on an analytical scale have not been entirely successful, but the following simplified procedure is of value when small amounts only of esters are available.[447]

"The methyl esters (100 mg) are dissolved in hexane (4 ml) and urea (1.5 g) moistened with methanol (15 drops) added. After standing overnight, the solid is filtered off and thoroughly washed with hexane; the washings and the hexane filtrate are combined, washed with water, dried over anhydrous sodium sulphate and evaporated yielding the branched-chain or polyunsaturated fraction. The straight-chain fraction can be recovered by dissolving the urea in water, extracting with diethyl ether, drying the solvent layer and evaporating as before."

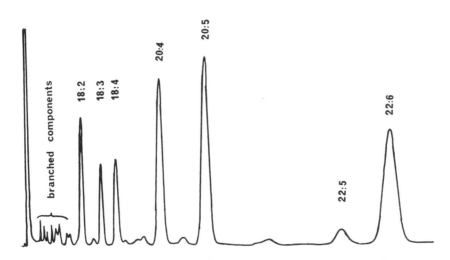

FIG. 5.12. GLC recorder trace (the EGSS-X column of table 5.1 and Fig. 5.1) of methyl esters, that did not form urea adducts, from the natural mixture illustrated in Fig. 5.1.

The main advantages of methods using urea are that large quantities of esters can be separated and that with care there is little chance for harm to come to polyunsaturated esters. Comprehensive reviews of urea fractionation procedures in general[662] and applications to unsaturated fatty acids in particular[741] have been published.

(IV) HIGH CAPACITY ADSORPTION CHROMATOGRAPHY

When the starting material is plentiful so that the efficiency of separation is not important, adsorption chromatography on large columns of silicic acids can be used to purify fatty acids. Generally, in conventional adsorption chromatography, only those components held least strongly are eluted in a pure fraction, but concentrates of the more strongly held components can be obtained by using silicic acid for displacement chromatography, i.e. those components held most strongly serve as internal displacers for those held less strongly. Privett *et al.*[534, 595] have described an elegant application of this method for the preparation of 100 g quantities of methyl linolenate (>99 per cent pure).

(V) SEPARATION AS MERCURIC ACETATE ADDUCTS

Concentrates rich in polyunsaturated fatty acids (as the methyl esters) can be obtained from suitable starting materials, by preparing the mercuric acetate adducts (see Chapter 4 for practical details of the preparation of adducts and regeneration of the original double bonds) and partitioning them between methanol and pentane; the methanol layer retains the adducts of the more unsaturated esters which can be regenerated unchanged. Methyl linoleate of 95 per cent purity and methyl linolenate of 90 per cent purity can be produced on the 50–100 g scale by this method from the esters of safflower and linseed oils respectively.[730] The author has prepared methyl linoleate of greater than 99 per cent purity simply by partitioning the adducts of safflower oil esters between methanol–water (95:5, v/v) and hexane in separating funnels (unpublished work), and others[825] have obtained methyl linolenate of similar purity by partitioning the adducts of linseed oil esters between methanol–water (90:10, v/v) and diethyl ether in a liquid–liquid continuous extractor.

The method warrants much wider recognition. Although mercuric acetate is not cheap, large quantities of high purity unsaturated esters can be obtained in a relatively short time. If these compounds are to be used in biochemical studies however, particularly in nutritional experiments, great care must be exercised to remove all traces of mercury. This aspect of the process requires further study.

CHAPTER 6

The Analysis of Simple Lipid Classes

LIPID extracts from animal tissues commonly contain triacylglycerols as the predominant simple lipid class, together with cholesterol esters, cholesterol, partial glycerides and free fatty acids in addition to complex lipids, which are considered as a single class in this chapter. As well as the more common simple lipids, trace amounts of hydrocarbons, methyl esters, wax esters, and glycerol ethers and vinyl ethers (neutral plasmalogens) may be found and, on occasion, one or more of these may assume major proportions. Plant lipid extracts may contain similar simple lipids except that sterols other than cholesterol are likely to be present. Most of these lipid classes can be separated from each other by adsorption chromatography on thin layers or on columns. Where the fatty acid constituents of lipids contain polar functional groups in the alkyl chain, however, simpler molecular species of these will be separated by adsorption chromatography and the elution pattern may be rather complicated; separations of this kind are discussed further in Chapter 8 and are not considered here. Methods are available for the analysis of some lipid classes without prior separation (e.g. cholesterol, triacylglycerols). In general, precautions should be taken to minimise the effects of autoxidation (see Chapter 3).

A. CHROMATOGRAPHIC SEPARATION OF THE COMMON SIMPLE LIPID CLASSES

1. Thin-layer chromatography

(I) SOLVENT SYSTEMS

A variety of solvent systems have been used to separate simple lipids by TLC on silica gel G in a single dimension. Those used most frequently contain hexane, diethyl ether and acetic (or formic)

acid in various proportions; for example, with these solvents in the ratio 80:20:2 (by vol.), the separation illustrated in Fig. 6.1 (plate A) is achieved, in which most of the common simple lipids are separated — cholesterol esters near the solvent front followed by triacylglycerols, free fatty acids, cholesterol, diacylglycerols (1,3- and 1,2-), monoacylglycerols and phospholipids. If diacylglycerols are important components of the sample, more distinct separations can be achieved with a solvent system consisting of benzene–diethyl ether–ethyl acetate–acetic acid (80:10:10:0.2 by vol.) as illustrated in Fig. 6.1 (plate B). Unfortunately, slight changes in the amount of acetic acid in the system can greatly alter the nature of the separation (in the interests of safety, toluene should perhaps be used instead of benzene). It should be recognised that isomerisation of diacylglycerols occurs on TLC plates, and it is generally advisable to combine the two bands for quantification purposes. Other useful high-performance TLC systems have been developed for the separation of individual neutral lipids and phospholipids in a single direction (see Chapter 7).

As little as 0.5 μg of lipid can be applied as a spot to a TLC plate if the separated components are detected by the most sensitive of the charring procedures. For preparative purposes, 20 mg of lipid may be applied with ease as a band on a 20 × 20 cm plate coated with a layer 0.5 mm thick of silica gel and, in some circumstances, for example when the lipids consist largely of triacylglycerols, up to 50 mg of lipid can be chromatographed on a single plate in this way. Simple lipids should be applied to the plate in a solvent such as chloroform, that does not evaporate too quickly, depositing material on the outside of the syringe.

As cholesterol esters migrate very close to the solvent front with these solvent systems, where a

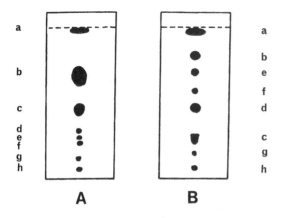

FIG. 6.1. TLC separations of simple lipids on silica gel G layers. a, cholesterol esters; b, triacylglycerols; c, free fatty acids; d, cholesterol; e, 1,3-diacylglycerols; f, 1,2-diacylglycerols; g, monoacylglycerols; h, complex lipids. Developing solvents; plate A, hexane–diethyl ether–formic acid (80:20:2 by vol); plate B, benzene–diethyl ether–ethyl acetate–acetic acid (80:10:10:0.2 by vol).[739]

lipid-like contaminants which may be in the silica gel will also appear, it is sometimes recommended that plates be developed first right to the top in a solvent system of chloroform–methanol (4:1, v/v), then dried immediately before the sample is applied, so that any impurities in the adsorbent will lie ahead of the earliest running component of the mixture. Although this treatment is quite effective, the retention characteristics of the adsorbent may be slightly altered.

Triacylglycerols (and other lipids) that contain fatty acids differing widely in chain-length such as are found in milk fats, will migrate as a more diffuse band on TLC adsorbents, as they are partially separated into molecular species that contain either mainly long-chain ($>C_{16}$) or mainly short-chain components (slower running).

(II) DETECTION AND IDENTIFICATION OF SIMPLE LIPIDS ON TLC PLATES

Lipid classes can be detected by any of the destructive or non-destructive non-specific reagents described in Chapter 3, and identified by their migration characteristics relative to authentic st‍‍‍ chromatographed simultaneously along‑ ples under investigation. Unfortunately, ecific spray reagent that can be used to e glyceride derivatives, but cholesterol holesterol can be positively identified

by means of various spray reagents (including charring reagents, see Chapter 3), of which the most useful is acidic ferric chloride solution.[458]

"Ferric chloride ($FeCl_3.6H_2O$, 50 mg) is dissolved in water (90 ml) with acetic (5 ml) and sulphuric (5 ml) acids. The developed TLC plate is sprayed with the reagent then heated at 100°C for 2–3 min, when the presence of cholesterol and cholesterol esters is indicated by the appearance of a red-violet colour. The colour for cholesterol appears slightly before that of its ester."

Free fatty acids can be identified by means of a specific sequence of spray reagents, although the chemical reactions involved are not understood.[188]

"The developed plate is sprayed to dampness in turn with a 2′,7′-dichlorofluoroescein spray (see Chapter 3), a solution of 1 per cent aluminium chloride in ethanol and finally with 1 per cent aqueous ferric chloride, warming the plate briefly to about 100°C after each spray. Free fatty acids give a rose-violet colouration."

Esterified fatty acids can also be recognised by a specific spray sequence.[826]

"Two reagents are necessary. (a) Hydroxylamine hydrochloride (10 g) is dissolved in water (25 ml)–ethanol (100 ml), and added to a saturated aqueous solution of sodium hydroxide (26 ml) and ethanol (200 ml). The mixture is filtered before use to remove any sodium chloride precipitated. (b) Ferric chloride ($FeCl_3.6H_2O$; 10 g) and concentrated hydrochloric acid (20 ml) are ground together in a mortar, then are shaken with diethyl ether (300 ml). The developed plate is sprayed with the first reagent, dried, then sprayed with the second reagent, when the fatty acid-containing lipids appear as purple spots on a yellow background."

(III) ESTIMATION OF SIMPLE LIPIDS SEPARATED BY TLC

Three basic types of method are in common use for the estimation of simple lipids separated by TLC (they are discussed in some detail in Chapter 3). In the first approach, all the lipids are charred by oxidising agents on or after elution from the plates, and the amount of carbon formed is determined by photodensitometry; in the second, a dye is sprayed on the plate and the lipids are determined by fluorometry; in the third approach, lipids are detected and recovered from the absorbent, and are estimated individually by methods that are appropriate to each lipid class. Procedures of the first type are rapid, but are destructive to samples and are sometimes of limited accuracy only, those of the second type are not destructive to the sample but require a costly dedicated instrument, while those of the third type are very time-consuming but are capable of greater accuracy and can yield more information.

For rapid routine analysis of large numbers of samples, the author favours the charring procedure developed by Shand and Noble[685], in which bands of lipid, charred under controlled conditions, are suspended in an emulsifier-scintillator, chosen according to whether the counter has a cooling unit[550], and are quantified by measuring the amount of quenching on counting in the external standard mode in a liquid scintillation counter. In common with all such methods, it must be calibrated for each lipid class, but the calibration is very stable and is not in general affected appreciably by the degree of unsaturation of the fatty acid moieties. It can be applied to any lipid class for which standards are available, and is capable of high accuracy down to 10 μg of lipid; it is probably as precise as most other methods at lower lipid levels.

Simple lipids can be quantified by charring and photodensitometry in solution following elution from the TLC adsorbent with chloroform or chloroform–methanol (9:1, v/v), or by precipitating the silica gel by centrifugation following charring of a suspension. In the most commonly-used procedures of this type, the lipids are oxidised with potassium dichromate solution and the reduction in absorbance of the dichromate ion at 350 nm is measured (see Chapter 3)[25, 215, 702], or the lipids can be charred at 200°C with concentrated sulphuric acid and the absorbance at 375 nm measured[488]. Again it is necessary to have a separate calibration curve, obtained with an appropriate standard of a similar degree of unsaturation to the samples, for each lipid class. Such procedures are capable of greater accuracy than those involving photodensitometry of charred spots on a TLC plate, as calibration is easier and the Beer–Lambert laws are more likely to hold when substances are in solution in the absence of

solid material. Also, the absorbance of the solutions can be measured on a general purpose laboratory UV spectrophotometer rather than with a specialised scanning photodensitometer.

Fluorometric methods are now being applied in a number of laboratories for the estimation of simple lipids separated by TLC (see also Chapter 3), but the specialised instrument that is needed is unlikely to be available to many analysts[547, 631].

Cholesterol and cholesterol esters may be estimated by the methods described below (section B). The amounts of tri-, di- and monoacylglycerols can be determined by quantitative IR spectrophotometry after they have been eluted from TLC plates[217]. After removal of the solvent, the lipid classes are dissolved in a known volume of carbon tetrachloride, and the absorbance due to the characteristic frequency of the carbonyl function at 5.85 μm is measured. The amount of each component present is then obtained by reference to calibration curves, which must be regularly checked, prepared from tri-, di- and mono-olein. The hydroxamic acid colorimetric reaction for acyl-ester groups is also used to estimate simple glycerides, as are methods involving analysis of the glycerol moiety (section B). Similarly, a variety of methods are available for the analysis of free (unesterified) fatty acids (see also section B).

The spray reagents used to detect lipids may interfere with colorimetric or spectrophotometric determinations if they are applied too liberally to the plates.

Complex lipids, considered as a single class, can be estimated by phosphorus determination (assuming that non-phosphorus containing glycolipids are not major components of the sample) by the procedure described in Chapter 7. The approximate amount of phospholipid is obtained by multiplying the weight of phosphorus found in the sample by the factor 25.

If, in addition to estimating the amount of each lipid class in a natural mixture, it is intended that the fatty acid composition of each be determined, then both quantities can be obtained in a single analysis by gas chromatography of the methyl ester derivatives of the fatty acid components with an added internal standard (see Chapter 3 for a description of the underlying principle). A known amount of the standard, which should be the methyl ester or other derivative of a fatty acid that does not occur in the sample to be analysed, is added to each lipid class when it is eluted from the TLC adsorbent, before it is transesterified and the methyl esters subjected to GLC analysis (see Chapter 5). The fatty acid composition of the sample is in this way determined, and the amount of the sample is found by relating the total area of the fatty acid peaks to the area of the peak for the standard ester[141]. It is necessary to allow for the weight of non-fatty acid material (e.g. glycerol or cholesterol) in each lipid class by multiplying each result by appropriate arithmetic factors, calculated by dividing the molecular weight of the heptadecanoic acid derivative (assuming 17:0 is the internal standard) of the lipid class by the molecular weight of methyl heptadecanoate. These factors are listed in Table 6.1[141]. Alternatively, an average molecular weight for each lipid class can be determined from the fatty acid compositions, and more accurate correction factors then calculated for particular samples.

Methyl heptadecanoate and pentadecanoate are chosen most often as internal standards[141] but to ensure accurate quantification of polyunsaturated fatty acids, as are found in phospholipids, for example, it may be advisable to use 20:3(n-3)[230] or 22:1[147] fatty acids for the purpose. They are usually added in methanol solution. Suitable transesterification procedures have been described in Chapter 4; methanolic hydrogen chloride can be used to esterify all lipid classes (remembering that an inert solvent must be added to effect solution of triacylglycerols and cholesterol esters), or free fatty acids alone may be esterified by this reagent and the other lipid classes by the more rapid and milder alkaline transesterification procedures. Losses can be minimised by carrying out the reaction in the vessel used to collect

the eluant, so that no transfers are carried out before all the fatty acids are methylated, or by transesterifying lipids without prior elution from the adsorbent. In the latter circumstance, the solution of the internal standard, the transesterifying reagent and a solvent (if necessary) are added together to the lipid on the adsorbent in a test-tube (see Chapter 4 also). With these precautions, the method is at least as accurate as any other in current use. Free cholesterol must of course be determined separately if this method is used for lipid analysis.

The Iatroscan analyser (see Chapter 3) can be used to separate and quantify neutral lipids[137, 701, 789] but the reproducibility of the procedure is not nearly as good as the charring method of Shand and Noble[685], for example.

2. Column chromatography

Larger amounts of simple lipids can be separated preparatively by column chromatography with silicic acid, acid-washed Florisil or Florisil as adsorbents (see Chapter 3), eluted in a stepwise sequence with hexane containing increasing proportions of diethyl ether[107]. The best resolutions of lipid classes are obtained with fine mesh silicic acid[305], which yields separations of the kind illustrated in Fig. 6.2. Hydrocarbons, cholesterol esters and triacylglycerols are all well separated (though not illustrated, free fatty acids elute between triacylglycerols and free cholesterol and may overlap with the latter), cholesterol and diacylglycerols overlap slightly, but monoacylglycerols are well-separated. The precise amount of diethyl ether that must be added to the hexane to elute each lipid class will vary from batch to batch of the adsorbent, so will not always be the same as that given in Fig. 6.2 and must be determined by experiment. Complex lipids can be recovered as a single class by elution with methanol, or individual complex lipids can be separated by eluting with choloroform containing increasing proportions of methanol (see Chapter 7). Acid-washed Florisil has similar adsorptive properties to silicic acid, but larger amounts of material may be chromatographed at much faster flow rates, although some loss of resolution is inevitable[106].

Florisil itself is a useful adsorbent with somewhat different elution characteristics to silicic acid[105]. The

TABLE 6.1. CORRECTION FACTORS TO CONVERT THE TOTAL AMOUNT OF FATTY ACIDS IN A LIPID CLASS (DETERMINED BY GLC ANALYSIS) TO WEIGHT OF LIPID[141]

Lipid class	Factor
cholesterol esters	2.246
triacylglycerols	0.995
diacylglycerols	1.049
monoacylglycerols	1.211
free fatty acids	0.951
phospholipids*	1.340

* As an approximation, it is assumed that the phospholipids consist of phosphatidylcholine only.

degree of hydration of the adsorbent can be important, and it may be necessary to add up to 7 per cent water to it for optimum resolution and to minimise "tailing". Fig. 6.3 illustrates a typical separation. Cholesterol esters and triacylglycerols are well separated, and cholesterol, diacylglycerols and monoacylglycerols can be resolved with care, though some positional isomerisation occurs with the last two lipids. Free fatty acids are eluted with diethyl ether–acetic acid (98:2 v/v), but complex lipids are not easily recovered quantitatively. When large amounts of cholesterol esters, triacylglycerols or methyl esters must be isolated from natural mixtures or from synthetic preparations for further study, and complete recovery or resolution of other lipids is unnecessary, Florisil chromatography is

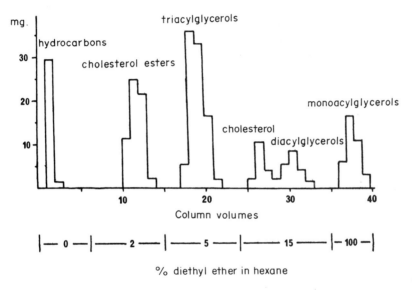

FIG. 6.2. Chromatography of simple lipids on a column (1.5 × 20 cm) containing silicic acid (30 g) eluted with hexane–diethyl ether.

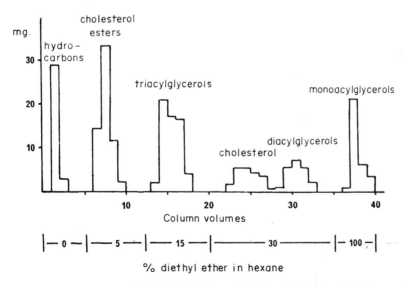

FIG. 6.3. Chromatography of simple lipids on a column (1.5 × 20 cm) containing Florisil (30 g) eluted with hexane–diethyl ether mixtures.

especially recommended, as up to 1 g of lipid can be separated rapidly on columns containing 20 g of adsorbent. Though the adsorbent is commonly thrown away after use, Florisil and acid-washed Florisil can apparently be regenerated and re-used repeatedly, by washing the columns with methanol, followed by chloroform and then by hexane[107].

Simple lipids as a class can be separated from complex lipids on silicic acid or acid-washed Florisil columns (30 mg/g of adsorbent), as chloroform (10 volumes) elutes simple lipids quantitatively, while the complex lipids are recovered by elution with methanol (10 volumes).

Procedures suitable for quantifying lipids separated by preparative column chromatography are described in Chapter 3. In addition, the specific chemical and spectroscopic procedures discussed later in this chapter can be used.

There have been a few reports of separations of simple lipids by HPLC[19], generally using microparticulate silicic acid as adsorbent, and of these the most promising appears to be that of Stolyhwo and Privett[737]. The technique is not likely to be widely used because of the lack of a suitable commercial detector.

3. High temperature gas-liquid chromatography

Kuksis and coworkers have adapted the technique of high-temperature GLC (see Chapter 8 for a discussion of the technical problems) to the separation and quantification of mixtures of simple lipids, phosphoglycerides and sphingomyelin[415, 421, 422, 424]. The method was developed primarily for the routine analysis of plasma lipids in clinical laboratories, but undoubtedly could be used much more widely. Lipids are first digested with phospholipase C (see Chapter 9), which converts lysophosphatidylcholine, phosphatidylcholine and sphingomyelin to monoacylglycerol, diacylglycerol and ceramide respectively. The hydrolysis products are converted to the volatile trimethylsilyl ether (preferred) or acetate derivatives (cholesterol is similarly derivatised, while unesterified fatty acids can be converted to the trimethylsilyl esters), and are injected directly into a gas chromatographic column containing a highly thermostable stationary phase.

By means of temperature-programmed analysis (200° to approx. 340°C) and with tridecanoin as an internal standard, it is possible to separate and quantify triacylglycerols, cholesterol esters, cholesterol, free fatty acids and the choline-containing phospholipids. The separations are effected according to the combined chain-lengths of the fatty acid components of each lipid class, as in the representative analysis in Fig. 6.4, so information is obtained on both the composition and content of individual lipids in a sample. Although more samples could be analysed in a clinical laboratory with automatic colorimetric analysis procedures, the GLC method is more accurate over a range of lipid concentrations and gives more information, while the analysis of the data can be subjected to a high degree of automation[421]. Small amounts of diacylglycerols or monoacylglycerols endogenous to the sample would be swamped by the same compounds derived from phospholipids, but the procedure could undoubtedly be adapted to take this into account should it be necessary. For example, by eliminating the phospholipase C digestion and derivatization steps, Mares et al.[486, 487] were able to effect excellent quantitative analyses of simple lipids only, in particular of cholesterol, cholesterol esters and triacylglycerols (hydrogenation prior to analysis sharpened the GLC peaks and improved quantification).

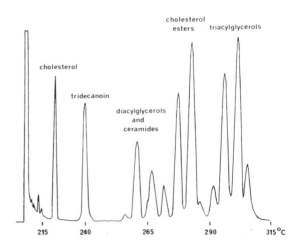

FIG. 6.4. Gas chromatographic trace (temperature-programmed) of plasma lipids after treatment with phospholipase C and acetylation. Tridecanoin was added as an internal standard. (Redrawn from Kuksis et al.[424])

B. THE ANALYSIS OF INDIVIDUAL SIMPLE LIPIDS AND OF THEIR HYDROLYSIS PRODUCTS

Methods are available for the analysis of some simple lipid classes without prior separation from other lipids. Often these methods are also applicable to the analysis of hydrolysis products of simple lipids. For the unequivocal identification of a simple lipid, it is necessary to isolate the hydrolysis products to ascertain their relative molar proportions.

1. Unesterified (free) fatty acids

Numerous methods have been developed for the analysis of the metabolically important unesterified fatty acid fraction of tissue lipids, especially of plasma. Any of the procedures described above can be used for the purpose (e.g. TLC separation followed by GLC of the methyl ester derivatives with an internal standard), but more specific methods of greater sensitivity and precision have been developed. Titrimetric methods were used for many years, but procedures that involve spectrophotometric measurement of highly-coloured copper complexes are now favoured.

Although in at least one modification the method is applied directly to a plasma sample without prior extraction of the lipids[337], most workers stress the importance of extracting the free fatty acids quantitatively from tissues, as they may be very strongly bound to albumin or other proteins. For example, it has been recommended[190] that plasma samples buffered with phosphate be extracted by refluxing with chloroform–methanol (2:1, v/v) three times under reflux for 30 min each. The lipids in chloroform are then reacted with a solution of aqueous 1M triethanolamine–1M acetic acid (9:1, v/v) containing $Cu(NO_3)_2.3H_2O$ (6.45% w/v)[189] to form the copper complexes, which are extracted into the chloroform layer before the colour is developed by adding diethyldithiocarbamate[189] or the more sensitive diphenylcarbazide-diphenylcarbazone[337] for spectrophotometric determination. Sensitive enzymic[503, 692] and radiochemical[172] methods for the assay of free fatty acids have recently been described.

2. Bound fatty acids

The estimation of the total amount of fatty acids in simple lipids by means of gas chromatography with an internal standard has been discussed above. They can also be determined chemically by the ferric hydroxamate procedure.[181, 613]

"Two reagents are necessary and must be prepared immediately before use. The first is prepared by dissolving hydroxylamine hydrochloride (100 mg) in water (0.17 ml) and adding absolute alcohol (3.2 ml). To this is added aqueous sodium hydroxide solution (0.2 ml; 5 g per 10 ml) in absolute alcohol (3.2 ml). Sodium chloride formed is removed by centrifugation. The second reagent consists of ferric perchlorate (100 mg) and perchloric acid (2.6 ml) in 95 per cent ethanol (97.4 ml). The samples and a set of standards containing 0.5 to 5.0 μeq. of fatty esters are dried thoroughly in a vacuum oven at 65°C (in particular, all traces of chloroform must be removed). The basic hydroxylamine reagent (0.1 ml) is added to each sample in diethyl ether solution (3 ml), and the whole is heated at 65°C at atmospheric pressure until the ether evaporates and then in a vacuum oven until all the remaining solvent is gone. The samples are cooled, ferric perchlorate reagent (5 ml) is added and the mixture shaken and left to stand for 30 min, when the absorbance of the solution at 530 nm is compared with that of a reagent blank. The amount of ester in each unknown is determined from the curve obtained with the standards."

3. Glycerides and glycerol

Simple glycerides can be estimated by measuring the amount of the glycerol moiety by chemical, enzymic or chromatographic means. Chemical determination of glycerol is time-consuming and subject to a number of errors. A more sensitive procedure utilises a coupled enzyme assay. The simple glycerides are hydrolysed with tetraethylammonium hydroxide under conditions such that phosphate ester bonds are unaffected. Glycerol kinase is then used to convert glycerol into glycerol-

3-phosphate, which is in turn oxidised by α-glycerophosphate dehydrogenase in the presence of NAD. The amount of NADH produced, which is measured by spectrophotometry or fluorometry, is directly proportional to the amount of glycerol present.[122] The procedure may be used with as little as 0.05 μmoles of glycerol. It is possible to purchase all the necessary enzymes and reagents separately for the method, but a number of biochemical suppliers offer kits containing the required reagents in the correct proportions for the assay together with full instructions. The method is suitable for automatic analysis. When applied to total lipid extracts, the combined simple glycerides are in fact assayed, although results are often quoted simply as "triglyceride" or "triacylglycerol". sn-1,2-Diacyl-glycerols specifically can be estimated by an enzymatic method that utilises diacylglycerol kinase;[391] it is essential that the method of extraction of the lipids should not cause isomerisation in this instance.

Gas chromatographic procedures for estimating glycerol have been reviewed by Roberts.[629] The following method is suitable for the determination of glycerol in simple glycerides; the glycerides are subjected to hydrogenolysis with lithium aluminium hydride yielding glycerol and fatty alcohols, which are acetylated for GLC analysis.[312]

"The glyceride (3–10 mg) and a known amount of a suitable methyl ester as internal standard (e.g. methyl pentadecanoate) are dissolved in dry diethyl ether (2 ml), and a solution of lithium aluminium hydride (20 mg) in dry diethyl ether (3 ml) is added in 0.1 ml portions until the boiling stops. A one volume excess of the lithium aluminium hydride solution is added, and the mixture refluxed for an hour. Acetic anhydride is added dropwise with cooling to decompose excess reagent, followed by further acetic anhydride (2.5 ml) and xylene (3 ml). The ether is evaporated off and the solution refluxed for 6 hr (the reflux temperature should be at least 110°C if no ether remains), when the reagents are removed on a rotary evaporator and the products taken up in dry diethyl ether for GLC analysis."

The procedure can be scaled down considerably. Triacetin elutes just before octadecanyl acetate on an EGS column at 170°C, but very much earlier (near

octanyl acetate) on non-polar stationary phases such as ApL or SE-30.[329] It is necessary to determine experimentally the molar response factor of the GLC detector for triacetin relative to the alcohol acetates, but with care the ratio of glycerol to fatty alcohols and, therefore, to fatty acids can be accurately determined. The analysis of fatty alcohol acetates is discussed further below (Section C3).

4. Cholesterol

Free cholesterol (and cholesterol esters) in lipid extracts can be quantified by a number of GLC procedures, including that described in Section A3 above. In most, a known amount of lipid extract is separated by TLC using one of the solvent systems described above, the cholesterol band is located and the compound is eluted from the adsorbent with chloroform. A known amount of cholestane (or octadodecane or desmosterol) is added as internal standard, the solvent is removed and the compounds are taken up in a little diethyl ether for GLC analysis. Columns (50–100 cm long) containing 3–5 per cent SE-30 as lipid phase operated at 200–240°C are generally satisfactory, but more symmetrical peaks may be obtained, and a slightly lower column temperature used, if the cholesterol is first converted to the acetate or trimethylsilyl ether derivative (see Chapter 4). The amount of cholesterol in the sample is calculated by relating the area of the cholesterol peak on the GLC trace to that of the internal standard. Cholesterol in cholesterol esters can also be estimated by this method if the latter are first hydrolysed or transesterified. The procedure has been thoroughly evaluated.[772] The most widely used chemical method is based on the Liebermann–Burchard reaction.[334]

"Sulphuric acid (10 ml) is added dropwise with stirring to acetic anhydride (60 ml), taking care that the temperature of the mixture does not rise above 10°C. Glacial acetic acid (30 ml) and anhydrous sodium sulphate (0.6 g) are then dissolved in the mixture with care. The reagent (5 ml), which is stable for about two weeks, is added to the lipid in acetic acid solution (0.2 ml) at 25°C and, after 20 min, the absorbance of the unknown and a series of standards and blanks are determined at 550 or 610 nm. The amount of free cholesterol in the sample is read from a calibration curve prepared from the standards."

Cholesterol esters must again be hydrolysed before the cholesterol residue can be determined; the procedure works best with pure compounds isolated by chromatography. An alternative procedure is available that utilises less hazardous reagents, and is dependent on a reaction between cholesterol and ferric salts.[856]

An enzymatic method has been described that uses a coupled reaction with cholesterol oxidase.[776] It has found favour with clinicians, as it lends itself to automatic analysis and with care can be applied directly to plasma samples without prior extraction of the lipids.[868] Kits containing all the reagents necessary for the purpose are available from a number of commercial sources.

C. SOME SPECIAL CASES

1. Preparative separation of isomeric mono- and di-acylglycerols

Partial glycerides can be recovered from tissues with minimal isomerisation, if this is necessary, by extracting the tissues with non-alcoholic solvents such as diethyl ether or chloroform, taking care not to heat extracts at any stage, although it should be recognised that other lipid constituents will not be recovered completely. In addition, partial glycerides are formed as essential steps in chemical or enzymatic degradation of other lipids (e.g. in stereospecific analysis of triacylglycerols or on phospholipase C hydrolysis of phosphoglycerides; see Chapter 9) or by chemical synthesis, and in such circumstances they must be purified by chromatographic procedures in which the isomerisation permitted is minimal. There should be no delay between the extraction and purification steps. The simplest method of obtaining pure positional isomers is to chromatograph the partial glycerides on TLC plates coated with silica gel G impregnated with boric acid (see Chapter 3 for details of preparation), using a solvent system of chloroform (alcohol-free)–acetone (96:4, v/v).[767] 1,2- and 1,3-Diacylglycerols and 1- and 2-monoacylglycerols can

be cleanly and safely separated by this procedure, as illustrated in Fig. 6.5. Lipids are detected by means of the 2′,7′-dichlorofluorescein spray and recovered from the adsorbent by elution with chloroform–acetone (9:1, v/v) or with diethyl ether (diacylglycerols only), which should be washed with ice water to remove any boric acid and dried rapidly over anhydrous sodium sulphate before removal of the solvent at ambient temperature. Diacylglycerols will isomerise slowly on standing in inert solvents or in the dry state even at low temperatures, and if they are required for any synthetic or biosynthetic purpose, they should be used as soon as possible after preparation.

Up to 20 mg of partial glycerides can be purified preparatively on a 20 × 20 cm plate coated with a layer 0.5 mm thick of silica gel impregnated with 10 per cent boric acid, but larger amounts can be purified by column chromatography on acid-washed Florisil impregnated with boric acid.[684]

Individual isomers of partial glycerides have distinctive NMR spectra[684] that can be used to aid their identification. Separations of mono- and diacylglycerols into individual molecular species are described in Chapter 8.

2. Alkyldiacylglycerols and neutral plasmalogens

(I) SEPARATION AND IDENTIFICATION

Alkyldiacylglycerols (glyceryl ethers) and neutral plasmalogens are generally minor components of animal tissues, but may occasionally assume greater

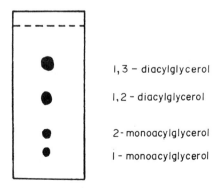

FIG. 6.5. Thin-layer separation of partial glycerides on silica gel G containing 5% (w/w) boric acid. Solvent system, chloroform–acetone (96:4 v/v).[767]

significance. Methods of analysis of these compounds have been reviewed.[665, 722, 803]

Adsorption TLC on silica gel layers can be used to separate simple alkyl and alkenyl lipids; neutral plasmalogens tend to migrate ahead of alkyldiacylglycerols, which in turn migrate just in front of triacylglycerols as illustrated in Fig. 6.6. Complete resolution of each is not always easy but a double development in a single direction with toluene, toluene–methanol (199:1, v/v)[619] or hexane–diethyl ether (95:5, v/v)[667] as solvent system can give satisfactory results, although it may be necessary to try various proportions of the more polar component in the mixture. A TLC procedure has been described for separating trace amounts of alkyl glycerides from large quantities of triacylglycerols,[667] but column chromatographic procedures do not in general possess sufficient resolving power.

Neutral plasmalogens may be detected by spraying the TLC plates with 2,4-dinitrophenylhydrazine (0.4 per cent) in 2 M hydrochloric acid; aldehydes are released which show up as yellow-orange spots on warming the plates. Alternatively, the TLC plate can be exposed to fumes of concentrated hydrochloric acid to split the vinyl ether bond,[668] and the released aldehydes detected as purple spots by spraying with a 2% aqueous solution of 4-amino-5-hydrazino-1,2,4-triazole-3-thiol in 1 M NaOH.[605] A two-dimensional reaction TLC technique can be used; the sample is applied in the bottom left-hand corner and the plate run in the first direction with hexane–diethyl ether (95:5, v/v) as

solvent system, the plate is exposed to hydrochloric acid fumes to liberate the aldehydes, and then turned anticlockwise and run in the second direction with hexane–diethyl ether (80:20, v/v) as developing solvent. The reaction products are easily distinguished from the unchanged compounds after detection with suitable reagents (see Fig. 6.7),[668] and all can be recovered from the plate for further analysis (although there are doubts as to whether the recovery of aldehydes is quantitative).

Spectroscopic aids to identification are also useful with plasmalogens; the ether-linked double bond exhibits a characteristic band in the IR spectrum at 6.1 μm and the olefinic protons adjacent to the ether bond produce a doublet centred at 4.11 τ in the NMR spectrum.[666]

There is no simple spot test for the identification of alkyldiacylglycerols and the ether linkage is not easily disrupted. They must be identified by their chromatographic behaviour on TLC adsorbents relative to an authentic standard or by the chromatographic behaviour of the hydrolysis products — free fatty acids and 1-alkoxy glycerols. The latter compounds migrate close to monoacylglycerols on TLC adsorbents, but cannot be hydrolysed further, and react with Periodate–Schiff reagent (see Chapter 7). The ether bond in alkyldiacylglycerols produces a sharp characteristic band at 9 μm in their IR spectrum.[59, 109] In the NMR

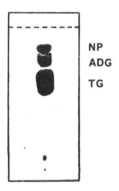

FIG. 6.6. Thin-layer separation of triacylglycerols (TG), alkyldiacylglycerols (ADG) and neutral plasmalogens (NP) on silica gel G layers. Solvent system; hexane–diethyl ether (95:5 v/v; double development in same direction).

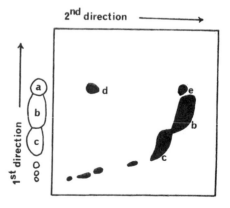

FIG. 6.7. Two-dimensional reaction TLC of neutral plasmalogens and related lipids on silica gel G. Solvent system, 1st direction: hexane–diethyl ether (95:5 v/v). Plate then exposed to hydrochloric acid fumes. Solvent system, 2nd direction: hexane–diethyl ether (80:20 v/v).[668] a, neutral plasmalogens; b, alkyldiacylglycerols; c, triacylglycerols; d, diacylglycerols; e, aldehydes.

spectra of the compounds, the protons on the carbon atom adjacent to the ether group give rise to a characteristic signal in the form of a triplet centred on 6.6τ.[684]

As cautioned earlier (Chapter 2), lipid samples containing plasmalogens should not be stored for long periods in solvents containing acetone, methanol or glacial acetic acid, as some rearrangement may occur.

(II) QUANTIFICATION AND THE ANALYSIS OF THE ALKYL MOIETIES

When they have been adequately resolved by TLC, alkyldiacylglycerols, neutral plasmalogens and triacylglycerols can be quantified by procedures such as charring followed by densitometry, or by GLC of the methyl esters of the fatty acid components of the compounds with a suitable internal standard. Basic transesterification procedures should be used to avoid disruption of the vinyl ether bond, which would result in contamination of the methyl esters by free aldehydes or dimethyl acetals. The methyl esters should be separated from the other hydrolysis products, which can also be recovered for analysis by the procedures described below, by preparative TLC on silica gel G layers with hexane–diethyl ether (95:5, v/v) as solvent system; methyl esters migrate close to the solvent front and the other products remain near the origin. Ultimately the calculation of the results should take into account the differing molar proportions of fatty acids in each lipid class.

Alternatively, such lipids can be quantified by analysis of the alkyl or alk-1-enyl moieties and indeed in a complete analysis, these should be characterised. The total amount of plasmalogenic material in a lipid sample can be determined by preparing the p-nitrophenylhydrazone derivative (see Chapter 4 for details of preparation) of the aldehydes released under acidic conditions and estimating these spectrophotometrically at 395 nm relative to a suitable blank. The amount present is obtained from a standard calibration curve prepared at the same time.[208, 613]

If it is intended that individual isomers of the fatty aldehydes be determined, they must first be liberated quantitatively from the neutral plasmalogens. Anderson et al.[28] have critically examined some of the acidic hydrolysis procedures that have been described, and recommend the following:

"The plasmalogens (0.2–2 mg) in diethyl ether (1.5 ml) are shaken vigorously for 2 min with concentrated hydrochloric acid (1 ml). The ether layer is removed and the aqueous phase is extracted once more with ether and once with hexane. The combined extracts are washed twice with distilled water before the solvent is removed in a stream of nitrogen. The free aldehydes are obtained by preparative TLC on silica gel G layers, with hexane–diethyl ether (90:10, v/v) as solvent system. Aldehydes migrate to just below the solvent front, and can be recovered from the adsorbent for further analysis by elution with diethyl ether. 2,3-Diacyl-sn-glycerols, the other product of the reaction, are found much further down the plate."

The hydrolysis procedure can be carried out in the presence of silica gel if necessary. It is now known that, although complete hydrolysis of the vinyl ether bond occurs with this method, only 80 per cent recovery of aldehydes can be attained, although they are probably representative in composition of those originally present in the natural compound.[843] It is possible that generation of aldehydes with 1% trichloroacetic acetic acid containing mercuric chloride (5mM)[169] will have fewer side-effects, but a critical comparison with more widely used methods does not appear to have been made.

The alkenyl moieties of neutral plasmalogens can be converted directly to dimethylacetal derivatives of aldehydes by reaction with anhydrous methanolic hydrogen chloride (see Chapter 4 for details). The methyl ester derivatives of fatty acids are formed simultaneously, but the two classes of compound can be separated for analysis by means of preparative TLC with toluene[480] or dichloroethane[833] as solvents for development (esters migrate ahead of acetals).

Gray[239] and Mahadevan[475] have reviewed GLC procedures for the analysis of aldehydes and their derivatives. Aldehydes can be analysed by GLC, without prior conversion to other derivatives, on similar columns to those used for methyl ester analysis, and they can be identified and estimated by analogous procedures. Standard aldehyde mixtures are available commercially or they can be synthesised easily from the corresponding fatty acids by a number of methods.[475] Because of the tendency of free aldehydes to polymerise on standing, especially in the presence of traces of alkali, it is more usual to convert them to more stable derivatives. Nonetheless, aldehydes have been reported to be stable for long periods, if stored at −20°C in solution in carbon disulphide[842] or other inert solvents such as pentane or diethyl ether. They should not be stored in contact with other lipids, especially those containing ethanolamine, which catalyses a condensation reaction in which 2,3-dialkylacroleins are formed.[670]

Of the derivatives, acetals are the most popular, especially dimethylacetals which are easy to prepare, although cyclic acetals (of 1,3-propanediol in particular) are also used because of their stability. Dimethyl acetals may decompose to form alk-1-enyl methyl ethers, if aluminium columns or stationary phases containing acid are used in the gas chromatograph.[475] If glass columns are used with packing materials that contain catalyst-free stationary phases, artefact formation is neglible, and dimethyl acetal derivatives are particularly recommended for the analysis of aldehydes by Gray[239]. They can be separated according to chain-length and number of double bonds, under similar GLC conditions as are used with the analogous methyl esters, and can be identified by their retention times relative to authentic standards or by using equivalent chain length (ECL) values, as illustrated earlier (Chapter 5) for methyl esters. In addition, individual components can be isolated and characterised by similar procedures to those used to determine the structures of fatty acids.

The IR spectra of free aldehydes are similar to those of the related esters, except that the characteristic frequency of the carbonyl function is at 5.9 μm with an additional band at 3.7 μm.[479] A triplet at 0.3τ is characteristic of the proton on the carbonyl group in NMR spectra of aldehydes.[475] Mass spectra of many aldehydes have been recorded; though the parent molecular ion is rarely seen, there are characteristic fragments at m/e = M-18, M-28 and M-44.[123] More distinctive spectra are obtained from the alcohols prepared from the aldehydes by lithium aluminium hydride reduction.[516] The melting points of most aldehydes are too low to be of assistance in their identification, but those of the 2,4-dinitrophenylhydrazone derivatives of fatty aldehydes are often sufficiently distinct.[479]

1-Alkylglycerols are released from alkyldiacylglycerols by saponification or transesterification,

but better yields are obtained if hydrogenolysis with lithium aluminium hydride is used.[771, 849] With the latter procedure, alk-1'-enylglycerols are released from plasmalogens. Although the technique has also been used to remove the phosphorus group from phosphoglycerides, hydrogenolysis with vitride reagent (70% sodium bis-(2-methoxyethoxy) aluminium hydride in benzene) reportedly gives better results.[723, 742] The following method is recommended for neutral lipids.[849]

"Lithium aluminium hydride (15 mg) in diethyl ether (3 ml) is added to the lipid (5 mg) in diethyl ether (0.5 ml) in a test-tube, and the solution is refluxed for 30 min. On cooling, water (3 ml) is cautiously added, followed by 4 per cent acetic acid (3 ml) and diethyl ether (3 ml) and the mixture is thoroughly shaken. The ether layer is removed by Pasteur pipette (after centrifugation to break any emulsions if necessary), and the aqueous layer is extracted twice more with diethyl ether (6 ml portions). The solvent is removed from the combined extracts in a stream of nitrogen (the last traces of water are removed *in vacuo*), and the samples are dissolved in a little chloroform and applied to a silica gel G TLC plate, which is then developed in diethyl ether — 30 per cent aqueous ammonia (400:1, v/v). The products are identified by their R_f values relative to standards and are eluted from the adsorbents with several volumes of diethyl ether."

The alkyl- and alk-1'-enylglycerols can be estimated on the TLC plate by charring and photodensitometry,[849] or they can be recovered from the adsorbent and analysed separately by appropriate procedures. Alk-1'-enyl-1-glycerols are not normally analysed in this form, but are converted to aldehydes as described above. The fatty acid components of the sample are reduced to fatty alcohols during the reaction, but can be recovered for analysis as described below.

Alkyl-1-glycerols must be converted to less polar volatile derivatives such as isopropylidene compounds,[275, 570, 742] trimethylsilyl ethers,[846] trifluoroacetates[846] or acetates[23] for GLC analysis (see Chapter 4 for details of preparation). Isopropylidene derivatives appear to be generally favoured. They may be separated both according to chain-length and to the number of double bonds in the alkyl chain, on GLC columns packed with similar polyester liquid phases, e.g. EGS, EGSS-X or EGSS-Y, as are used to separate methyl esters, except that somewhat higher column temperatures are necessary. By using appropriate internal standards, the amount of alkyl ether lipid in a neutral lipid sample can be estimated,[570, 742] or indeed the amounts of alkyl- and alk-1'-enylglycerols present and the nature of the aliphatic moieties can be determined[742]. Individual isomers of natural alkyl-1-glycerols or their derivatives can be isolated by procedures related to those described for fatty acids[846], and the positions of double bonds can be determined by permanganate-periodate oxidation[275] or by ozonolysis[610], with GLC identification of the fragments.

Of the many other methods of analysing alkyl-1-glycerols that have been described, one other merits further consideration; the preparation under basic conditions of thionocarbonate derivatives, which are estimated by their absorbance at 235 nm and which are also suited to GLC analysis (it has been suggested that alk-1'-enyl-1-glycerols might be analysed in this form).[609]

IR and NMR spectra of alkyl-1-glycerols or their derivatives are similar to those of the parent alkyldiacylglycerols, except that the free hydroxyl groups or specific functional groups in the derivatives introduce additional features.[848] Individual isomers can be identified by mass spectrometry.[266]

3. WAX ESTERS AND RELATED COMPOUNDS

Natural waxes may contain a wide variety of simple lipid components including hydrocarbons, wax esters, fatty alcohols and ketones, sterols and sterol esters, as well as the more common simple lipids such as triacylglycerols. Thin-layer chromatography on layers of silica gel has proved to be the best method of separating these compounds; silicic acid column chromatography in the conventional mode lacks resolution, but some potentially useful HPLC separations have been described.[20] High-temperature gas chromatography also has some value. Methods for the analysis of waxes have been reviewed in some detail.[346, 779]

Holloway and Challen[313] have systematically studied TLC methods for separating wax con-

stituents and have developed several valuable solvent systems, although no single one will adequately resolve all compounds. Benzene–chloroform (ethanol-free) (7:3, v/v) was a good general solvent, but others may be preferred for specific purposes.

Hydrocarbons tend to elute near the solvent front with even the least polar of solvent systems, but normal paraffins are eluted just ahead of squalene if carbon tetrachloride[313] or hexane–benzene (9:1, v/v)[260] are the developing solvents. Individual hydrocarbon isomers can be separated by gas chromatography; for example on a capillary column linked to a mass spectrometer, 93 different compounds from wool wax have been separated and identified.[504] Homologous series of hydrocarbons differing in degree of unsaturation have been separated by silver nitrate TLC.[727]

Of the compounds generally classed as wax components, wax esters, which consist of long-chain fatty acids esterified to primary alcohols of analogous structure, and their constituents have been most studied as they are major components of the lipids of many marine animals. They are only slightly more polar than cholesterol esters from which they are not easily separated. Double development in carbon tetrachloride[313] or hexane–diethyl ether (98:2, v/v) on layers of silica gel G may be successful, however, although there is frequently some band spread because of partial separation into molecular species. It is also claimed that these compounds can be separated on thin layers of magnesium oxide[545], but this cannot always be achieved reproducibly. Methods of analysis of molecular species of wax esters are discussed in Chapter 8. The fatty acids and alcohols of wax esters are liberated by alkaline hydrolysis, and may be separated from each other by solvent extraction of the alkaline solution (see Chapter 4) or by preparative TLC (transesterification can also be used in this instance). Methods for the analysis of fatty alcohols have been reviewed by Mahadevan.[476] Primary fatty alcohols migrate just ahead of free cholesterol on TLC with the solvent systems shown in Fig. 6.1, while secondary alcohols migrate just before the primary compounds. Individual isomers of the fatty alcohols, usually in the form of less polar derivatives to minimise tailing, can be separated by chain-length and degree of unsaturation, and estimated by GLC on similar polyester liquid phases as are used with methyl esters. Acetate derivatives are generally preferred as the resolution of individual isomers on polyester columns is somewhat better than that obtained with trifluoroacetate or trimethylsilyl ether derivatives, although these are eluted earlier than acetates and are sometimes used also.[352] It should be noted that the acetate derivatives of polyunsaturated fatty alcohols do not necessarily elute in the same order as the corresponding methyl esters on polyester columns.[352]

Methyl esters of fatty acids are occasionally found as constituents of tissues, but as they may be formed as artefacts during extraction of tissues with solvents containing methanol, the natural origin of such compounds must be established (see Chapter 2), when their presence is detected.

It is also possible to analyse waxes by high temperature GLC[779] (see Chapter 8 for a discussion of the technical problems). Separation of each lipid class is achieved according to the combined chain-lengths ("carbon numbers") of the aliphatic constituents; although the chromatographs obtained are highly complex, components can often be identified by comparisons of the results obtained before and following acetylation or methanolysis.

CHAPTER 7

The Analysis of Complex Lipids

COMPLEX lipids from natural sources may contain such a multiplicity of components that complete separation, identification and estimation of each one is an extremely difficult task. The nature of the problem is dependent upon the types of lipid in a given sample, and this in turn is dependent on its source i.e. whether it is of animal, plant or bacterial origin. The preferred approach to the problem will depend on the amount of information required. Simple methods may be available for the isolation and estimation of particular complex lipid classes, but if a detailed knowledge is needed of the amounts of all the lipid classes present in a sample together with their fatty acid compositions say, it is necessary to use a range of chromatographic methods. It is not possible to describe a definitive procedure that can be used with confidence in all circumstances.

Commonly, the total lipid extract from a tissue is subjected to a preliminary separation by column chromatography on silicic acid and the complex lipids separated from simple lipids. It is often convenient to separate complex lipids into phospholipid and glycolipid fractions, and to treat these separately. Single and two-dimensional TLC systems have been developed that permit separation of most complex lipids on an analytical or semi-preparative scale, but when larger amounts of individual components are required for more detailed analysis, column chromatography on DEAE cellulose, for example, must be used to obtain discrete fractions that can be further subdivided by column or one-dimensional TLC techniques. When only one compound or group of compounds is required, short cuts may be available. Unequivocal identification of complex lipids requires that the various hydrolysis products from each compound be identified and estimated. Precautions should always be taken to minimise the effects of autoxidation (see Chapter 3).

A. PRELIMINARY SEPARATION AND PREPARATION OF LIPID SAMPLES

1. Simple group separations

The following procedures apply to extracts of complex lipids that do not contain significant amounts of gangliosides, i.e. samples obtained by the conventional extraction-washing procedures described in Chapter 2 (and also below).

Complex lipids can be separated from simple lipids by column chromatography on silicic acid or acid-washed Florisil (30 mg lipid/g adsorbent) as described in Chapter 6. Briefly, chloroform or diethyl ether (10 column volumes) will elute the simple lipids, and methanol (10 column volumes) the complex lipids. For simple fractionations of this nature, short wide columns are used so that rapid flow rates are achieved. If necessary further subdivision of natural mixtures is possible, as glycolipids can be recovered as a class by elution with acetone after the simple lipids have been washed from the column. In particular, mono- and digalactosyldiacylglycerols and plant sulpholipid may be separated from the phospholipids of plant tissues[810] and glycosyldiacylglycerols can be isolated from bacterial lipids[688] by elution with this solvent (10 column volumes). Acetone (40 column volumes) will elute ceramide monohexosides, ceramide polyhexosides and sulphatides from lipid extracts of animal tissues,[639] and tetrahydrofuran–methylal–methanol–water (10:6:4:1 by volume) has been used with equal effect[237]. Phospholipids, virtually free of glycolipids, are then recovered from the column by elution with methanol.

The principal disadvantage of the procedure is that small amounts of the less-polar acidic phospholipids, such as cardiolipin and phosphatidic acid, occasionally accompanied by phosphatidyl-ethanolamine, may contaminate the acetone fraction. With some batches of adsorbent, greater amounts of these may be eluted with acetone than is acceptable, but when this is seen to occur it can be minimised by eluting with chloroform–acetone mixtures rather than with acetone alone. In addition, acetone has a strong dehydrating effect on the adsorbent causing the phospholipids to be retained more strongly than otherwise would be the case, so that recoveries are sometimes less than ideal.

With care, individual glycolipids can be further resolved by eluting with chloroform containing various proportions of acetone, while individual phospholipids are recovered subsequently by eluting with chloroform containing methanol; this is discussed in greater detail below.

Simple and complex lipids can be effectively separated from each other by solvent partition procedures, and that described by Galanos and Kapoulas[223] is especially recommended as a semi-quantitative preparative method.

"Hexane is equilibrated with 87 per cent ethanol. The hexane layer is contained in two separating funnels (45 ml in each), and up to 10 g of lipid in the ethanolic phase (15 ml) is added to the first funnel. After thorough shaking, the bottom layer is run into the second funnel and the two phases are again mixed, after which the bottom layer is run off. A further portion (15 ml) of fresh ethanolic phase is added to the first funnel, and the extraction procedure is repeated. Six further portions of fresh ethanolic phase are shaken with the hexane layer in this way. The combined ethanolic layers then contain all the complex lipids, and the combined hexane layers most of the simple lipids."

The method is particularly suited to lipid samples that contain a high proportion of simple lipids, for example adipose tissue extracts or milk fats, which may contain less than 1 per cent of the total lipid in the form of complex lipids. Emulsion formation can be troublesome on occasion, but the only other disadvantage of the method is that free fatty acids partition fairly evenly between the two phases. These and traces of other simple lipids can be removed later by column chromatography as described above.

Simple and complex lipids can also be separated by dialysing lipid extracts in hexane solution through a rubber membrane against fresh hexane. Simple lipids, including free fatty acids in this instance, pass through the membrane but complex lipids do not.[788]

2. Cations associated with complex lipids

It is not always realised that the solubility and chromatographic properties of acidic phospholipids are somewhat dependent on the cations with which they are associated. In particular, divalent ions, for example those of calcium or magnesium, confer quite different elution characteristics than those of the sodium or potassium salts, on acidic lipids such as phosphatidic acid, phosphatidylserine, phosphatidylinositol and cardiolipin on adsorption chromatography. Single phospholipid classes may on occasion be split into two or three bands by TCL, according to the types of cation associated with the molecules. Before complex lipids are separated into their constituents, it may therefore be advantageous to ensure that all the phospholipid classes are in a single salt form — preferably as the sodium or potassium salts. This can best be achieved by passing the lipid extracts through a chelating resin as described below.[110] Washing the lipid in an appropriate solvent with a solution containing the chosen cation is only partially effective.

"The resin (Chelex 100, 50–100 mesh, Biorad Laboratories) is prepared in the proper ionic form by washing it twice with two volumes of 2 M hydrochloric acid, then thoroughly with deionised water. The cycle is repeated three times with 2 M sodium hydroxide solution (or potassium hydroxide depending on which cation is required) and then the resin is washed with deionised water until its pH is approximately 12. The pH is adjusted to 8.0 by washing with 2 M acetic acid, when the solvent is changed to chloroform–methanol–water (5:4:1 by volume) by thorough washing. The lipids (up to 10 g) are added to a column (4 × 50 cm) of the resin in this solvent, which is also used for elution (3 column volumes). Complete conversion to the required salt form occurs."

Although this procedure may not be necessary with all lipid samples and is not practical when large numbers of samples have to be analysed routinely, it should be accepted as general practice when highly precise analysis is intended. The benefits of this treatment will of course be negated if the lipids are subsequently chromatographed on adsorbents containing appreciable amounts of metal ions (e.g. silica gel G). Chromatography with solvents containing ammonia will effectively convert all the lipids to the ammonium salt form and eliminates the need for the pretreatment of lipid samples.

B. GROUP SEPARATIONS BY COLUMN CHROMATOGRAPHY

Complex lipids eluted from columns can be determined by the general procedures described in Chapter 3. Methods more specific for complex lipids are also available, for example determination of phosphorus in glycerophosphatides or of nitrogen in glycosphingolipids, and these are discussed in detail below. The choice of a particular method will be dependent on the nature of the lipid sample, the degree of purity required and the scale of the operation.

1. Preparative chromatography on silicic acid columns

Chromatography on columns of silicic acid is a useful preparative procedure for subdividing large amounts of complex lipids into simpler fractions. Pure single compounds are not easily obtained from natural mixtures by this technique, however, and it is customary to use the method in conjunction with preparative TLC or with DEAE cellulose chromatography for complete separations. The preparation of suitable columns has been described in Chapter 3, and the application of the method to the analysis of complex lipids has been reviewed elsewhere.[640, 755]

The principal advantages of the method lie in the ease of preparation of the column and in the comparatively large amount of lipid that can be separated (30 mg lipid per g of adsorbent). On the debit side, the properties of the adsorbent can change with slight changes in its physical state, e.g. in the size of the particles or in the degree of hydration. As a result, the acidic lipids such as phosphatidylserine or phosphatidylinositol can change their elution

characteristics with respect to non-acidic lipids. With chloroform containing various proportions of methanol, acidic lipids are eluted in the order:

cardiolipin
  phosphatidic acid
    phosphatidylglycerol
      phosphatidylserine
        phosphatidylinositol
          di- and tri-phosphoinositides

and non-acidic complex lipids in the order:

ceramide
  ceramide monohexoside and glycosyldiacyl-
    glycerols
    phosphatidylethanolamine
      phosphatidylcholine, lysophosphatidyl-
        ethanolamine, ceramide dihexoside
        sphingomyelin
          lysophosphatidylcholine

The elution characteristics of the acidic lipids are also influenced by the cations with which they are associated. Both groups of lipids are always eluted in a constant order, but the regions where the two groups overlap may vary somewhat and it is not always possible to obtain clear-cut fractions. In consequence, it is usually necessary to determine the elution characteristics of a fresh batch of adsorbent before proceeding. As a rough guide, phosphatidylethanolamine and phosphatidylserine tend to elute together while phosphatidylinositol elutes just ahead of phosphatidylcholine.

More useful separations can be obtained if chloroform–acetone mixtures are used to elute glycolipids, before phospholipids are recovered by elution with chloroform–methanol mixtures. With plant lipids[810], for example, monogalactosyldiacylglycerols are recovered free of most other complex lipids by elution with chloroform–acetone (1:1 v/v; 5 column volumes); acetone alone (10 column volumes) will then elute digalactosyldiacylglycerols and plant sulpholipid together with some of the less polar phospholipids such as cardiolipin or phosphatidic acid, although the latter are not often major components of plant lipids. Chloroform–acetone mixtures are occasionally used with animal lipids but are generally less useful, as glycolipids tend to be minor components (with the important exception of brain lipids), but acetone alone (40 column volumes) can be usefully employed to elute cerebroside, cerebroside sulphate and ceramide dihexoside as discussed above.

Phospholipids can then be recovered from the silicic acid column by elution with chloroform containing increasing amounts of methanol. In a typical separation, phosphatidic acid and cardiolipin are eluted first with the solvents in the ratio 95:5, v/v (10 column volumes) followed by phosphatidylethanolamine and phosphatidylserine (80:20, v/v; 20 column volumes), phosphatidylcholine and phosphatidylinositol (50:50, v/v; 20 column volumes), and finally sphingomyelin and lysophosphatidylcholine are eluted with methanol alone. If care is taken and very slow flow rates, low lipid loads and gradient elution techniques are used, much finer separations are possible, but these are usually attainable with less effort by other methods. The nature of the fractions obtained can be monitored by means of micro-TLC plates (see Chapter 3).

Acid-washed silicic acid (prepared by washing the adsorbent with 6 M hydrochloric acid followed by water and activation at 120°C) can sometimes be used to obtain separations that are not possible with the untreated adsorbent. For example, phosphatidylserine is eluted before rather than with

phosphatidylethanolamine on acid-washed silicic acid.[640]

Silicic acid-silicate columns also are a useful alternative to the untreated adsorbent. A column prepared in the usual manner is washed with chloroform–methanol (4:1, v/v; 1 column volume) containing 1 per cent ammonia. Phosphatidylethanolamine is eluted with chloroform–methanol (4:1, v/v; 8 column volumes), and phosphatidylserine with methanol (3 column volumes) from such a column.[640, 643] Phosphatidylcholine and sphingomyelin are also more easily separated on silicic acid prepared in this way than on the unmodified adsorbent.[638, 640] Table 7.1 contains a summary of some useful procedures.

2. Preparative chromatography on Florisil and alumina columns

Florisil itself is little used for the separation of phosphatides, as they are very strongly retained by the adsorbent and are recovered only with difficulty, especially when the water content of the adsorbent is low. This phenomenon can be utilised, however, in the isolation of certain glycolipids, if very dry solvents (obtained by incorporating 5 per cent 2,3-dimethoxypropane) are used. For example, cera-

TABLE 7.1. CHROMATOGRAPHY OF COMPLEX LIPIDS ON SILICIC ACID COLUMNS

Fractions	Compounds eluted	Solvents*	Column volumes
Procedure A	Animal lipids on silicic acid		
1	simple lipids	chloroform	10
2	glycolipids with traces of acidic phospholipids	acetone	40
3	phospholipids	methanol	10
Procedure B	Fraction 3 from procedure A on silicic acid		
1	any remaining phosphatidic acid and cardiolipin	chloroform–methanol (95:5 v/v)	10
2	phosphatidylethanolamine + phosphatidylserine	,, ,, (80:20 v/v)	20
3†	phosphatidylinositol + phosphatidylcholine	,, ,, (50:50 v/v)	20
4	sphingomyelin and lysophosphatidylcholine	methanol	20
Procedure C	Fraction 2 from procedure B on silicic acid-silicate[643]		
1	phosphatidylethanolamine	chloroform–methanol (80:20 v/v)	8
2	phosphatidylserine	methanol	3
Procedure D	Plant lipids on silicic acid[810]		
1	monogalactosyldiacylglycerol	chloroform–acetone (50:50 v/v)	8
2	digalactosyldiacylglycerol and sulpholipid	acetone	10
3–6	as procedure B		

* The solvent mixtures described should be taken as an approximate guide only.

† Fractions 3 and 4 of procedure B can be better resolved on occasion with silicic acid-silicate as adsorbent.[638]

mide is eluted from Florisil with chloroform–methanol–dimethoxypropane (95:5:5 by vol.; 10 column volumes) and cerebroside and cerebroside sulphate with chloroform–methanol–dimethoxypropane (70:30:5 by vol.; 20 column volumes).[636] Glycosyldiacylglycerols from plant lipids can be separated from phosphoglycerides by a related procedure.[558] Glycolipids of animal origin are most easily recovered as a class for further analysis by Florisil chromatography, following acetylation (see below).[651] Such methods should be considered primarily as a means of obtaining concentrates of particular glycolipids rather than as precise analytical techniques, as the subsequent recovery of phospholipids is poor. Short wide columns and rapid flow rates are then used, as resolution is less important.

Alumina is rarely used as an adsorbent for column chromatography of complex lipids because of its basic nature; hydrolysis of glycerophosphatides occurs and significant amounts of lysophosphatides, not originally present in the sample, are found. Nonetheless, neutral alumina (Brockman grade IV) is occasionally made use of in a rapid method for obtaining concentrates of specific lipids. For example, neutral alumina (Merck, A.G., Germany; heated at 110°C for 24 hr then deactivated by adding 10 per cent water) provides a rapid means of obtaining phosphatidylcholine[699] and phosphatidylinositol[462] from lipid samples. Phospholipids are eluted in a quite different order from that described above for silicic acid, i.e.

phosphatidylcholine
lysophosphatidylcholine and sphingomyelin
cardiolipin and phosphatidylinositol
phosphatidylserine

The non-acidic lipids are eluted with chloroform, methanol and water in various proportions, but it is necessary to add ammonium salts to the solvents to recover the acidic lipids.[462]

Batches of alumina may vary much more markedly in their physical properties than those of silicic acid, so solvent systems quoted in the literature can only be taken as an approximate guide. However, smaller volumes of eluting solvents can be used than is required with silicic acid and comparatively rapid flow rates (up to 10 ml/min) are permissible. Although some hydrolysis of lipids

inevitably occurs even with deactivated adsorbents, the lipids recovered appear to be identical in fatty acid composition to the original compounds (i.e. hydrolysis of acyl groups occurs randomly), and they are suitable for analyses of molecular species.

3. DEAE, TEAE and CM-cellulose chromatography

DEAE cellulose chromatography is a valuable alternative column procedure to conventional adsorption chromatography for the separation of complex lipids. The principle of the method, the preparation of the adsorbent and packing of the columns are discussed in Chapter 3. The nature of the separations is very different from that of the procedures just described: choline- and ethanolamine-containing lipids are separately eluted from the columns with chloroform–methanol mixtures; weakly acidic lipids such as phosphatidylserine are eluted with acetic acid, and more highly acidic or ionic lipids, such as phosphatidic acid or phosphatidylinositol, are eluted with solvents to which inorganic salts or ammonia are added. Table 7.2 lists some useful general elution schemes for animal and plant lipids (bacterial lipids including glycophospholipids can also be separated efficiently by such procedures).[96, 211] Clear-cut fractions are obtained with virtually no cross-contamination and, when more than one compound is present in any given fraction, these are usually separated comparatively easily by subsequent thin-layer chromatography or by other column procedures. Different batches of DEAE cellulose do not vary markedly in their properties so that the elution schemes described in Table 7.2 are reproducible. As a result, bulk fractions can be taken with almost no danger of fraction overlap, even when the volumes of solvent used are increased.

DEAE cellulose procedures can also be used on a comparatively large scale for isolating acidic lipids. For example, 0.5–0.6 g of phosphatidylserine was isolated from extracts of brain tissue (50 g) on a 2 × 14 cm column of DEAE cellulose in a very short time.[657] CM cellulose may be even more useful in this particular instance.[146]

With plant lipids, a preliminary fractionation of glyco- and phospholipids on silicic acid columns, as described above, is unnecessary with DEAE

Lipid Analysis

TABLE 7.2. ELUTION SCHEMES FOR DEAE CELLULOSE COLUMN CHROMATOGRAPHY

Fraction	Lipids eluted	Solvents	Column volumes
A	*Animal lipids*[640]		
1	phosphatidylcholine lysophosphatidylcholine sphingomyelin cerebroside	choloroform–methanol (9:1, v/v)	10
2	phosphatidylethanolamine ceramide di- and poly-hexosides lysophosphatidylethanolamine	chloroform–methanol (1:1, v/v)	10
3	—	methanol	10
4	phosphatidylserine	glacial acetic acid	10
5	—	methanol	4
6	cardiolipin phosphatidic acid phosphatidylglycerol phosphatidylinositol cerebroside sulphate	chloroform–methanol (4:1, v/v) made 0.05 M with respect to ammonium acetate to which is added 20 ml of 28 per cent ammonia/litre	10
7	—	methanol	10
B	*Plant lipids*[541]		
1	monogalactosyldiacylglycerol	chloroform	20
2	phosphatidylcholine cerebroside sterol glycoside	chloroform–methanol (95:5, v/v)	10
3	digalactosyldiacylglycerol	chloroform–methanol (90:10, v/v)	10
4	phosphatidylethanolamine	chloroform–methanol (60:40, v/v)	10
5	cardiolipin phosphatidic acid phosphatidylglycerol phosphatidylinositol plant sulpholipid	solvent as for fraction 6 above	10
6	—	methanol	10

cellulose chromatography, as the glycolipids are eluted in separate distinct fractions.[541] Indeed, no preliminary separation of simple and complex lipids is necessary as simple lipids, with the exception of free fatty acids, are eluted with chloroform (10 column volumes) before the elution sequences in table 7.2 are commenced (with plant lipids, simple lipids are eluted in the first 10 column volumes of chloroform and monogalactosyldiacylglycerols in the second 10 column volumes). Free fatty acids are eluted with the weakly acidic lipids (fraction 4 in scheme A and fraction 5 in scheme B), but if they are major components of the sample, a preliminary separation of the simple lipids on a silicic acid column is advantageous, otherwise the retention power of the DEAE cellulose adsorbent for the other acidic lipids is diminished.

Care should be taken not to change back from polar to non-polar solvents too quickly, as the column packing may be disturbed and the quality of the separations impaired. For this reason, methanol washes are inserted as fractions 5 and 7 in elution scheme A (Table 7.2). If the column is not allowed to run dry, it can be regenerated for further use after the final methanol wash by elution with glacial acetic acid (3 column volumes), methanol (3 column volumes), chloroform–methanol (1:1, v/v; 3 column volumes) and chloroform (4 column volumes). With comparatively uncomplicated lipid mixtures, each solvent is run rapidly (approximately 10 ml/min) through the column, but with more difficult samples, a slower flow rate (3 ml/min) may be preferable. Salts in column fractions are removed by Sephadex chromatography or by a "Folch" wash (see Chapter 2).[640]

DEAE cellulose in the borate form has somewhat different properties; it is prepared by washing the DEAE cellulose column, prepared as described earlier (Chapter 3), with a saturated solution of sodium tetraborate in methanol (3 column volumes), after which excess borate is removed by washing with methanol then chloroform–methanol mixtures. The chief virtue of such columns is that phosphatidyl-ethanolamine is more strongly retained so that ceramide polyhexosides are obtained as a separate class, free of glycerophosphatides, by elution with chloroform–methanol (2:1, v/v; 10 column volumes). After washing the column with methanol, pure phosphatidylethanolamine is recovered by elution with chloroform–methanol–acetic acid (63:33:1 by vol.; 8 column volumes). Accordingly, such columns are a useful supplement to conventional DEAE cellulose chromatography, as they permit the subdivision of the second fraction (scheme A, Table 7.2) into simpler components. Columns must be regenerated with sodium tetraborate in methanol before they are reused. In addition, DEAE-Sephadex and DEAE-silicic acid preparations have found favour, when glycosphingolipids are the compounds of principal interest (see Section E.1.).

TEAE cellulose in the hydroxyl form is a useful alternative to DEAE cellulose as it has a much higher capacity for acidic lipids and, in contrast to the latter, ceramide polyhexosides and phosphatidylethanol-amine are eluted in quite different fractions. Columns are prepared as follows:[640]

"The TEAE cellulose adsorbent is prepared and packed into columns in a slurry of acetic acid, as described earlier (Chapter 3) for DEAE cellulose. It is then washed with methanol (3 column volumes) and converted to the hydroxyl form by washing with 0.1 M potassium hydroxide in methanol (4 column volumes). After further washes with methanol (8 column volumes), chloroform–methanol (1:1, v/v; 4 column volumes) and chloroform (4 column volumes), the column is ready for use."

Approximately 10–30 mg of lipid per g of adsorbent can be separated on this column, if a solvent flow rate of 3 ml/min is not exceeded. The recommended elution sequence is outlined in Table 7.3. In this instance, fraction 2 contains only ceramide di- and polyhexosides, while phosphatidyl-ethanolamine is eluted cleanly in a later fraction. Therefore, when the former group of compounds is of particular interest to the analyst, TEAE cellulose chromatography is recommended.[782, 783]

Carboxy-methyl (CM) cellulose in the sodium form has been introduced comparatively recently as column packing for separations of complex lipids,[146] and has not yet been much used. It appears to offer a

TABLE 7.3. ELUTION SCHEME FOR TEAE CELLULOSE CHROMATOGRAPHY

Fraction	Lipids eluted	Solvents	Column volumes
1	As in fraction A-1 (Table 7.2)		8
2	ceramide polyhexosides	chloroform–methanol (2:1, v/v)	8
3	—	methanol	8
4	phosphatidylethanolamine phosphatidyl-N-methylethanolamine phosphatidyl-NN-dimethylethanolamine	chloroform–methanol–glacial acetic acid (63:3:1, by vol.)	6
5	phosphatidyl serine	glacial acetic acid	6
6	—	methanol	3
7	As in fraction A-6 (Table 7.2)		6

number of advantages, however. Acidic phospholipids are isolated by elution with solvents that are much less polar than are required with other adsorbents (chloroform–methanol mixtures only are needed), and lyso-phosphatides can be separated from the corresponding diacyl-compounds, for example. Elution conditions for some of the more common complex lipids are listed in Table 7.4.

Column chromatography on ion-exchange celluloses are among the more useful preparative methods for simplifying complex lipid mixtures. The only disadvantages are the rather lengthy conditioning necessary before columns can be used, some oxidation of phosphatidylethanolamine may occur and any phosphatidylserine plasmalogens present are destroyed by the acidic conditions necessary to elute them from DEAE-cellulose columns. The excellent review articles of Rouser et al.[640, 642] are recommended for reading before the technique is used. CM-cellulose does not suffer from some of the above disadvantages and is likely to find widespread use.

4. HPLC of complex lipids

Excellent separations of individual classes of phospholipids by means of HPLC have been obtained in a number of laboratories. Generally, the absorbent is microparticulate silicic acid, and complex lipids are eluted by gradients of polar solvents. Some of the more impressive separations have been made use of flame-ionisation detection systems (custom-built),[596, 737, 738] but others have obtained satisfactory results with UV detection at 206nm (isolated double bonds),[231, 289, 364] both on an analytical and a preparative scale, although quantification can present a problem, especially with more saturated lipids. Unfortunately, it appears that polar solvents such as methanol, water or ammonia, which are necessary for the elution of phospholipids with a minimum of tailing, rapidly deactivate the adsorbent and reduce the useful life-span of the rather costly columns. This is less of a problem with ion-exchange packing materials and some preliminary results appear promising,[243, 396] but further

TABLE 7.4 CONDITIONS FOR THE ELUTION OF COMPLEX LIPIDS FROM
CM-CELLULOSE COLUMNS[146]

Methanol (Vol. %)	Lipids eluted
0	Free fatty acids, cholesterol, di- and triacylglycerols
3	Diphosphatidylglycerol
4	Phosphatidylcholine
5	Sphingomyelin, phosphatidyl(m)ethanol, monoglucosyldiacylglycerol, trinitrobenzene sulphonate-labelled phosphatidylethanolamine
9	Phosphatidylethanolamine
12.5	Phosphatidic acid
15	Diglucosyldiacylglycerol
20	Phosphatidylglycerol
23	Lysophosphatidylcholine
30–35	Phosphatidylserine
35–50	Phosphatidylinositol
100 followed by 0	Regeneration

development work is necessary before such methods are recommended for general use. HPLC methods have been used to obtain particularly effective separations of glycolipids and these are discussed below (section E.1.).

C. THIN-LAYER CHROMATOGRAPHY OF COMPLEX LIPIDS

TLC is by far the most widely used method for the analytical scale separation of individual complex lipid classes. It can often be used with little modification and a minimum loss of resolution for small-scale preparative purposes, e.g. for further resolution of compounds separated by column chromatography. Similar separations to those described below can be obtained on silicic acid-impregnated paper,[207, 383] but for the reasons given in Chapter 3, TLC is generally preferred. TLC of complex lipids has been reviewed.[626, 627, 702] Rouser has concisely described the principles behind the selection of particular solvents for specific separations.[635]

One-dimensional TLC procedures are preferred for rapid group separations or for small-scale preparative purposes; two-dimensional TLC procedures tend to be preferred for resolution of the maximum number of distinct components. The choice of a particular method will also be influenced by the nature of the sample.

1. Single-dimensional TLC systems

Non-acidic phospholipids may be separated on plates coated with thin layers of silica gel G, developed in a solvent system of chloroform–methanol–water (25:10:1 by vol.). Lipids migrate in a similar order to the elution sequence outlined above for silicic acid columns, and a typical separation is illustrated in Fig. 7.1 (plate A). Satisfactory separations are achieved as long as acidic lipids are not present in appreciable amounts, otherwise phosphatidylserine will contaminate phosphatidyl-ethanolamine, and phosphatidylinositol will contaminate phosphatidylcholine. It is a particularly useful TLC system for isolating individual

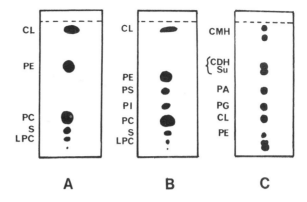

FIG. 7.1. TLC separations of complex lipids from animal tissues. Plate A, Silica gel G. Plates B and C, Silica gel H. Solvent systems: plate A, chloroform–methanol–water (25:10:1 by vol.); plate B, chloroform–methanol–acetic acid–water (25:15:4:2 by vol.);[704] plate C, 1st development–pyridine–hexane (3:1, v/v); 2nd development–chloroform–methanol–pyridine–2M ammonia (35:12:65:1 by vol.).[703] Abbreviations: CL, cardiolipin; PE, phosphatidylethanolamine; PC, phosphatidylcholine; S, sphingomyelin; LPC, lysophosphatidylcholine; PS, phosphatidylserine; PI, phosphatidylinositol; CMH, ceramide mono-hexosides; CDH, ceramide dihexosides; Su, sulphatide; PA, phosphatidic acid; PG, phosphatidylglycerol.

components from fractions 1, and occasionally 2, of the general DEAE cellulose elution scheme for animal lipids, outlined in Table 7.2, or for isolating the products of phospholipase A hydrolysis of phospholipids (see Chapter 9). Up to 10 mg of phospholipids can be applied to a 20×20 cm plate coated with a layer 0.5 mm thick of silica gel G. In most analytical applications, thinner layers (0.2–0.3 mm) are used, and much less phospholipid (<1 mg) is applied to the plate. Variations in the properties of the adsorbent are countered by raising or lowering the amount of water in the solvent system. When simple lipids are present in the sample, they migrate with the solvent front.

There is no single TLC system that can be used to separate all known complex lipids in one dimension, but Skipski et al.[704] have developed a useful procedure for the more common acidic and neutral glycerophosphatides. Plates are coated with a layer of silica gel H (without calcium sulphate as binder), made in a slurry of 1mM sodium carbonate solution to render it basic. After the plates are activated, lipids are applied in the usual way and the plates are developed in a solvent system consisting of chloroform–methanol–acetic acid–water (25:15:4:2 by vol.). Phospholipids migrate in the order:

cardiolipin and phosphatidic acid
phosphatidylglycerol and phosphatidylethanolamine
phosphatidylserine
phosphatidylinositol
phosphatidylcholine
sphingomyelin
lysophosphatidylcholine

When the acidic lipids, phosphatidylserine and phosphatidylinositol, are not completely separated from the non-acidic lipids because of slight differences in the properties of the adsorbent, small changes in the proportions of acetic acid and water in the solvent mixture will generally rectify the situation. Figure 7.1 (plate B) illustrates the nature of the separations that can be obtained with a natural mixture (pig liver lipids), which contains these components. Simple lipids run with the solvent front, but better results are obtained if they are first removed to the top of the plate by a prewash with acetone–hexane (1:3, v/v)[702], and the plate dried at room temperature in a vacuum oven before the phospholipids are separated as above. It has been claimed[388] that the addition of 0.4 per cent ammonium sulphate to the silica gel H layer and a slight change in the proportions of the solvents (to 50:25:8:1 by vol.) improves the reproducibility of the system.

The principal disadvantages of the procedure are that phosphatidic acid and cardiolipin, and phosphatidylglycerol and phosphatidylethanol-amine are not separated and, in addition, glycosphingolipids may contaminate phosphatidyl-ethanolamine. These difficulties can be partly overcome by a two-step single-dimensional TLC system.[703] Lipids are applied to a silica gel H plate, prepared as above, that is first developed in pyridine–hexane (3:1 v/v) to the top, when it is removed from the solvent tank and dried in a vacuum oven at room temperature. It is then redeveloped in the same direction with chloroform–methanol–pyridine–2M ammonia (35:12:65:1 by vol.) as solvent system, until this is about 2 cm below the previous solvent front. Lipids migrate in the order:

cerebrosides
ceramide dihexosides and sulphatides
phosphatidic acid
phosphatidylglycerol
cardiolipin
phosphatidylethanolamine

as illustrated in Fig. 7.1 (plate C). Fractions 2 and 6 of the DEAE cellulose elution scheme (Table 7.2A) can therefore be separated into individual components with this system. With care, good separations of individual phospholipids and glycolipids from tissues, such as brain, can be obtained by one-dimensional TLC[806], but in general better separations of complex glycosphingolipids are attained, if these compounds are first separated from glycero-phosphatides by an appropriate procedure. Such methods have been mentioned briefly and are discussed in greater detail below (Section E.1.).

Fractions enriched in acidic lipids, such as phosphatidylserine or phosphatidylinositol obtained from DEAE cellulose columns, are best purified on layers of silica gel containing magnesium silicate (10 per cent, w/w) as binder;[637] chloroform–meth-anol–7 M ammonia (60:35:4, by vol.) is a suitable developing solvent in this instance. Separations of polyphosphoinositides, phosphonolipids and ether-containing phospholipids from animal tissues are discussed below (Section F).

Single-dimensional TLC systems are used less with plant lipids, as glycolipids tend to overlap with phospholipids when chloroform–methanol solvent systems are used. Nonetheless, solvents containing acetone may on occasion offer useful separations. For example, acetone–acetic acid–water (100:2:1 by vol.) has been used to separate mono- and digalactosyldiacylglycerols from other complex lipids on layers of silica gel G (as illustrated in Fig. 7.2, plate A)[225] and, with some modification, the corresponding monoacyl compounds can also be separated.[521] More comprehensive separations have been described by Nichols (Fig. 7.2, plate B)[539] on silica gel G layers with diisobutyl ketone–acetic acid–water (40:25:3.7 by vol.) as the developing solvent, and further useful systems have been described by others.[521, 582]

High performance TLC (HP-TLC) has not yet been widely used in lipid analysis, but one valuable application has been to separate the main simple and

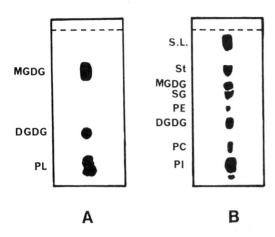

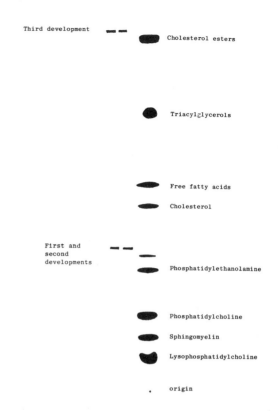

FIG. 7.2. TLC separations of plant complex lipids on silica gel G. Solvent systems: plate A, acetone–acetic acid–water (100:2:1 by vol.);[225] plate B, diisobutyl ketone–acetic acid (40:25:3.7 by vol.).[539] Abbreviations: MGDG, monogalactosyldiacylglycerols; DGDG, digalactosyldiacylglycerols; PL, phospholipids; St, sterols; SG, sterol glycosides; PE, phosphatidylethanolamine; PC, phosphatidylcholine; PI, phosphatidylinositol.

FIG. 7.3. High-performance TLC of plasma lipids. First and second development with chloroform–methanol–water (60:30:5 by vol.) and third development with hexane–diethyl ether–acetic acid (80:20:1.5 by vol.). (Redrawn from Kupke and Zeugner[427].)

complex lipid classes of human plasma in one dimension in a single analysis as shown in Fig. 7.3.[427] To reduce the number of steps in the analysis, aliquots of plasma, rather than of the lipid extracts, are applied directly to the plate and this is first given a double development with a polar solvent (chloroform–methanol–water, 65:30:5 by vol.) to about one quarter of the way up the plate in order to free the lipids from their association with plasma proteins and to effect a separation of the phospholipid classes. The plate is then developed to its full length with hexane–diethyl ether–formic acid (80:20:1.5 by vol.) to separate the simple lipid classes. In spite of the three developments, the combined elution time is only 30 min. Unfortunately, only small amounts of lipid (in 0.5 μl of plasma) can be applied as a spot to the plate and this can lead to problems in quantification. Commercial precoated HP-TLC plates are expensive.

Methods of recovery of complex lipids from TLC adsorbents have been discussed earlier (Chapter 3 and 6) and methods of detecting, identifying and estimating complex lipids are described in detail below. If single-dimensional TLC systems are to be used for routine separations of complex lipids, the

identity and purity of components should be checked periodically by two-dimensional TLC to confirm that single spots do not conceal more than one compound.

2. Two-dimensional TLC systems

Many more complex lipid components can be separated on a single TLC plate if two-dimensional systems are used (see Chapter 3). Very many such systems have been devised, and Rouser et al.[640] in particular have described a variety of solvent combinations suited to the analysis of animal lipids. The most successful separations are achieved when contrasting solvents are used for development in each direction; for example, a neutral or basic solvent mixture in the first direction may be followed by development with an acidic solvent mixture in the second direction, or the second system may contain acetone to retard the migration of phospholipids relative to glycolipids, with which they might otherwise overlap. Four solvent systems are favoured by Rouser:

   (a) chloroform–methanol–water (65:25:4 by vol.).

   (b) n-butanol–acetic acid–water (60:20:20 by vol.).

   (c) chloroform–methanol–28 per cent aqueous ammonia (65:35:5 by vol.).

   (d) chloroform–acetone–methanol–acetic acid–water (10:4:2:2:1 by vol.).

In Fig. 7.4, plate A illustrates the type of separation possible with solvent a in the first direction and solvent b in the second; plate B has solvent c in the first direction and solvent d in the second. Many other permutations of these solvent systems are possible, and more than thirty different complex lipids have been resolved in this way. Silica gel H is usually the favoured adsorbent, but Rouser et al.[637] preferred silica gel containing magnesium silicate (10 per cent w/w) as binder, as this apparently gave more compact spots, particularly with acidic lipids. Gray[238] has described a useful TLC system in which solvent mixture a above is used in the first direction and tetrahydrofuran–methylal–methanol–2M aqueous ammonia (10:5:5:1 by vol.) is used in the

second direction (plate C, Fig. 7.4). Unfortunately solvents such as butanol are not easy to remove prior to charring, and the intermediate drying step can sometimes lead to deactivation of the plates, especially under humid atmospheric conditions, so that the second development does not produce the desired separations.

Innumerable two-dimensional TLC systems for the separation of lipids from animal tissues have been described in the literature, some designed to effect better than usual separations of particular phospholipids e.g. phosphatidic acid[633] or phosphatidylglycerol[589]. Many published systems represent minor adjustments only of earlier ones to suit local conditions (e.g. of temperature or humidity), or particular commercial brands of solvents or adsorbents. It is rarely easy for an independent observer to judge whether they represent real advances.

Smaller than usual two-dimensional HP-TLC plates (10 × 10 cm) have been used to obtain good separations of brain lipids.[586]

Excellent separations of complex lipids of plant origin can be achieved with two-dimensional TLC systems as shown in Fig. 7.5.[540, 698] Similar principles to those described above for animal lipids apply. Bacterial lipids are frequently separated by similar solvent systems to those used for plant lipids.

The amount of lipid that can be separated by two-dimensional TLC varies with the thickness of the

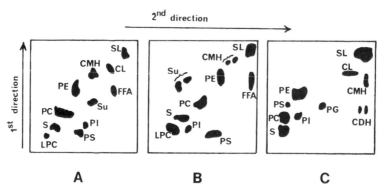

FIG. 7.4. Two-dimensional TLC separations of complex lipids from animal tissues on silica gel H containing 10 per cent (w/w) magnesium silicate (plates A and B) or silica gel H alone (plate C). Solvent systems: plate A,[640] 1st direction, chloroform–methanol–water (65:25:4 by vol.); 2nd direction, n-butanol–acetic acid–water (60:20:20 by vol.); plate B,[640] 1st direction, chloroform–methanol–28 per cent aqueous ammonia (65:35:5 by vol.); 2nd direction, chloroform–acetone–methanol–acetic acid–water (10:4:2:2:1 by vol.); plate C,[238] 1st direction, chloroform–methanol–water (65:25:4 by vol.); 2nd direction, tetrahydrofuran–methylal–methanol–2M aqueous ammonia (10:5:5:1 by vol.). Abbreviations as in Fig. 7.1; SL, simple lipids; FFA, free fatty acids.

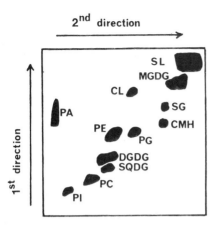

2nd direction

1st direction

FIG. 7.5. Two-dimensional TLC separation of plant complex lipids[540] on layers of silica gel G. Solvent system, 1st direction, chloroform–methanol–7M ammonium hydroxide (65:30:4 by vol.); 2nd direction, chloroform–methanol–acetic acid–water (170:25:25:6 by vol.). Abbreviations: SL, simple lipids; MGDG, monogalactosyldiacylglycerols; CL, cardiolipin; PA. phosphatidic acid; SG, sterol glycosides; CMH, ceramide monohexosides; PE, phosphatidylethanolamine; PG, phosphatidylglycerol; DGDG, digalactosyldiacylglycerols; SQDG, sulphoquinovosyldiacylglycerols; PC, phosphatidylcholine; PI, phosphatidylinositol.

layer of adsorbent. Up to 3 mg can be applied as a spot or a small streak (1 cm or so long) on layers 0.5 mm thick. Better resolution is obtained with layers 0.25 mm thick, but less material can be applied to the plates. The principal disadvantage of the technique lies in the fact that, although lipids migrate in a constant order in each direction, their precise orientation on the plate may vary greatly with small changes in the properties of the adsorbent or indeed in the atmospheric conditions. The double development is time-consuming and despite the lack of resolution, single-dimensional TLC systems are often preferred for routine analyses. For preliminary examination of lipid samples and for checks on the purity of lipid components isolated by other techniques, two-dimensional TLC offers considerable advantages.

3. Location and identification of complex lipids on TLC plates

Complex lipids can be located on TLC plates by most of the non-specific destructive or nondestructive reagents described in Chapter 3. If the latter are used, care must be taken to remove polar

solvents, such as water or glacial acetic acid, otherwise lipid spots will be obscured. This is particularly important if the fatty acid components are required for further analysis and, in this instance, solvents should be evaporated in a vacuum oven at close to room temperature or by blowing nitrogen on to the surface of the plate. Excess acetic acid can be removed by neutralisation with an ammonia spray.[521] Such precautions minimise the risk of autoxidation, and care and patience at this stage are generally rewarded.

Iodine vapour can be used to distinguish between glycolipids and phospholipids, as only the latter are stained significantly, but a number of spray reagents are available that are specific for the presence of certain functional groups in complex lipids and, therefore, aid in the identification of the latter. Spray reagents should *always* be used in fume cupboards.

(I) PHOSPHORUS

The "Zinzadze" reagent in one or other of various modifications is commonly used to detect and identify lipids containing phosphorus. The following reagent, developed from that of Dittmer and Lester[180], offers several advantages and has a shelf-life of about six months.[648]

"Solution 1. Molybdic oxide (8 g) is dissolved in 70% H_2SO_4 (200 ml) by boiling gently, with magnetic stirring, then cooled.
Solution 2. Powdered molybdenum (0.4 g) is added to solution 1 (100 ml) and the mixture refluxed for 1 hr, then cooled.
Solution 3. Solutions 1 and 2 are mixed with water (200 ml) and filtered. The final spray reagent is prepared by adding solution 3 (1 vol.) to water (2 vol.) and glacial acetic acid (0.75 vol.); it is left for 3–4 days before use."

When TLC plates are sprayed lightly with this reagent, phospholipids appear within 10 min as blue spots on a white background. With slight modification, the reagent can be utilised with reversed-phase TLC plates. The Dittmer–Lester reagent can be used to differentiate between phospho- and phosphonolipids.[735]

(II) GLYCOLIPIDS

A number of specific reagents are available for the detection of the carbohydrate moieties of glycolipids. That most widely used is an orcinol-sulphuric acid mixture.[750]

"The reagent is prepared by dissolving orcinol (200 mg) in 75 per cent sulphuric acid (100 ml). The whole surface of the plate is wetted by a fine spray of the solution, and then is heated in an oven at 100°C for 10–15 min. Glycolipids appear as blue-violet spots on a white background."

The solution is stable for about one week if refrigerated and kept in the dark.

α-Naphthol[696] or diphenylamine[356] sprays also react specifically with glycolipids. The former reagent is prepared and used as follows.

"α-Naphthol (0.5 g) is dissolved in methanol–water (1:1, v/v), before it is sprayed on the plate until the surface is wet. After air drying, the plate is sprayed lightly with 95 per cent sulphuric acid, then is heated at 120°C when glycolipids appear as purple-blue spots and other complex lipids as yellow spots."

(III) COMPOUNDS WITH VICINAL DIOL GROUPS

All compounds of this type including phosphatidylinositol, phosphatidylglycerol, 1-monoacylglycerols and glycolipids can be detected with the periodate-Schiff's reagent.[687]

"Two reagents are necessary: (a) 0.2 per cent aqueous sodium periodate. (b) Schiff's reagent; pararosaniline–fuchsin (0.2 g) is dissolved in water (85 ml), a 10 per cent solution of sodium bisulphite (5 ml) is added and the mixture allowed to stand overnight, when it is decolorised with charcoal and filtered. The TLC plate is sprayed with the periodate solution, and left at room temperature for 15 min after which it is treated with sulphur dioxide to destroy excess reactant. The plate is then sprayed with the Schiff's reagent, and treated once more with sulphur dioxide."

After a short time glycol-containing lipids appear as blue-purple spots, while most other complex lipids appear as yellow spots. Plasmalogens and free aldehydes are also detected by the procedure, and the presence of such compounds should be separately confirmed by an independent analysis (see Chapter 6 and below). Unfortunately, autoxidised lipids also react, so particular care must be taken to prevent the formation of lipid peroxides. When the glycolipids have been visualised, phospholipids can be detected on the same plate with the reagent described above.

(IV) FREE AMINO GROUPS

Phospholipids such as phosphatidylethanolamine, phosphatidylserine and the related lyso compounds, having free amino groups, can be detected with the aid of a ninhydrin spray.

"The plate is sprayed with a solution of 0.2 per cent ninhydrin in butanol saturated with water. Lipids having free amino groups appear as red-violet spots when the plate is heated in an oven at 100°C in a water-laden atmosphere."

(V) CHOLINE

Phosphatidylcholine, lysophosphatidylcholine and sphingomyelin give a positive reaction with the "Dragendorff" reagent in the following modification.[811]

"Two solutions are necessary: (a) potassium iodide (40 g) in water (100 ml), (b) bismuth subnitrate (1.7 g) in 20 per cent acetic acid (100 ml). Immediately before use, the first solution (5 ml) is mixed with the second solution (20 ml) and diluted with water (75 ml). When the plate is sprayed with the mixture, choline-containing lipids appear within a few minutes, especially if the plate is warmed gently, as orange-red spots."

An alternative procedure has been described.[797]

(VI) SPHINGOLIPIDS

The amide group of sphingolipids can be detected by means of a sodium hypochlorite–benzidine spray. However, benzidine is such a potent carcinogen, capable of being absorbed through the skin as well as by inhalation, that its use is prohibited in the United Kingdom. No procedure can be recommended for the detection of sphingolipids at present, therefore.

(VII) GANGLIOSIDES

The sialic acid residues of gangliosides form a distinct and specific coloured complex with resorcinol.[751]

"A solution of resorcinol (2 g) in water (100 ml) is prepared and stored at 4°C. This stock solution (10 ml) is added to concentrated hydrochloric acid

(80 ml) and 0.1 M copper sulphate solution (0.25 ml), and made up to 100 ml immediately before use. The plate is sprayed with this reagent and heated at 110°C for a few min. A violet-blue colour is obtained with gangliosides, while other glycolipids appear as yellow spots."

(VIII) IDENTIFICATION OF PHOSPHOLIPIDS BY IR SPECTROSCOPY

The nitrogenous bases of phospholipids can be identified by their IR absorption spectra in the regions 9–11 μm.[536] Free amine groups, such as are found in phosphatidylserine or phosphatidyl-ethanolamine, exhibit a single sharp band with a maximum at 9.3 μm, while monomethylamines (e.g. in phosphatidyl-N-methylethanolamine) display a strong maximum at 9.5 μm with a weaker band at 9.2 μm. The spectra of dimethylamines exhibit a pronounced doublet with maxima at 9.2 and 9.5 μm, while quaternary amines, such as choline in phosphatidylcholine or sphingomyelin, have, in addition to this doublet, a sharp band at 10.4 μm. In addition, the spectra of these compounds in the IR region 5.5 to 7.0 μm exhibit features useful as diagnostic aids.[51] Spectra are normally obtained with the phospholipids in chloroform solution.

D. DETERMINATION OF COMPLEX LIPIDS SEPARATED BY CHROMATOGRAPHIC PROCEDURES

(1) Universal procedures

Many methods are available for quantifying particular lipids separated by chromatography e.g. phospholipids can be determined by phosphorus assay, and glycosphingolipids by assaying the long-chain base components (see below). An enzymic method is available for phosphatidylcholine[42], while lipids containing free amine groups, e.g. phosphatidylethanolamine and phosphatidylserine, can be determined spectrophotometrically following conversion to the trinitrobenzenesulphonic acid derivatives.[695] No single universal method has become established as the standard for phospholipids in general, but a variety of methods are available for use in differing circumstances.

Charring followed by densitometry has not been widely used for the purpose possibly because of the lack of availability of suitable standards for all classes. Shand and Noble[685] obtained uniform results with phosphatidylcholine, phosphatidylethanolamine and sphingomyelin, and a modification of the method has been applied to glycolipids.[682] With fluorometric methods, there may be difficulties in the availability of standards. Also, the fluorescence phenomenon requires the presence of two or more double bonds in the fatty acid molecule, so such methods are of little use for comparatively saturated lipids such as sphingolipids or bacterial lipids. Nevertheless, excellent results have been obtained with phospholipids of animal tissue by several workers.[281, 303, 427, 661]

It is also possible to determine the amounts of complex lipids, particularly those separated by TLC, by gas chromatography of the methyl ester derivatives of the fatty acid constituents with an added internal standard.[128, 141, 521] The amounts and fatty acid compositions of all lipid classes are thereby obtained simultaneously in a single analysis. As fatty acids are the only hydrolysis product common to all glycerophosphatides and glycolipids, the method is particularly apposite. The principle of the procedure has been described in Chapter 3, and application to the analysis of simple lipids in Chapter 6. Glycerophosphatides can be transesterified by acid-catalysed methanolysis (anhydrous) in the presence of the TLC adsorbent (see Chapter 4), but sphingolipids must be transesterified by methanol containing hydrochloric acid (see Chapter 4 and below). If the internal standard (15:0 or 17:0) in methanol solution (about 0.2 mg/ml) is added to the lipid and the adsorbent, along with the transesterifying reagent, and the reaction is taken to completion, there will be virtually no loss of lipid relative to the internal standard. The procedure has been tested with a wide range of phospholipids, including sphingomyelin, and with glycosyldiacylglycerols of plant origin. In calculating the molar proportion of each lipid class, it must be remembered that some contain only one fatty acid per molecule and others two or more.

2. Phospholipid determination by phosphorus assay

Phospholipids separated chromatographically can be estimated by determining the amount of phosphorus in each fraction, and a detailed procedure suitable for the purpose is described below. Components should preferably be eluted from TLC adsorbents with chloroform–methanol–water (5:5:1 by vol.), after detection with 2′,7′-dichlorofluorescein spray or with iodine vapour. The estimation can be carried out in the presence of the adsorbent if necessary, but if this is not washed thoroughly with solvent before the plates are prepared (see Chapter 3), the blank values may be high and the accuracy of the procedure diminished. In calculating molar amounts of phospholipids from phosphorus analyses, it should be remembered that certain of these, e.g. cardiolipin, contain 2 moles of phosphorus per mole of total lipid. One disadvantage of the technique is that it is destructive, and when only small amounts of material are available for analysis, it may not be possible to obtain any more information from the sample.

Most methods for the determination of phosphorus involve digestion of phospholipids with the release of inorganic phosphate, which is reacted with ammonium molybdate to form phosphomolybdic acid; this is reduced and the product determined spectrophotometrically. Innumerable modifications to the procedure have been described, but the following is well-tried and is essentially that of Bartlett.[56, 181]

"The ammonium molybdate reagent is prepared by dissolving the compound (4.4 g) in water (500 ml), adding concentrated sulphuric acid (14 ml) and making up to 1 litre. The reducing reagent consists of sodium bisulphite (2.5 g), sodium sulphite (0.5 g) and 1-amino-2-naphthol-4-sulphonic acid (0.042 g), which are dissolved in water (250 ml) and allowed to stand in the dark for several hours. The solution is filtered into a brown bottle and, when refrigerated, is stable for about one month. A standard solution of 0.5 mM sodium dihydrogen phosphate is also prepared for calibration purposes. Perchloric acid (0.4 ml; at least 70 per cent) is added to the dry lipid sample (containing up to 1 μmole of phosphorus) in a test-tube, and the lipid is digested by gentle refluxing for 20 min on a heating block at a temperature such that the tube acts as an air condenser, in a fume cupboard (take care!). The mixture is cooled and ammonium molybdate reagent (2.4 ml) is added, followed by the reducing reagent (2.4 ml). The

solutions are mixed thoroughly and are heated on a boiling water bath for 10 min for colour development. On cooling, the absorbance of the solution is measured at 830 nm (some dilution may be required). A blank sample is analysed simultaneously. The amount of phosphorus in the unknown sample is read from a calibration curve, prepared at the same time by performing the reaction on known amounts of the standard phosphate solution."

When TLC absorbent is present, the solution is centrifuged at approximately $800 \times g$ and decanted for measurement following colour development. It should be noted that there is a slight danger of explosions occurring with perchloric acid digestions, especially if all traces of solvents are not removed. A safety screen should always be used. Because of this hazard, efforts have been made to devise safer digestion procedures[186], or to avoid the need for digestion[648, 734, 792].

3. Glycerophospholipid determination by glycerol assay

A gas chromatographic procedure for the determination of glycerol in simple lipids has already been described (Chapter 6). The following method, which involves acetolysis with acetic anhydride and trifluoracetic acid, saponification of the resulting diacylglycerol acetates and preparation of triacetin for GLC analysis, is more suited to the determination of glycerol in complex lipids.[311]

"The phospholipid (up to 6 mg) is refluxed with acetic anhydride–trifluoracetic acid (4:1, 2 ml) for 10 hr in the presence of a known amount of hexadecanylacetate, which serves as an internal standard. Reagents are then removed *in vacuo*, 0.3 M sodium methoxide in methanol (2 ml) is added, and the mixture refluxed for 10 min. The methanol is removed by evaporation, then water (0.3 ml) and 1 M sodium hydroxide (0.2 ml) are added, and the mixture refluxed for 2.5 hr to complete the hydrolysis. Acetic anhydride (3.5 ml) and xylene (3.5 ml) are added, and triacetin is prepared as described earlier (Chapter 6)."

The conditions for gas chromatographic analysis of the product are also described in Chapter 6.

4. Nitrogenous bases of phospholipids

Five bases commonly occur in phospholipids — serine, ethanolamine, N-methylethanolamine, NN-dimethylethanolamine and choline. Chromatographic procedures have been developed for their analysis as they can have related metabolic origins. They are eluted in the above order from columns of Dowex-50 ion-exchange resin with 1.5 M hydrochloric acid.[156] This, and other, chromatographic methods for the analysis of these compounds have been reviewed by McKibbin.[473] Specific chemical methods have also been developed for the analysis of choline[822], and serine and ethanolamine[179].

5. Determination of glycosphingolipids and sphingomyelin

Glycolipids can be determined by analysis of the carbohydrate residues after reaction with anthrone (see section E.5.)[859], but as the molar proportions of these in a given lipid may not be known, the procedure is not definitive. On the other hand, sphingomyelin and all glycosphingolipid molecules contain one mole of long-chain base, and these can be assayed, following hydrolysis of phosphate, glycosidic and amide bonds, by a variety of methods to provide an accurate measure of the amount of glycolipid present in a sample or in a chromatographic fraction. Long-chain bases can be determined by nitrogen assay, for example.[224]

"The lipids (up to 4 mg) are heated under reflux for 3 hr in 2 M methanolic hydrogen chloride, when the mixture is taken to dryness. 3 M Aqueous hydrochloric acid (1 ml) is then added, and the products are heated in a stoppered tube at 100°C for 1 hr. On cooling, chloroform–methanol (2:1, v/v; 1 ml) is added, the aqueous layer, which contains the liberated carbohydrates, is removed carefully by means of a Pasteur pipette and the chloroform layer is washed twice with fresh upper phase (prepared by partitioning chloroform–methanol (2:1, v/v) with an equal volume of water). The chloroform layer contains only long-chain bases and free fatty acids, and the amount of the former can be determined by nitrogen analysis by standard micro-Kjeldahl procedures."

Alternatively the bases can be determined spectrophotometrically and with greater sensitivity as the trinitrobenzenesulphonic acid derivatives.[859]

"The chloroform extract obtained as above is taken to dryness, and 4 per cent aqueous sodium bicarbonate solution (1 ml) and 1 per cent aqueous trinitrobenzenesulphonic acid solution (1 ml) are added. The mixture is left in the dark for 1 hr at 40°C. Methanolic hydrogen chloride (1 M, 1 ml) is added, and the solution extracted three times with hexane (2 ml portions). The hexane is removed and the sample dissolved in 95 per cent ethanol (4 ml), when the absorbance of the solution at 340 nm is obtained relative to an appropriate blank. The amount of glycolipid is read from a calibration curve prepared with pure standards."

Related methods use the reaction of long-chain bases with methyl orange[566, 567] or fluorescamine[394] to produce coloured or fluorescent complexes, as the basis for assay.

E. THE ANALYSIS OF GLYCOSPHINGOLIPIDS, SPHINGOMYELIN AND THEIR HYDROLYSIS PRODUCTS

As sphingolipids, with the exception of sphingomyelin, are only trace components of most tissues, it is usual to separate glycosphingolipids from phospholipids and to analyse the glycolopids separately. As the hydrolysis products tend to be very different from those obtained with phospholipids, the chemical problems involved in their analysis also differ. Methods of analysis of glycosphingolipids (reviewed elsewhere[431, 757, 834]) are especially important because of the association of these compounds with certain physiological and pathological disease states, and because of their involvement in the immune-defence system.

1. Cerebrosides and oligoglycosylceramides

Cerebrosides and oligoglycosylceramides as a class can be obtained by a variety of methods, some of which are discussed above (Section 2). For example, they are isolated in fairly pure form by elution with acetone or tetrahydrofuran–methylal–methanol–water (10:6:4:1 by vol.) from silicic acid columns, or with super-dry solvents from Florisil. In addition, they are eluted in specific fractions from DEAE and TEAE cellulose columns (Section B.3.). Distinct fractions containing either neutral or acidic glycosphingolipids can be obtained by column chromatography on DEAE-sephadex[437, 865] or DEAE-silicic acid[425, 426] preparations, and such methods appear to be replacing earlier ones. However, glycerolipid contaminants must still be removed by chemical or chromatographic means.

The favoured chemical procedure consists in subjecting complex lipid fractions to mild alkaline transesterification, by means of which those lipids containing O-acyl bound fatty acids are converted to methyl esters and water-soluble products, while sphingomyelin and other glycosphingolipids are unaffected. The following method is suitable:

"The complex lipids (100 mg) are heated at 50°C for 10 min in 0.5 M sodium methoxide in methanol (3 ml). Acetic acid (0.15 ml) is added followed by chloroform (6 ml) and water (2 ml). After thorough shaking, the upper layer is removed by Pasteur pipette, and the lower layer is washed twice more with methanol–water (1:1, v/v; 2 ml portions). After removal of the solvent, the crude products must be purified by chromatography."

The hydrolysis products of the glycerophosphatides, including any deacylated ether derivatives, are easily removed by column chromatography and if need be, the glycosphingolipids can be separated from sphingomyelin. Chloroform elutes the methyl esters, then chloroform–methanol (1:1 by vol.) elutes the pure glycosphingolipids from silicic acid columns.[233] With animal lipids, the acetone fraction from silicic acid columns should be treated in this way to remove small amounts of the acidic phospholipids, although traces of glycosyldiacylglycerols, that may be present in some tissues, will also be eliminated by such a step.

More thorough separations of glycolipids from phospholipids can be obtained if the carbohydrate moieties of the former are first acetylated with pyridine–acetic anhydride (see Chapter 4 for practical details).[651] Acetylated total lipid extracts are chromatographed on Florisil; 1,2-dichloro-ethane elutes neutral lipids, while 1,2-dichloro-ethane–acetone (1:1, v/v) elutes all the acetylated glycolipids completely free of phospholipids. The

latter remain on the column, from which they can be recovered (although yields may be poor) if necessary. The glycolipids are deacylated with sodium methoxide in methanol as described above, before they are analysed further. Gangliosides are also obtained in the glycolipid fraction if crude unwashed lipid extracts are taken to dryness, acetylated and fractionated in this way, although the deacetylated product must later be purified by dialysis against ice-water to remove non-lipid contaminants. No detectable structural changes occur in the process. In an ingenious modification of the procedure, glycolipids were acetylated with ^{14}C-acetate, separated by TLC in this form and quantified by radioactivity measurement, in amounts corresponding to only 10^{-9} g.[75]

Two-dimensional TLC systems that separate individual glycolipids along with phospholipids are described above, and a one-dimensional system for the separation of mono-, di- and triglycosyl-ceramides from phospholipids has been described (silica gel H-magnesium silicate layers with tetrahydrofuran–water (4:1 by vol.) as developing solvent).[194] When phospholipids are absent, glyco-lipids are easily resolved by TLC in a single dimension. For example, Svennerholm and Svennerholm[754] separated four main classes of glyco-lipids, i.e. cerebrosides, di- and triglycosylceramides and ceramide trihexoside-N-acetyl-galactosamine, on silica gel G layers with solvent system chloroform–methanol–water (65:25:4 by vol.) as illustrated in Fig. 7.6. The first three of these components can under some conditions appear as double bands, since those molecules containing normal fatty acids are separated to some extent from those containing hydroxy fatty acids.[236, 754] In fact, the separation is not as complete as it is sometimes claimed to be since, under optimum conditions, three bands should be separable, i.e. fractions containing normal fatty acids with dihydroxy bases, hydroxy fatty acids with dihydroxy bases together with normal fatty acids with trihydroxy bases, and finally hydroxy fatty acids with trihydroxy bases. The pattern can be even more complicated as compounds containing very long-chain fatty acids are occasionally partially separated from those having a more normal range of chain-lengths; sphingomyelin frequently chromatographs as a double spot for the same reason. Cerebroside

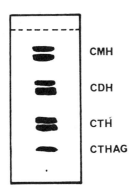

FIG. 7.6. TLC separation of glycosphingolipids on layers of silica gel G.[754] Solvent system, chloroform–methanol–water (65:25:4 by vol.). Abbreviations: CMH, CDH and CTH, ceramide mono-, di- and trihexosides respectively; CTHAG, ceramide trihexoside-N-acetylgalactosamine.

sulphate migrates just ahead of ceramide dihexosides with the above TLC system. Glycolipids containing more than four hexose units are not easily separated unless they are first acetylated.[265, 861]

The nature of the hexose unit affects the chromatographic behaviour of glycolipids on silica gel only slightly, but galactosylceramide is clearly separated from glucosylceramide on silica gel impregnated with sodium tetraborate (approx. 5 per cent w/w). With the solvent system used above, the glucoside migrates just ahead of the galactoside.[864]

Perhaps the most successful applications of HPLC in lipid analysis have been in the separation and quantification of glycolipids. Benzoylated glycolipids (picomole quantities) are readily separated on HPLC columns of pellicular silica gel, and are detected and quantified by means of UV detectors at 230 nm. Such procedures are still in a state of active development, but already offer a number of advantages in terms of ease of separation and of sensitivity. The technique most widely used to date consists of benzoylation of glycolipids with benzoylchloride in pyridine at 60°C for 1 hr.[469] Glycolipids are both N-acylated and O-acylated by this procedure, except for those containing 2-hydroxy fatty acids which are O-acylated only. The excess reagents can be removed by various methods, and the derivatives injected onto the HPLC column from which they are eluted with either single solvents or solvent gradients e.g. hexane–ethyl acetate or hexane–dioxane mixtures.

By this means, glycolipids with one to four glycosyl moieties can be separated, in addition to cerebrosides containing glucose and galactose, and those containing hydroxy- and nonhydroxy-fatty acids.[343, 365, 469, 470, 552, 785, 786] The sulpholipid, seminolipid, has been analysed by HPLC following benzoylation[746], but better results were obtained with cerebroside sulphate, when it was desulphated prior to analysis[469, 552].

More recently, it has been shown that no N-benzoylation occurs if benzoic anhydride and 4-dimethylaminopyridine in pyridine are used for derivatization[244], and the native glycolipids are more easily recovered for structural analysis. A separation of benzoylated glycolipids, prepared in this way, is shown in Fig. 7.7. Others have preferred to prepare O-acetyl-N-p-nitrobenzoyl derivatives.[749, 860] HPLC has also been used to separate molecular species of glycolipids (see Chapter 8).

Glycolipids may be recognised by the specific spray reagents described earlier, but a complete

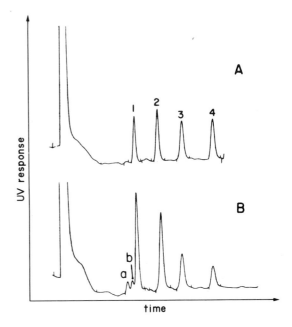

FIG. 7.7. HPLC separation of benzoylated standard (A) and plasma (B) glycosphingolipids on a Zipax column (2.1 × 500 mm), and eluted with a 13 min linear gradient of 2.5–25% dioxane in hexane with detection at 230 nm.[244] 1, glucosylceramide; 2, lactosylceramide; 3, globotriaosylceramide; 4, globotetraosylceramide; a, unidentified; b, hydroxy-fatty acid-containing galactosylceramide. (Reproduced by kind permission of Dr. R. H. McCluer and *Analytical Biochemistry*.)

analysis requires that the molar amounts of each hexose be determined relative to the amount of ceramide, and that the fatty acid and long-chain base compositions be ascertained by the methods discussed below. In addition, the sequence of hexose units should be determined, but this is the province of the carbohydrate chemist and cannot be discussed further here.

2. Gangliosides

Gangliosides, although they are only minor components of most tissues, are becoming increasingly the subject of study because of their role in various lipidoses and because of their antigenic properties. They are found in the aqueous phase, along with non-lipid contaminants, after a "Folch" wash of lipid extracts, and they can be recovered from this, after dialysis, by lyophilisation (discussed earlier in Chapter 2).[97, 111, 368, 468] Alternatively, a pure ganglioside fraction can be obtained by Sephadex chromatography[696], or they can be obtained, along with other glycosphingolipids, by Florisil chromatography of acetylated lipid extracts (as discussed above)[651].

Individual gangliosides are most often isolated by means of preparative TLC in a single dimension from such concentrates, although with conventional TLC plates, very long development times may be necessary because of the highly-polar nature of the compounds and of the developing solvents. Two groups of developing solvents have been widely used — basic, e.g. chloroform–methanol–2.5 M aqueous ammonia (60:40:9 by vol.)[434, 436], or neutral, e.g. propanol–water (70:30, v/v)[407]. Fig. 7.8 illustrates the kind of separations that are attainable. The purity of fractions isolated with a neutral solvent system should be checked by using a basic one; certain components which co-chromatograph with one solvent system are separable with another of a different type. Acidic solvents must be avoided as they may cause degradation of gangliosides. More recently, it has been shown that very much faster separations can be achieved with improved resolution, if high-performance TLC is used[35, 214, 869], especially if the developing solvents contain chloride ion e.g. tetrahydrofuran–0.1% aqueous KCl (5:1, v/v)[195] or methyl acetate–isopropanol–0.25% KCl

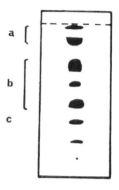

FIG. 7.8. TLC separation of gangliosides on layers of silica gel G. Solvent system, chloroform–methanol–2.5M ammonia (60:40:9 by vol.). a, mono-, b, di-, and c, trisialogangliosides.[434]

(45:30:20 by vol.)[869]. Gangliosides, separated in this way, can be detected by the specific resorcinol reagent described above (Section C.3.).

Individual gangliosides are identified partly by their Rf values in various solvent systems relative to compounds isolated from tissues of known ganglioside composition, e.g. bovine brain.[435] More positive identification requires that the hexose and sialic acid residues be identified, and the molar amounts of each determined as well as the fatty acid and long-chain base compositions. Methods of analysing for all these components are discussed below. Complete identification requires that the mode of linkage of each of the hexose and sialic acid residues be ascertained, but structural studies of this kind are outside the scope of this text.

Excellent preparative-scale separations of gangliosides have been obtained by column chromatography on anion-exchange resins. Various media have been tried, but the best results seem to have been achieved on Spherosil-DEAE-Dextran[214] and DEAE-Sepharose[344]. With these columns, fractions are obtained differing in the number of sialic acid residues, and further subfractionation by TLC is possible. Perbenzoylated monosialogangliosides have been successfully subjected to HPLC.[84]

N-Acetylneuraminic acid and other sialic acids in lipids can be determined by spectrophotometric estimation of the complex formed with resorcinol.[751]

"A stock solution of resorcinol (2 g) in water (100 ml) is prepared, and stored in a dark bottle at 4°C. Immediately before use, this solution (10 ml), concentrated hydrochloric acid (80 ml) and 0.1M copper sulphate solution (0.25 ml) are mixed, and made up to a volume of 100 ml with water. Samples containing up to 1 μmole of N-acetylneuraminic acid are dried under vacuum (*iso*-amyl alcohol may be added to prevent foaming). Water (1 ml) and the resorcinol reagent (1 ml) are added, the mixture is heated on a boiling water bath for 15 min and then chilled in ice. n-Butyl acetate and n-butanol (85:15, v/v; 2 ml) are added and, after thorough mixing and centrifugation, the absorbance of the upper layer of the sample and an appropriate reagent blank at 580 nm are determined. The amount of N-acetylneuraminic acid in the sample is calculated from a calibration curve prepared from the pure compound."

The procedure can be applied to gangliosides containing N-glycolylneuraminic acid, which has a greater molar extinction coefficient, but as a suitable standard may not be available for calibration purposes, values obtained from the N-acetyl-neuraminic acid calibration curve must be corrected by multiplying by a factor of 0.77. Related procedures have been developed for quantifying gangliosides in the presence of thin-layer absorbents.[35, 712, 862]

In an alternative procedure, gangliosides are determined in the presence of other lipids without prior isolation, by a sensitive thiobarbituric acid assay of the sialic acid residues.[177]

3. Long-chain bases

In order to obtain the long-chain bases from sphingolipids, it is first necessary to hydrolyse any glycosidic linkages in glycolipids or the phosphate bond in sphingomyelin, as well as the amide bond, before the various products are isolated and purified. The chemistry and methods of analysis of long-chain bases have been reviewed by Karlsson[375] and by Weiss[819]. Methods of obtaining the total long-chain base content of lipids are discussed above (Section D.5.).

The amide, glycosyl and phosphate bonds in glycosphingolipids, sphingomyelin and related phosphonolipids can be cleaved simultaneously by reaction with either acid or base under appropriate conditions. With acidic conditions, much degradation and modification of the bases occurs, especially in those with double bonds allylic to hydroxyl groups. For example, 3- or 5-methoxy and 5-hydroxy-compounds are formed, some dehydration may occur and there is liable to be some inversion of configuration of the hydroxyl and amine groups. The most widely used acidic hydrolysis procedure is that of Gaver and Sweeley[228], in which by-product formation is less than in other methods; only the trihydroxy-bases are greatly altered but the main by-products are 5-methoxy compounds, that are not easily separated from the native bases. Karlsson[375] recommends an alternative method, in which somewhat more artefact formation occurs, though in this instance, the artefacts are 5-hydroxy compounds and are easily removed chromatographically, especially when the amino-group is stabilised by conversion to the dinitrophenyl (DNP) derivative. The hydrolysis procedure is as follows.

"The sphingolipids (4 mg) are refluxed for 6 hr with 2M aqueous hydrochloric acid (3 ml). On cooling, chloroform (8 ml) and methanol (4 ml) are added and, after thorough mixing, the lower phase is evaporated to dryness and the products are added in chloroform to a silicic acid column (1 g). Chloroform–methanol (98:2, v/v; 10 ml) elutes the free fatty acids, and chloroform–methanol (1:3, v/v; 10 ml) elutes the bases."

The DNP derivatives are prepared and purified by the following procedure.

"The bases (5 mg) are dissolved in methanol (1 ml) containing 1-fluoro-2,4-dinitrobenzene (5 μl). 2M Potassium borate buffer (pH 10.5, 4 ml) is added dropwise, and the solution is heated at 60°C for 30 min. On cooling, the mixture is partitioned between chloroform, methanol and water in the ratio 8:4:3 by vol., the lower phase is evaporated and the material chromatographed on a silicic acid column (1 g). Hexane–diethyl ether (7:3, v/v; 20 ml) elutes the less polar impurities, hexane–ether (1:1, v/v; 20 ml) elutes the DNP derivatives of the natural bases and hexane–ether (1:3, v/v) elutes more polar artefacts."

The bases can be further analysed in DNP form; the derivatives are yellow in colour and are easily seen on a TLC plate.

Basic hydrolysis of sphingolipids with barium hydroxide[518] permits isolation of long-chain bases with very much less degradation than is found with acidic procedures, but surprisingly the method does not appear to have been much used. The recommended procedure is as follows.

"The sphingolipids (5 mg) are dissolved in warm dioxane (2.5 ml), a similar volume of 10% aqueous barium hydroxide is added and the mixture is heated at 110°C for 24 hr in a sealed tube. On cooling, the reagents are washed into a separating funnel with water (2 × 5 ml), and the whole is extracted with chloroform (2 × 15 ml). The required bases are obtained after drying and removing the solvent, and are purified by TLC as described below. The fatty acid components of the sphingolipids can be recovered, if necessary, by re-extraction of the aqueous extract following acidification."

Even this procedure causes some degradation of trihydroxy bases and if these are present in small amounts only, it may be advisable to convert the bases directly to aldehydes for analysis, by reaction of the intact lipids with periodate (a suitable method is described below), before generating the remaining bases for further analysis.[517]

DNP derivatives of long-chain bases are readily separated by TLC on layers of silica gel G impregnated with 2 per cent (w/w) boric acid, into three groups, i.e. saturated dihydroxy-, unsaturated dihydroxy- (with a *trans*-double bond in position 4) and trihydroxy- components, with chloroform–hexane–methanol (50:50:20 by vol.) as developing solvent as illustrated in Fig. 7.9.[375, 377] When the bases are prepared by acidic hydrolysis methods, unnatural *threo*-isomers of the unsaturated dihydroxy-bases are found just below the natural *erythro*- compounds on the TLC plate, but they need not interfere with subsequent GLC analyses. Components are recovered from the adsorbent by elution with chloroform–methanol (2:1, v/v), but the eluate should be washed with 1/5 volume of water to remove any boric acid, which may also be eluted. DNP derivatives of bases separated in this way can be further resolved according to degree of unsaturation by silver nitrate chromatography[373], or into critical pairs by reverse phase TLC[372]. By using these techniques in sequence, individual com-

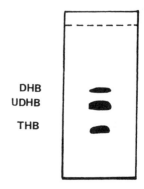

Fig. 7.9. TLC separation of dinitrophenyl-derivatives of long-chain bases on layers of silica gel G containing 2 per cent (w/w) boric acid.[375, 377] Solvent system, chloroform–hexane–methanol (50:50:20 by vol.). Abbreviations: DHB, saturated dihydroxy bases; UDHB, Δ^4-*trans*-unsaturated dihydroxy bases; THB, trihydroxy bases.

ponents of natural mixtures can be isolated in sufficient quantity for structural studies.

The free bases or other derivatives of the bases can also be separated by TLC, although the resolutions attainable are not quite as good as those with the more stable DNP derivatives[375, 819], but as steps involving preparation and hydrolysis of derivatives are avoided, many workers prefer to work with the free bases.

Isomers of the various classes of bases can be separated by gas chromatography if they are first converted to non-polar derivatives, in particular to the trimethylsilyl ethers. Components differing in chain-length or in the number of hydroxyl groups in the molecule can then be separated on silicone liquid phases such as SE-30.[228] The N-acetyl, O-TMS ether derivatives of long-chain bases are also useful for gas chromatography. The N-acetyl derivative is prepared by reacting the free base with acetic anhydride in methanol (1:4, v/v) at room temperature overnight (under these conditions no O-acetylation occurs).[229] The TMS ether derivatives are prepared from the N-acetyl compound as described earlier (Chapter 4), and may be subjected to GLC on silicone liquid phases such as OV-1 or OV-17 (3 per cent stationary phase at column temperatures of 220 to 240°C). Although isomers differing in degree of unsaturation cannot be separated on such columns, this can be achieved if the double bonds are first hydroxylated and then converted to the TMS ether derivatives (see Chapter 4 for practical details).

The facility to chromatograph intact long-chain

base derivatives is especially important when they are isotopically labelled, but much better resolutions are obtained if the bases or their DNP derivatives are oxidised by periodate to form long-chain aldehydes. Such compounds are readily separated both by chain-length and degree of unsaturation by GLC on conventional polyester liquid phases, as discussed in Chapter 6. Di- and trihydroxy bases, whether they are of the *threo-* or *erythro*-configuration, give similar aldehydes (although aldehydes of trihydroxy bases will be one carbon atom shorter than those of dihydroxy bases), so it is essential to isolate the three major classes of bases by preparative TLC as described above prior to GLC analysis. Various procedures utilising sodium metaperiodate have been described for cleaving the bases, but the mildest reagent is probably sodium metaperiodate in methanol, used as follows.[756]

"0.2M aqueous sodium metaperiodate (1 ml) is added to the long-chain bases (5 mg) in methanol (5 ml). The mixture is stirred in the dark at room temperature for 1 hour when dichloromethane (10 ml) and water (5 ml) are added and the whole is shaken and centrifuged. The aqueous layer is extracted twice more with dichloromethane and the combined extracts are dried over anhydrous sodium sulphate before the solvent is removed."

The aldehydes derived from bases are more easily identified than the parent compounds, as a wide range of aldehydes are available for use as standards. Aldehydes can also be reduced to long-chain alcohols with sodium borohydride and converted to the TMS ether or acetate derivatives for GLC analysis[516], and this is preferred by some workers.

When bases are prepared by acidic hydrolysis it should be remembered that, although the relative amounts of the various isomers from each class of base determined by GLC analysis reflect reasonably accurately the amounts present in each class in the native state, the measured amounts of the three groups of bases relative to each other may be in error. Saturated dihydroxy-bases are not affected by the hydrolysis reaction, but as little as one-half the original amount of the allylic bases may be recovered, and there will also be losses of trihydroxy bases. As stated earlier, no completely satisfactory method of circumventing this problem is yet available.

Karlsson[376] has reviewed procedures for determining the structures of long-chain bases. The most powerful tool for identifying them is mass spectrometry, and they are generally converted to less polar derivatives for the purpose. For example, Karlsson[375] converts the bases to the DNP derivatives before methylating the free hydroxyl groups by refluxing them overnight with methyl iodide and silver oxide. These derivatives give characteristic fragmentation patterns in a mass spectrometer, so that the molecular weight and primary structure of the bases can be ascertained. The positions of double bonds in the chain can be established if they are first hydroxylated and then converted to the TMS ethers, in a manner similar to that discussed earlier for unsaturated fatty acids (see Chapter 5). Definitive mass spectra are also obtained from N-acetyl, O-TMS ether derivatives of long-chain bases (prepared as described above).[584, 585] In addition, these compounds are sufficiently volatile for GLC–mass spectrometer combinations to be used for the identification of components of natural mixtures.

The long-chain aldehydes produced by periodate oxidation of the bases of alcohols prepared from the aldehydes can be identified by mass spectrometry as well as by their GLC retention characteristics, and the positions of any double bonds can be established by permanganate–periodate oxidation (see Chapter 5 for a description of the method) with GLC identification of the fatty acid fragments.[516] The number of hydroxyl groups in the intact molecule cannot of course be obtained if the aldehydes alone are examined by this technique, but this feature can be established from the chromatographic behaviour of the intact bases or their derivatives on thin-layer adsorbents or by gas chromatography, if authentic standards are available.

Infrared absorption spectroscopy can be used to advantage to detect the presence of *trans* double bonds in the aliphatic chain of a long-chain base or its derivative, by means of the characteristic absorption band at 10.3 μm.

4. The fatty acid components of glycolipids

Although the problems of analysis of fatty discussed at length in Chapter 5, a remind appropriate that sphingolipids contain two unique groups of fatty acid constituents — very long-chain and odd- and even-numbered saturated and monoenoic fatty acids (up to C_{28}) and a group of related compounds with hydroxyl groups in position 2, both of which are linked by amide bonds to the long-chain bases. In addition, there may occasionally be fatty acids linked to hydroxyl groups.

The problems involved in the preparation of methyl esters are discussed in Chapter 4; O-acyl esters are transesterified with sodium methoxide and N-acyl with methanol containing hydrochloric acid.

If the original sphingolipids have not already been separated by TLC into hydroxy- and non-hydroxy-fatty acids, the methyl esters of the two types are readily separated by TLC (see Chapter 5). Alternatively, as 2-hydroxy fatty acids form copper chelates, they can be isolated as such.[393] Normal fatty acids are identified by the standard procedures described earlier (Chapter 5). The hydroxy esters can be analysed by GLC after conversion to the acetate or TMS ether derivatives.

5. The analysis of hexoses

The high sensitivity and resolving power of GLC make it an ideal technique for the separation, identification and estimation of the carbohydrate moieties of complex lipids.[165, 430] Two basic methods are in widespread use. In one, the glycolipids are hydrolysed with methanolic acid, cleaving the glycosidic bonds and releasing the hexoses, which are recovered and silylated for analysis by GLC; during hydrolysis, methylglycosides are produced from the various anomeric forms of the hexoses, so several peaks are obtained on the GLC trace for each component.[628, 758] In the second procedure, the glycolipids are hydrolysed with aqueous acid, before the hexoses are reduced to the corresponding alditols and converted to acetates prior to GLC analysis. During the reduction step, the anomeric centre is destroyed so that each component chromatographs as a single peak. Aqueous acid can bring about disruption of some hexoses, so care is necessary during the hydrolysis step to keep such effects to the minimum. It is desirable to calibrate the method with glycolipids of known structure, rather than by using pure hexoses only.[369, 774] The following alditol–acetate procedure is suitable for galactose, glucose and the corresponding hexosamines.[369]

"The glycolipid sample (50–100 μg hexoses) is hydrolysed by reaction with aqueous 1M hydrochloric acid (1 ml) at 100°C for 12 h. On cooling, arabinose (50 μg) is added as internal standard, and the fatty acids released are removed by extraction with hexane (3 × 2 ml) and then chloroform (2 × 2 ml). The aqueous layer is dried down in a stream of nitrogen at 35°C, and the residue dried overnight in a vacuum desiccator. 0.8M Sodium borohydride in water (0.1 ml) is added and the mixture left at room temperature for 2 h, when reaction is stopped by addition of acetic acid (2 drops). The solution is taken to dryness by evaporation with repeated additions of methanol–acetic acid (200:1, by vol.), and finally by being left in a vacuum desiccator overnight. The alditols are acetylated by reaction with acetic anhydride (0.3 ml) at 100°C for 2 h, after which excess reagent is removed by evaporation with addition of toluene (3 × 2 ml). The product in chloroform (2 ml) is washed with water (2 ml), solvent is removed and the derivatives dissolved in acetone for analysis."

A GLC recorder trace of a standard mixture of hexoses prepared in this way is shown in Fig. 7.10. The method is not suitable for sialic acids, although it can be used for the other hexose moieties in gangliosides. On the other hand, if the methanolysis-silylation procedure is used, N-acetylneuraminic acid is converted to the 2-O-methylketal derivative of methylneuraminate and this, on silylation, gives a single peak on a GLC recorder trace.[758] Hexosamines, however, must be N-acetylated prior to silylation. With both procedures, accurate calibration of the detector response is necessary.

Inositol in complex lipids can be determined by means of related gas chromatographic procedures.[823]

The total amount of hexose units in a glycolipid fraction can be determined by converting the hexoses to the furfural derivatives in strong acid and reacting them with anthrone, with which they form coloured complexes that can be estimated spectrophotometrically. The following method is essentially that of Yamamoto and Rouser.[859]

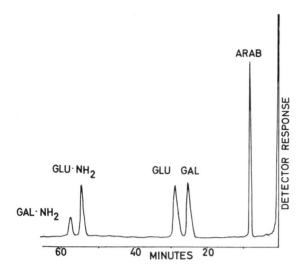

FIG. 7.10. GLC trace of the alditol acetates prepared from a standard mixture of glucose, galactose, glucosamine and galactosamine with arabinose as internal standard. The column contained 1% OV-225 and was temperature-programmed from 190–226°C. (Reproduced by kind permission of Dr. H. Debuch and the *Journal of Chromatography*).[369]

"A stock solution of 2 per cent anthrone in 98 per cent sulphuric acid (10 ml), which is stable for 3 weeks if refrigerated, is diluted with 87.5 per cent sulphuric acid (90 ml). This reagent (4 ml) is added to the glycolipid in dimethylformamide (1 ml), and the sample is heated in a boiling water bath for 4 min. The sample is cooled rapidly and its absorbance, together with that of a reagent blank, is measured at 625 nm. The amount of hexose present is determined from a calibration curve, obtained ideally with a pure glycolipid standard or, if this is not available, with pure glucose or galactose."

The reaction can be carried out in the presence of thin-layer adsorbents, provided they are precipitated by centrifugation prior to the absorbance measurement. The extinction coefficients of the anthrone derivatives of the common sugars are similar, so no allowance has to be made for the nature of the hexose components in the lipid.

Hexosamine[752], inositol[18] and sulphate[181, 729] can be determined by chemical procedures, as can sialic acid (see Section E.2.).

F. SOME SPECIFIC PROBLEMS IN THE ANALYSIS OF PHOSPHOLIPIDS

1. Alkyl- and alkenyl-ether forms of phospholipids

Phospholipids isolated by the procedures described above may include alkyl–acyl- and alkenyl-acyl-forms of the glycerophosphatide, in addition to the more common diacyl derivatives. While such ether-containing lipids are not often major components of tissues (other than heart muscle or nervous tissue), they are almost always present in trace amounts. Although alkyl ether and plasmalogen forms of simple lipids can be separated with care from the analogous acyl derivatives (see Chapter 6), related phosphoglycerides of the three types can only be separated after conversion to non-polar derivatives or following dephosphorylation. For example, phosphatidylcholine can be reacted with phospholipase D with the formation of phosphatidic acid, which is in turn converted to a non-polar dimethylphosphatide by reaction with diazomethane (see Chapter 9 for further practical details). By means of TLC on silica gel layers with diethyl ether or hexane–chloroform (1:1 v/v) as developing solvent and repeatedly running the plate in the same direction, this compound can be separated into alkenyl-, alkyl- and diacyl-forms.[624] Unfortunately, the alkenyl groups may form a by-product with diazomethane, limiting the potential of the method. Phosphatidylethanolamine has been converted to the N-dinitrophenyl, O-methyl derivative, which again is sufficiently non-polar to be separable into the three groups of components by TLC.[622] The derivatives are prepared as follows:

"The phosphatidylethanolamine (1 mg) is dissolved in benzene (2 ml), and triethylamine (40 μl) and fluoro-2,4-dinitrobenzene (4 μl) are added. After 2 hr the reagents are removed *in vacuo* (< 1 mm at 90°C). The product in chloroform (1 ml) is methylated with excess diazomethane in diethyl ether at 20°C for 30 min."

Repeated TLC on silica gel G layers in a single dimension (up to 8 times) is necessary to separate effectively the diacyl and ether-containing forms of the compound, with hexane–chloroform (4:6, v/v) and/or toluene–chloroform (4:6, v/v) as solvent systems. The principal advantage of a procedure of this type is that the phosphorus atom is retained, and metabolic studies of the various subgroups of phospholipids with ^{32}P as marker are possible.

The three subgroups of each phospholipid class can also be separated if the polar head group is removed by reaction with phospholipase C, (see Chapter 9 for practical details) and the resulting partial glycerides rendered less polar by acetylation. Alkenyl–acyl-, alkyl–acyl- and diacyl–glycerol acetates prepared in this way have similar mobilities on thin layers of silica gel G to the related simple lipids described earlier (Chapter 6). Acceptable separations are possible if the plate is developed first half-way with hexane–diethyl ether (1:1, v/v) as solvent and then fully in the same direction with toluene.[625–627] Some particularly good separations have been achieved, on a sufficient scale for further analysis, by means of HPLC.[158]

Concentrates of ether-containing lipids can be obtained by enzymatic hydrolysis of phospholipids. Under appropriate conditions, most phospholipases will react preferentially, if not exclusively, with diacylphosphatides hydrolysing them while leaving the corresponding lipids with ether bonds comparatively unchanged, provided that the reaction is not allowed to proceed too far (see Chapter 7 for a discussion of these enzymes). Concentrates of plasmalogens and alkyl-phosphatides have also been prepared by treating the mixed phospholipids briefly with mild alkali; the diacyl forms of the phospholipids are transesterified much more rapidly than are the ether-containing analogues enabling concentrates of the latter to be obtained. Pure alkyl-, acyl-phosphatides are then prepared by selectively hydrolysing the plasmalogens with acid. Although such procedures have proved of value for some preparative purposes, their value is limited because yields of alkyl and alkenyl lipids are not quantitative, and there may be some selective loss of particular molecular species. A better approach consists in hydrolysing the phospholipids with highly-purified pig pancreatic lipase[835] or the lipase from *Rhizopus arrhizus*[571]. The latter is now available commercially, and removes only those fatty acids in position 1 of diacylphosphatides. Ether lipids are thus not affected and can be recovered quantitatively (see also Chapter 9).

The analysis of the various hydrolysis products of simple plasmalogens and alkyl ethers, principally aldehydes and 1-alkoxy-glycerols respectively, are discussed in Chapter 6 and similar procedures are used to detect these moieties in complex lipids. For example, plasmalogens can be detected by spraying TLC plates with the 2,4-dinitrophenylhydrazine reagent described earlier; aldehydes are released by the action of aqueous acid or of mercuric chloride, and are analysed as described earlier. 1-Alkoxy-glycerols and 1-alkenyloxy-glycerols are liberated from phosphoglycerides by hydrogenolysis with lithium aluminium hydride[849] or more effectively with vitride reagent[723, 742], and are also analysed as described in Chapter 6. A comprehensive procedure for the analysis of alkyl- and alkenyl-phospholipids has been described, in which the phospholipids are first hydrolysed by means of the vitride reagent, then alkenyl moieties are converted to alkyl-substituted dioxanes in the presence of 1,1-dimethoxyheptadecane as internal standard for quantification by GLC; alkylglycerols are analysed as isopropylidene derivatives using 1-O-heptadecylglycerol as internal standard.[723] A spectrophotometric method for the analysis of ether-containing phospholipids has also been described.[69] The products of vitride reduction are acidified, and the liberated aldehydes measured colorimetrically after reaction with fuchsin reagent; the total alkyl and alk-1-enyl content is determined by oxidation of the hydrolysate with periodate to form alkylglycolic aldehydes, which are quantified similarly.

Plasmalogens can in addition be recognised and determined by reaction TLC.[568] The complex lipids (about 1 mg) are applied to the bottom left-hand corner of a TLC plate, which is developed in chloroform–methanol–water–acetic acid (65:43:3:1 by vol.). Volatile solvents are removed from the plate in a stream of nitrogen, residual acetic acid is neutralised by exposing the plate to ammonia vapour for some time and excess ammonia is removed in a desiccator or vacuum oven at room temperature, before the lipid lane is sprayed with 5 mM aqueous mercuric chloride solution. After 1 min, the plate is turned 90° anticlockwise and developed in chloroform–methanol–water (60:35:8 by vol.). The spots are visualised by standard procedures, and results similar to those illustrated in Fig. 7.11 are obtained. The aldehydes liberated along with the lysophosphatides formed simultaneously

2nd direction →

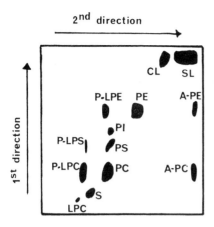

FIG. 7.11. Two-dimensional reaction TLC of phospholipids[568] on silica gel H layers. Solvent system, 1st direction, chloroform–methanol–water–acetic acid (65:43:3:1, by vol.). Lipid lane is then sprayed with 5 mM mecuric chloride. Solvent system 2nd direction, chloroform–methanol–water (60:35:8, by vol.). Abbreviations: SL, simple lipids, CL, cardiolipin; PE, phosphatidylethanolamine; PI, phosphatidylinositol; PS, phosphatidylserine; PC, phosphatidylcholine; S, sphingomyelin; LPC, lysophosphatidylcholine; prefix P-L and A-, lysophosphatide and aldehydes respectively derived from the appropriate plasmalogen.

are clearly separated from the unchanged diacyl- and alkyl–acyl-phosphatides so that, if non-destructive spray detection reagents are used, the aldehyde and fatty acid components of the plasmalogens can be obtained and analysed, while the relative amounts of related plasmalogens and non-plasmalogenic components can be determined by phosphorus analyses of the appropriate spots. There is a possibility that mercuric chloride may form adducts with poly-unsaturated fatty acids limiting the recovery of these; if this is found to be an important factor, hydrolysis of plasmalogens with 2M hydrochloric acid in methanol can be used.[796]

Useful reviews of methods of analysis of ether-containing phospholipids have appeared else-where.[665, 722, 803]

2. Analysis of glycerophosphoryl moieties

Methods of estimating glycerophosphatides after deacylation have been developed in a number of laboratories and are especially advocated by Dawson[167], who was the principal originator of the procedure. Lipids are deacylated by treatment with mild alkali, and the water-soluble phosphorus-containing compounds are isolated and separated by paper chromatography. These are more easily separated by this technique than the parent phosphatides by adsorption TLC and, in particular, the acidic phospholipids are especially well-resolved. In addition, alkyl- and alk-1-enyl-bonds are not disrupted by the procedure, and these forms of each lipid class can be separated and estimated. The procedure has been recommended for lipids labelled isotopically in the glycerol or the phosphorus moiety.

The principal disadvantages of the technique are that no information is obtained about the fatty acids attached to each phosphatide and that lysophosphatides (although they are usually comparatively minor components of tissues) cannot be distinguished from the diacyl compounds. It is also a somewhat lengthy and tedious procedure which must be performed under rigorously controlled conditions, and it is necessary to make certain corrections for side reactions. Somewhat unfairly, it is shunned by lipid chemists who tend to eschew the handling of organic compounds in aqueous media if at all possible. This "hydrophobia" would no doubt be overcome if a procedure were developed for complementary analysis of the intact lipids as well as the deacylated products, because of the extra information that might be obtainable.

3. Phosphonolipids

Phosphonolipids are not easily separated from the related phospholipids as they have very similar polarities and they are generally isolated together, the presence of the former being detected mainly by the recovery of hydrolysis products containing a carbon–phosphorus bond that is stable to prolonged acidic hydrolysis. Ceramide aminoethylphosphonate, for example, is eluted with phosphatidylethanolamine from DEAE cellulose columns, and most other phosphonate lipids migrate with their phosphoryl analogues with the more usual TLC systems. Kapoulas[370], however, has described chloroform–acetic acid–water solvent systems, which will separate phosphatidylcholine and phosphatidylethanolamine and their phosphonate analogues. For example, with the above solvents in the ratio 60:34:6 (by vol.), phosphonate components migrate ahead of the more common phosphate compounds as illustrated in Fig. 7.12. Berger and Hanahan[63] have used chloroform–acetic acid elution mixtures to separate the natural phosphonate derivative of phosphatidylethanolamine from its phosphoryl analogue on silicic acid columns, used in the conventional manner, while others have adapted ascending dry-column chromatography to the purpose[805]. Before such separations are attempted, it is probably advisable to isolate groups of related components from natural mixtures by the more usual procedures.

Phosphonate bases are liberated from phosphonolipids by acidic hydrolysis, and a variety of chromatographic methods for the isolation and identification of these have been described, of which the most useful are probably paper chromatography[170] and TLC[726]. Snyder and Law[726] have described a procedure for the determination of

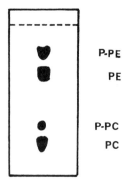

P-PE

PE

P-PC

PC

FIG. 7.12. TLC separation of phospho- and phosphono-analogues of phospholipids on silica gel G;[370] solvent system, chloroform–acetic acid–water (60:34:6 by vol.). Abbreviations: PC, phosphatidylcholine; PE, phosphatidylethanolamine; prefix "P", phosphono- analogue of each phosphatide.

phosphonate phosphorus, that offers a number of advantages over others available. Lipids are digested first with acid and then by reaction with the enzyme alkaline phosphatase (from *E. coli*) under conditions such that all the phosphorus in phosphate form is degraded to inorganic phosphate, although the phosphorus–carbon bond is not disrupted. The inorganic phosphate liberated is determined by the procedure described above, and the total phosphorus by the procedures of Ames[26]; the amount of phosphonate phosphorus present is obtained by subtracting the first result from the second.

4. Polyphosphoinositides

Incomplete recoveries of polyphosphoinositides are obtained and some degradation occurs unless suitable precautions are taken during extraction of the tissues with solvents (see Chapter 2). Polyphosphoinositides can be resolved on silica gel H layers to which 2.5 per cent potassium oxalate has been added (to sequester any calcium), with chloroform–methanol–4M ammonia (9:7:2 by vol.) as solvent system.[235] Alternatively, a two-stage one-dimensional TLC system has been described in which cyclohexanediamine tetraacetic acid is added to the second solvent.[286] Several column procedures have been described using DEAE-cellulose[299] (CM-cellulose might now be considered more suitable), silicic acid-oxalate[152] and neomycin reductively-coupled to glass beads[660]. The last utilises the property that neomycin binds specifically to polyphosphoinositides; most phospholipids are washed from the column with solvents of low ionic strength while di- then triphosphoinositides are eluted in solvents of increasing ionic strength. Such columns can be recycled and used repeatedly.

CHAPTER 8

The Analysis of Molecular Species of Lipids

A. INTRODUCTION

IN NATURE, lipid classes do not exist as single pure compounds, but rather as complex mixtures of related components in which the composition of the aliphatic residues varies from molecule to molecule. In some lipids, such as cholesterol esters, only the single fatty acid moiety will vary; in others, for example triacylglycerols, each position of each molecule may contain a different fatty acid. Sphingolipids contain a number of different long-chain bases which may be linked selectively via the amide bond to specific fatty acids. A complete structural analysis of a lipid therefore requires that it be separated into *molecular species* that have single specific alkyl moieties (fatty acids, alcohols, ether-linked aliphatic chains, long-chain bases, etc.) in all the relevant portions of the molecule. With lipids that contain only one or two alkyl groups, this is now often technically feasible. With lipids which have more than two alkyl groups, means have yet to be developed for physically separating all the possible species that may exist, although, if stereospecific enzymatic hydrolyses are performed on fractions separated by the available methods, it may be possible to at least calculate the amounts of all the molecular species that are present. The analyst must at the moment be content to isolate simpler rather than single molecular species in such instances. Alkyl-, alkenyl- and acyl- forms of a given lipid are *not* molecular species of it and can themselves be fractionated into molecular species and should therefore be isolated separately before analysis of this kind is begun.

Ideally, it would be preferable if lipids could be separated into individual molecular species without being modified in any way so that, for example, the biosynthesis or metabolism of each part of the molecule could be studied with isotopically-labelled components. The technical problems of the analysis can often be greatly reduced, however, if the polar parts of complex lipids are rendered non-polar by the formation of suitable derivatives, or are removed entirely by chemical or enzymatic means. Whatever approach is adopted, it is always necessary to apply combinations of different chromatographic procedures to achieve effective separations. These methods have been reviewed elsewhere.[413, 417, 620]

The chromatographic procedures used for the separation of molecular species of lipids differ little from those used for simpler aliphatic molecules (fatty acids, alcohols, etc), and they are described in general terms in Chapter 3. When the methods are applied to the isolation of molecular species of more complicated lipids, the separations achieved depend on the combined physical properties of all the aliphatic residues. If triacylglycerols are considered for illustrative purposes, adsorption chromatography will permit the separation of molecules containing three normal fatty acids from those containing two normal fatty acids and one fatty acid with a polar functional group in the chain, from those containing one normal fatty acid and two polar fatty acids, and so forth. Silver nitrate chromatography will separate those molecules containing three saturated fatty acids from those with two saturated fatty acids and one monoenoic acid, and these are in turn separable from a number of other distinct fractions containing molecules with fatty acids of a progressively higher degree of unsaturation. High temperature GLC is used mainly to separate molecular species according to the combined chain-lengths of the fatty acid moieties and complements the separations achieved by silver nitrate chromatography particularly well. Liquid-liquid chromatography (including TLC and HPLC) is used to separate triacylglycerols in which the combined chain-lengths of the fatty acid components differ by two carbon atoms (a double bond reduces the effective chain-length of a fatty acid by two carbon atoms). The principles of these methods as they apply to molecular species separations are described further below.

It should be remembered that it is almost always advisable to calculate the molar rather than weight proportions of any molecular species isolated. Precautions should be taken to minimise the effects of autoxidation (see Chapter 3).

B. GENERAL METHODS OF ANALYSIS

1. Liquid–solid and liquid–liquid chromatography

Thin-layer systems of liquid–solid chromatography have tended to be favoured for separations of molecular species containing differing numbers of polar substituents because of the high resolution that can be achieved and because the equipment necessary is comparatively inexpensive. Charring or internal standard-GLC methods can be used for quantification purposes. On the other hand, with silver nitrate TLC, in which species are separated by degree of unsaturation, charring techniques are rarely used because of high background values and since the amount of carbon formed is not in this instance necessarily related linearly to the amount of material present. Gas chromatography of the methyl esters of fatty acid constituents with an added internal standard is most often the method of choice, as fractions are generally similar in polarity and, after transesterification, they can be characterised and quantified in a single analysis. This method may be combined with gas chromatography of the intact lipid, again in the presence of an appropriate internal standard, so that a check is obtained on the results. As an additional check, the fatty acid composition of the unfractionated lipid ought to be computed by reconstituting the results from the individual fractions, and this should agree with the original analysis.

Liquid–liquid TLC has also been much used for separations of molecular species, but many workers find the procedure messy and there are other disadvantages e.g. separated fractions are not always easily detected. HPLC with stationary phases chemically-bonded to the support is, therefore, being increasingly used instead.

2. Gas-liquid chromatography of intact lipids

(I) INSTRUMENTS AND COLUMNS

The principles of gas chromatography and applications of the method to comparatively simple molecules have been discussed in earlier chapters. Because of the low volatility of intact lipids, gas chromatographic analysis is beset with a number of difficulties and, indeed for some time, it was thought that molecules such as triacylglycerols, with molecular weights up to 900, would pyrolyse at the temperatures required to elute them from GLC columns. Work, principally from the laboratories of Kuksis and of Litchfield, has shown that this need not occur and useful separations of such compounds can now be achieved. The procedure has been the subject of several review articles,[268, 410, 416, 449] and it is also discussed briefly in Chapter 6.

The gas chromatograph required is similar to that described earlier for fatty acid analysis, but it is essential that it has a flame ionisation detector for maximum sensitivity and facilities for accurate temperature programming up to a temperature of at least 350°C. In addition, it should be of a construction such that on-column injection is possible, although a preheater is necessary to warm up the carrier gas before it reaches the column packing. The dead volume between the end of the packing and the detector flame should be as small as possible, and the flame jet should preferably be wider than normal so that comparatively high carrier gas flow rates can be used. Automatic flow controllers for the carrier gas are a useful accessory as the flow rate in short columns can change markedly on temperature programming. As bleeding of even the most thermally-stable liquid phases occurs at high temperatures, better results are obtained with dual-compensating column than with single column instruments. The analyst should not be discouraged from using equipment that does not meet all these criteria, especially for separating compounds of intermediate molecular weight such as diacylglycerols or their derivatives, as patience and skill can compensate for many instrumental deficiences.

Helium or nitrogen may be used as the carrier gas and excellent recoveries of high molecular weight triacylglycerols have been obtained with both, although the former is to be preferred whenever feasible, as better resolutions are attainable at high flow rates. Nitrogen is used most often for reasons of cost, but it is essential that it be of very high purity as traces of oxygen or water will destroy liquid phases at elevated temperatures.

Better results are obtained with glass columns than with those of other materials, but there may be technical difficulties in obtaining effective seals at each end of the column; O-rings of Viton or other materials generally become brittle and crack if used at temperatures above 250°C for any length of time, but graphite seals will stand up to such conditions for long periods. It is also possible to avoid this difficulty by using columns with direct glass to metal seals, and these are available for specific instruments. All-metal columns are not recommended. Narrow bore columns give the best resolutions, and those of 2 mm (i.d.) diameter are preferable to those of 4 mm; the length of the column selected will vary with the nature of the sample but it is usually necessary to compromise resolution by using short columns (50–100 cm) in order that compounds are eluted in a reasonable time. Short wall-coated capillary columns are increasingly being used.

The most useful liquid phases are silicone elastomers of high thermal stability such as SE-30, JXR, OV-1 and Dexsil 3000, with which separations are achieved solely on the basis of molecular weight. However, separations of wax esters and diacyl-glycerol derivatives according to degree of unsaturation have recently been accomplished on the newer temperature-stable polyester phases. Silanised solid supports (80–100 or 100–120 mesh) are essential and they are generally coated with low levels (1–3 per cent) of stationary phase by the filtration technique described earlier (Chapter 3). Care is necessary in preparing the columns, which must be packed firmly to ensure adequate resolution, but not too tightly, otherwise higher temperatures than are advisable may be necessary to elute samples in a reasonable time, and losses of higher molecular weight components may occur. Finally, the column is sealed with a plug of silanised glass wool and conditioned at a temperature at least 25°C higher than that at which it is to be used for 4 hr. It may be necessary to attempt the preparation of suitable columns several times before success is achieved and the analyst should not be discouraged by initial failures.

(II) OPERATING CONDITIONS AND NATURE OF THE SEPARATIONS

The precise operating conditions and the resolutions attainable will vary with the nature of the samples to be analysed, and they are discussed in the appropriate sections below, but some general comments can be made. With high molecular weight compounds such as triacylglycerols, short columns are essential and the column temperature must be programmed in the range 250–350°C at 2–5°/min. Slow programming rates give improved resolution generally. The optimum carrier gas flow rate will vary with the dimensions of the column and the amount of stationary phase on the packing material, but will generally be 100 ml/min or greater. The sample, in solution, should be injected by means of a syringe, directly on to the column packing at a temperature about 40°C below that at which the first component emerges from the column. In this way, all the sample is vaporised but remains as a narrow band at the top of the column until temperature programming is under way. Samples containing about 20 μg of the largest component provide the optimum load. The entire analysis should be completed in 25–45 min. With samples of intermediate molecular weight such as diacyl-glycerol acetates, wax esters or cholesterol esters, longer columns can be used to improve the resolutions attainable and the temperature limits of the analysis will be lower than with triacylglycerols.

The silicone stationary phases, which must be used for GLC analysis of high molecular weight compounds, do not in general permit the separation of saturated from unsaturated components of the same chain length. Separations are then based solely on the approximate molecular weights of the compounds (for example, tripalmitin and myristo-palmitoolein elute together), and components that differ by two carbon atoms in the combined chain-lengths of the alkyl moieties must be separable before the columns are considered satisfactory. This can usually be achieved with column efficiencies of 500–1000 theoretical plates per foot of packing material, and indeed, with well-packed columns, components differing in molecular weight by one carbon atom can often be completely resolved. A shorthand nomenclature is in common use to designate simple glycerides separated in this way; the total number of carbon atoms in the aliphatic chains of the compounds (but not in the glycerol moiety) are calculated and this figure is used to denote the compound. As an example, tristearin, triolein and trilinolein are referred to as C_{54} triacylglycerols or as having a *carbon number* of 54. Distearin has a carbon number of 36 and the diacylglycerol acetate prepared from it has a carbon number of 38 (although the two carbon atoms of the acetate moiety are not counted by some authors in calculating carbon numbers).

During isothermal operation, as discussed in Chapter 3, there is a logarithmic relationship between the retention times and carbon numbers of components of a homologous series, but when linear temperature programming is used, there is a linear relationship between these parameters for short series of homologues. As a result, the elution temperature rather than the elution time is sometimes quoted to describe the retention characteristics of a given compound. With longer homologous series the relationship begins to break down, and improved resolutions are obtained with non-linear (concave) temperature programming profiles[410], but as few commercial gas chromatographs are equipped with this facility, the refinement has not been widely adopted.

When flame ionisation detectors are used, the detector response is, within limits, proportional to the weight of material eluting from the columns (see Chapter 3), and the amount of each component can be calculated from the areas of the peaks on the GLC recorder trace. There is no simple relationship between area, retention time and peak height for temperature programmed analyses, so it is necessary to measure the area of each peak manually by one of the methods described earlier or, better, by means of an electronic digital integrator. Because of the high temperatures necessary for gas chromatography of intact lipids, there is always a danger that losses will occur on the columns as a result of pyrolysis or of reaction with the column materials. It is, therefore, necessary to check that acceptable reproducible recoveries are obtained and, if necessary, to calibrate the columns to compensate for any losses. It is not easy to check absolute recoveries from columns as this requires preparative facilities or means of counting radioactive samples eluting from the columns; it is possible to check that recoveries are linearly related to the amount of material injected by inserting known quantities of standards (say 1–20 μg) into the columns and measuring the detector response. Alternatively, it can be assumed that in all but the very worst of columns, recoveries of tricaprin or trilaurin will be essentially complete, so that standard mixtures of these triacylglycerols and the higher molecular weight compounds can be analysed and relative losses determined. The losses that can be accepted will vary with the degree of difficulty of the analysis but, as a rough guide, recoveries of the highest molecular weight component of a mixture of triacylglycerols, for example, should be at least 90 per cent relative to tricaprin. When pyrolysis occurs on the column, peaks for pure standards are often preceeded by broad humps of decomposed material. Such effects can be minimised by adding to the sample a triacylglycerol of higher molecular weight than is normally present and this presumably decomposes preferentially[486]; triarachidin (C_{60}), for example, can be used in many circumstances. By this means, recoveries and quantification of minor components are greatly improved.

As a wide range of unsaturated compounds is unlikely to be available for standardisation purposes, there are a number of advantages to be gained by hydrogenating all samples prior to analysis. If this is done there are no selective losses of unsaturated relative to saturated components, peaks on the recorder trace are sharper, resolutions are improved and quantification of components is simplified. In the analysis of molecular species of lipids, it is necessary to know the molar proportions of all components separated, and the weight responses of the detector must be corrected by multiplying by appropriate arithmetic factors obtained from the molecular weights of the compounds, as described earlier for methyl esters of fatty acids (Chapter 5). Hydrogenation prior to analysis greatly simplifies the range of factors required and removes any dubiety about their numerical values. This is of course particularly important with samples containing polyunsaturated fatty acids.

Samples may be injected on to columns in carbon disulphide, diethyl ether, hexane or xylene solution;

chloroform has also been used on occasion but tends to strip the stationary phases from the packing and damages the flame ionisation detector. Hydrogenated lipids tend to be less soluble than the unsaturated compounds in most solvents but will usually dissolve on warming.

(III) PREPARATIVE GAS CHROMATOGRAPHY

As the lipid fractions separated by the gas chromatographic techniques described above may contain a wide variety of fatty acid components, it would be very useful if compounds separated could be collected for further analysis. Kuksis and Ludwig[419] succeeded in doing this with triacylglycerols by adapting conventional preparative GLC techniques (see Chapter 5). Columns (60 cm × 0.4 cm) containing higher amounts of stationary phase were used, and it was necessary to inject the same sample a number of times, pooling corresponding fractions, to obtain sufficient material for further analysis. It was also necessary to rechromatograph fractions to obtain material of satisfactory purity. Although there were selective losses of unsaturated components, the fractions recovered were apparently not degraded in any way. The procedure is tedious and does not appear to be sufficiently reliable to have been widely used.

C. MOLECULAR SPECIES OF SIMPLE LIPIDS

1. Cholesterol esters and wax esters

Cholesterol esters are readily separated according to the degree of unsaturation of the single fatty acid moiety, by TLC on adsorbents impregnated with silver nitrate. Separations are performed with the same TLC systems as were described earlier (Chapter 5) for methyl esters of fatty acids, and fractions that contain fatty acids with zero to six double bonds are obtained in an analogous manner. Some separation of a similar nature is obtained on normal silica gel G plates with carbon tetrachloride or hexane–benzene (9:1, v/v) as solvents for the development, although the principle involved is not understood.

Reverse-phase TLC systems can be used to separate cholesterol esters into fractions containing critical pairs of fatty acids. For example, Kaufmann et al.[385] used silica gel G impregnated with liquid paraffin (see Chapter 3 for practical details) as stationary phase, with methylethylketone–acetonitrile (7:3, v/v), 80 per cent saturated with liquid paraffin, as mobile phase and obtained resolutions comparable to those described earlier for methyl esters of fatty acids (Chapter 5). Column chromatography with hydroxyalkoxypropyl–Sephadex G25 as stationary phase has also been used for the purpose.[199]

In addition, cholesterol esters can be separated by high temperature GLC on columns similar to those used for triacylglycerol analysis (50 cm × 0.2 cm with 2.25 per cent SE-30 as stationary phase), and they elute at temperatures intermediate between those required for diacylglycerol acetates and triacylglycerols (see below).[414] With temperature programmed analyses, compounds differing in molecular weight by one carbon atom are separable and, with isothermal analyses some separation of unsaturated components is possible (e.g. cholesteryl arachidonate and cholesteryl arachidate may be resolved).[408]

Single pure components which may be required for radioactivity measurements for example, can be isolated preparatively by combining any two of the above procedures, although the proportions of the various molecular species are obtainable simply from a fatty acid analysis.

In wax esters there are long aliphatic chains in both the fatty acid and alcohol moieties, each of which may contain double bonds, and silver nitrate TLC may be used to separate components according to the number of double bonds in total in both parts of the molecule. Solvent systems similar to those described above for cholesterol esters and methyl esters (Chapter 5) are used.[1, 261] Wax esters have similar molecular weights to diacylglycerol acetates and are eluted under comparable conditions from GLC columns containing non-polar stationary phases.[1, 345] In this instance they are easily separated into components differing in carbon number by one or two units, where the carbon atoms of the alcohol and fatty acid moieties have equal value. Some separation of wax esters according to degree of unsaturation in addition to chain-length has been achieved on short columns of thermally-stable polyesters at elevated temperatures, i.e. on Silar 10C[760] and Apolar 10 C[762]. Also, wax esters of short-chain alcohols have been effectively resolved on a DEGS wall-coated capillary column.[16] Unfortunately, the gas chromatographic traces from such columns present a complex distribution of peaks that are not easily identified without the aid of mass spectrometry.[1] Reverse-phase column chromatography with hydroxyalkoxypropyl–Sephadex G25 as stationary phase has given excellent results.[200] Again, an integrated programme of analysis using two of the above procedures (as discussed below for diacylglycerol acetates) should ideally be applied to wax esters.

2. Monoacylglycerols

1-Monoacylglycerols can be separated from 2-monoacylglycerols without isomerisation by chromatography on adsorbents impregnated with boric acid (see Chapter 6), and it is generally advisable to separate the two forms before proceeding to more involved analyses. It is customary to prepare volatile derivatives such as acetates, trimethylsilyl (TMS) ethers or trifluoroacetates in addition to isopropylidene or benzylidene derivatives (see Chapter 4 for practical details), to reduce the polarity of the compounds and to prevent acyl migration. Indeed, trimethylsilyl ether derivatives of monoacylglycerols are sufficiently volatile to be analysed by GLC on polyester columns, such as EGS, PEGA, Silar 5CP or Silar 10C,[336, 529, 845] which permit the separation of compounds differing in the chain-length and degree of unsaturation of the fatty acid moieties. 1- and 2-Monoacylglycerol analogues can also be separated from each other on these columns and on non-polar stationary phases, the 2-acyl compound eluting first, but saturated and unsaturated compounds of the two types will overlap if an initial separation is not performed. Similar methods are used for the separation of 1-alkylglycerols (see Chapter 6).

Individual monoacylglycerol species could no doubt be separated by silver nitrate or reverse-phase TLC procedures. Mass spectrometry is of value in the identification of specific isomers.[361]

3. Diacylglycerols

Although diacylglycerols rarely occur in greater than trace amounts in tissues, they are important intermediates in the biosynthesis of other lipids. In addition, 1,2-diacyl-*sn*-glycerols are formed by the action of phospholipase C on glycerophosphatides (see Chapter 9) and, as these compounds or their derivatives are much more volatile and less polar than the parent complex lipids, it is common practice for molecular species of glycerophosphatides to be analysed in this form. As a result, chromatographic methods for the analysis of diacylglycerols are of some importance.

Methods of separating α,β- and α,α'-diacylglycerols on boric acid impregnated adsorbents are discussed above (Chapter 6), and it is advisable to carry out this separation before proceeding to more detailed analyses, to simplify identification and improve resolution of components. Methods of fractionating α,β-diacylglycerols are discussed at length below, but similar principles apply to the analysis of α,α'-compounds. They may be separated in the free state or as less polar stable derivatives, such as the acetates or TMS ethers, but if derivatives are not prepared it is advisable that the programme of analysis be performed as quickly as possible to minimise the risk of acyl migration. Acetates are the favoured derivatives as they are chemically stable on TLC adsorbents and during high temperature GLC, although the TMS ethers are also satisfactory for GLC purposes. t-Butyldimethylsilyl derivatives, which are comparatively stable chemically, can be subjected to both GLC and TLC analysis.[532] Trifluoroacetate derivatives of diacylglycerols decompose at the temperatures necessary to elute them from GLC columns.[411] Other derivatives with UV-absorbing groups are of value for HPLC analysis. A small amount of acyl migration occurs inevitably in the preparation of most derivatives.

Diacylglycerol acetates are readily separated by silver nitrate TLC into simpler molecular species according to the degree of unsaturation of the combined fatty acid moieties. They migrate in the order:

SS>SM>MM>SD>MD>DD>ST>MT>

STe>MTe>DTe>SP>SH

where S, M, D, T, Te, P and H denote saturated, mono-, di-, tri-, tetra-, penta- and hexa-enoic fatty acid residues respectively (they do not indicate the relative positions of the fatty acids on glycerol as positional isomers are eluted together).[420, 600] It is noteworthy that one linoleate moiety is more strongly retained than two monoenoic moieties, and one linolenate moiety is more strongly retained than two linoleate moieties. Figure 8.1 illustrates the separation of diacylglycerol acetates prepared from phosphatidylcholine of pig liver on silica gel G impregnated with 10 per cent silver nitrate; the solvent system in this instance was chloroform–methanol (99:1, v/v), but hexane–diethyl ether (95:15, v/v) also gives satisfactory results. (N.B. to ensure that the solvent system contains the stated amount of methanol, it is first necessary to free the chloroform of any alcohol added as stabiliser (see Chapter 3).) When the sample contains high proportions of polyunsaturated fatty acids, it may be necessary to repeat the fractionation with a solvent system containing an increased concentration of the more polar component. Diacylglycerol acetates with up to twelve double bonds in the combined fatty acid moieties have been separated on similar plates with chloroform–methanol–water (65:25:4, by vol.) as the solvent system.[623] It is possible with care to separate molecular species containing positional isomers of polyunsaturated fatty acids from each other. For example, diacylglycerol acetates contain-

ing 18:3 (n-6) migrate ahead of those containing 18:3 (n-3), and species containing 20:4 (n-6) migrate ahead of those containing 20:4 (n-3).[130]

Approximately 10 mg of diacylglycerol acetates can be separated preparatively on a 20×20 cm TLC plate coated with a layer of the silver nitrate-impregnated adsorbent 0.5 mm thick. Components are visualised under UV light after spraying with 2′,7′-dichlorofluorescein solution, and may be identified and estimated by GLC with an added internal standard, after recovery from the adsorbent and transesterification (see Chapters 3 and 6 for further discussion of the methods). Diacylglycerol acetates, free of dichlorofluorescein and silver compounds, can be recovered quantitatively from the adsorbent by the method of Akesson[21].

"Bands are scraped from the plate into test-tubes and a solution of the internal standard, say methyl pentadecanoate, in methanol (1 ml) added to each followed by a 20 per cent aqueous solution of sodium chloride (1 ml) and hexane–diethyl ether (1:1, v/v; 3.5 ml). The contents are mixed thoroughly by means of a vortex mixer and centrifuged to precipitate the solids. The top layer is pipetted off, and the bottom layer is washed twice more with similar volumes of hexane–ether. The combined extracts are washed with 0.05 M tris buffer (pH 9, 2 ml) and then with water–methanol (1:1, v/v; 2 ml) before being dried over anhydrous sodium sulphate."

After removal of the solvent, each sample is transesterified with sodium methoxide in the presence of dichloromethane or benzene, which ensures that it remains in solution and reacts (see Chapter 4 for practical details). Methyl heptadecanoate may also be used as the internal standard. The molar amounts of the fractions are calculated by relating the combined areas of the fatty acid peaks on the GLC recorder traces, corrected from weight to molar amounts, to the area of the internal standard peak. In addition, tridecanoin may be added as the internal standard (in the hexane layer in this instance), and the intact compounds subjected to high temperature GLC as discussed below.

Alternatively, bands from silver nitrate TLC can be scraped on to the top of a short column (2 cm) of carboxymethylcellulose packed in chloroform and the lipids recovered, quantitatively and free of

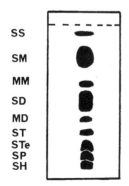

SS
SM
MM
SD
MD
ST
STe
SP
SH

FIG. 8.1. TLC separation of dyacylglycerol acetates prepared from the phosphatidylcholine of pig liver, on layers of silica gel G impregnated with 10 per cent silver nitrate. Developing solvent: chloroform–methanol (99:1, v/v). Abbreviations: S, M, D, T, Te, P and H denote saturated, monoenoic, dienoic, trienoic, tetraenoic, pentaenoic and hexaenoic fatty acid residues respectively esterified to glycerol.

contaminants, by elution with chloroform–methanol (98:2 by vol; 10–15 ml), containing an internal standard if necessary (Author, unpublished observations). The plates should be exposed to light for the minimum possible time while bands are being extracted.

Alternative methods of quantifying the fractions have been devised. For example, [1-^{14}C]-acetic anhydride has been used in the preparation of diacylglycerol acetates and fractions determined by radioactivity measurement.[53] Trityl as opposed to acetate derivatives can be estimated by measuring UV absorption following separation.[193] t-Butyldimethylsilyl derivatives are chromatographed in a similar way to acetates but have more distinctive mass spectra which facilitate characterization and estimation.[532]

Although free diacylglycerols can be subjected to gas chromatography some rearrangement and transesterification inevitably occurs, so it is more usual for the acetate derivatives to be prepared. The nature of the columns and the precise operating conditions that have been recommended vary somewhat from laboratory to laboratory but, as a rough guide, 1 m × 3 mm o.d. columns packed with 2 per cent SE-30 on a silanised support will give good results. Carrier gas flows of 100–200 ml/min and temperature programming from 220–280°C are commonly quoted. Some deviation from these optimum conditions is permissible, however, and the separations illustrated in Fig. 8.2 were obtained on 50 × 0.4 cm (i.d.) glass columns containing 1 per cent SE-30, temperature programmed from 250 to 300°C with a carrier gas flow of 50 ml/min of nitrogen. Components are identified by their retention times (or elution temperatures) relative to standard triacylglycerols having the same total number of carbon atoms. The resolutions obtained are sufficient to separate components with combinations of fatty acids differing in chain length by one carbon atom. Although the separations depend largely on the molecular weights of the components, some partial separation on the basis of degree of unsaturation may occur but may not be especially desirable. For example, in Fig. 8.2(A), the components of carbon number 38 consist of two

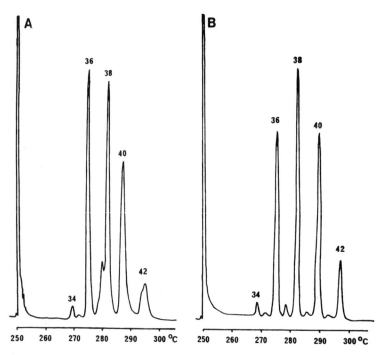

Fig. 8.2. GLC separation of diacylglycerol acetates prepared from the phosphatidylcholine of pig liver. 50 × 0.4 cm (6 i.d.) glass column packed with 1 per cent SE-30 on chromosorb W (acid-washed and silanised; 100–120 mesh); temperature programmed from 250–300°C at 2°/min; carrier gas — 50 ml/min nitrogen. Trace A before and trace B after hydrogenation. The numbers above each peak refer to the carbon number of the component.

partially resolved peaks; the first contains 16:0 together with 20:4 and the second consists of two C_{18} acids of varying degrees of unsaturation. On hydrogenation (a suitable procedure is described in Chapter 4) these components merge and the remaining peaks are distinctly sharper with improved resolution (Fig. 8.2B); in particular, trace amounts of components containing odd-chain fatty acids become apparent. With these particular columns and operating conditions, recoveries of a wide range of standards were essentially complete whether the compounds were hydrogenated or not, and similar results have been reported from many other laboratories.

t-Butyldimethylsilyl ether derivatives of diacylglycerols have been subjected to GLC analysis in an analogous manner to the above for mass spectral characterisation[532, 659], but acetates can also be analysed in this way if need be[282, 283].

Diacylglycerol acetate or trimethylsilyl ether derivatives are separable to some extent, both by the chain-length and degree of unsaturation of the combined fatty acid constituents, on short columns containing the more thermally-stable liquid phases. EGSS-X was used initially,[412] but improved resolution and column stability was obtained with Silar 5CP[530] and Silar 10C[336] as stationary phases. GLC traces obtained by this means are complex, and peaks are not easily identified, but internal standards help and it is also possible to use a series of separation factors similar to those used for fatty acid identification (reviewed by Myher[528]).

Reverse-phase TLC systems have been used to fractionate diacylglycerol acetates into molecular species containing critical pairs of fatty acid combinations; for example 16:0–16:0 migrated with 16:0–18:1, but both were separable from 16:0–18:0. Silicone (Dow Corning 200) liquid on silica gel G was the stationary phase and acetic acid–water mixtures, the mobile phase.[620] Reverse-phase column chromatographic methods (including HPLC) have also been used on various derivatives of diacylglycerols.[58, 157, 158]

A full analysis of molecular species of diacylglycerols requires that combinations of the above procedures be used. For example, diacylglycerol acetates may be analysed separately according to degree of unsaturation by silver nitrate chromatography, and then according to molecular weight by high-temperature gas chromatography. Fractions from silver nitrate plates should be identified and estimated by fatty acid analysis with an added internal standard, but should then be analysed as the intact compounds (hydrogenated if necessary) by high-temperature GLC with tridecanoin as internal standard. The proportions of the various fractions found by the two methods should agree. It is also advisable to check the accuracy of the analyses by summing the amounts of each fatty acid or diacylglycerol acetate of each carbon number in all the fractions, and comparing the results with similar analyses performed on the intact lipid. *Molar* proportions should always be quoted. As an alternative to GLC, reverse-phase chromatography has certain advantages especially in preparative applications.

The main limitation of all these procedures is that components with two given fatty acids in different positions of the glycerol moiety (e.g. 1-palmito-2-olein and 1-oleo-2-palmitin) are not separated. The amounts of various isomers of this type can, however, be calculated if each fraction is subjected to enzymatic hydrolysis with pancreatic lipase (see Chapter 9).

Alkyl- and alkenyl-analogues of diacylglycerols have been prepared, and analysed by similar techniques to those described above.[157, 158, 618, 619]

4. Triacylglycerols

As triacylglycerols are the most abundant lipid class in nature, and all the commercially important fats and oils consist almost entirely of this lipid, more effort has been applied to the determination of molecular species of triacylglycerols than of any other single lipid class. The problem can be extremely complicated; for example, a triacylglycerol with only five different fatty acid constituents may consist of 75 different molecular species (not including enantiomers). Methods for the analysis of triacylglycerols have been reviewed.[449, 528] Triacylglycerols from natural sources may contain a wide range of fatty acids, including those of very short chain-length (from acetic acid upwards) or those containing polar functional groups. In such instances, useful separations of molecular species can be obtained by adsorption chromatography. For example, triacylglycerols of the Beluga whale contain isovaleric acid, and species containing zero, one and two molecules of this acid have been isolated by TLC on silica gel G with a developing solvent of hexane–diethyl ether–acetic acid (87:12:1, by vol.);[450] the presence of a short-chain acid retards the migration of the molecule (Fig. 8.3, plate A). Similarly, the triacylglycerols of ruminant milks can be separated by related TLC systems into two rough fractions — one containing three fatty acids of normal chain-length and one containing two normal and one short-chain fatty acid. Polar fatty acid constituents also retard the migration of triacyl-glycerols on adsorbents, and triacylglycerols from seed oils that contain epoxy fatty acids, for example, can be separated into simpler species that contain zero, one, two and three epoxy-fatty acid residues per mole of glycerol.[149] A solvent system of hexane-diethyl ether (3:1, v/v) with silica gel G layers, for example, was used to isolate such fractions. (c.f. Fig. 8.3, plate B). Similar separations have been obtained of triacylglycerols that contain fatty acids with hydroxyl or other polar functional groups. Estolides can be separated in this way from normal triacylglycerols.[509]

GLC of the fatty acid constituents in the presence of a suitable internal standard (e.g. methyl pentadecanoate or methyl heptadecanoate) or charring-photodensitometry would normally be the methods of choice for estimating components isolated by adsorption procedures. The amounts that can be separated preparatively in this way vary with the nature of the fatty acid constituents, but generally, the more polar the functional groups in the fatty acid, the more that can be applied to a TLC plate. As a rough guide, 10–25 mg of material can be applied to a 20×20 cm plate coated with a layer 0.5 mm thick of silica gel G.

Countercurrent distribution has been used to separate triacylglycerol species containing polar fatty acids, but the same fractions can usually be obtained with very much less effort by adsorption chromatography, though admittedly on a smaller scale.

Silver nitrate chromatography has proved immensely useful in separating triacylglycerols, containing a more normal range of fatty acids with zero to three cis-double bonds, into simpler molecular species. A triacylglycerol of this type can contain species with up to nine double bonds in the fatty acid moieties per mole of glycerol. Components migrate in the order:

$$SSS>SSM>SMM>SSD>MMM>SMD>MMD>$$
$$SDD>SST>MDD>SMT>MMT>DDD>SDT>$$
$$MDT>DDT>STT>MTT>DTT>TTT>$$

where S, M, D, and T denote saturated, mono-, di- and tri-enoic acids respectively[255, 634] (they do not indicate the positions of the fatty acids on the glycerol moiety), although there may be some changes in this order depending on the nature of the solvent mixtures used for development. Some simplification of even more highly unsaturated triacylglycerols, such as fish oils, has also been attained by silver nitrate TLC.[78] 10–15% Silver nitrate is incorporated into the layers and the solvent systems generally employed consist of hexane–diethyl ether, benzene–diethyl ether or chloroform (alcohol-free)–methanol mixtures. As all the fractions listed above cannot be separated on one plate, it is common practice to separate the least polar fractions first with hexane–diethyl ether (80:20, v/v) or chloroform–methanol (197:3, v/v) as illustrated in Fig. 8.4 plate A, and then to separate the remaining fractions with more polar solvents such as

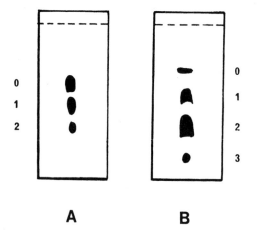

A   **B**

FIG. 8.3. TLC separation of triacylglycerols containing short-chain or polar fatty acids on layers of silica gel G. Plate A. Beluga whale oil triacylglycerols — the numbers refer to species with 0, 1 and 2 iso-valeroyl residues per mole. Solvent system: hexane–diethyl ether–acetic acid (87:12:1 by vol.)[149] Plate B. Seed oil containing vernolic acid — the numbers refer to species with 0, 1, 2 and 3 epoxy-fatty acid residues per mole. Solvent system: hexane–diethyl ether (3:1, v/v).

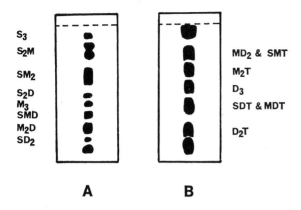

S_3
S_2M

SM_2

S_2D
M_3
SMD
M_2D
SD_2

MD_2 & SMT
M_2T
D_3
SDT & MDT
D_2T

A   **B**

FIG. 8.4. TLC separation of soyabean triacylglycerols on layers of silica gel G, impregnated with 10 per cent (w/w) silver nitrate. Developing solvent; Plate A. Chloroform–methanol (99:1, v/v). Plate B. Chloroform–methanol (96:4, v/v). Abbreviations: S, M, D and T denote saturated, monoenoic, dienoic and trienoic fatty acid residues respectively esterified to glycerol.

diethyl ether alone or chloroform–methanol (96:4, v/v), as illustrated in Fig. 8.4 plate B. Bands are detected under UV light after spraying with 2′,7′-dichlorofluorescein solution, components are recovered and they are identified and determined by gas chromatography of the fatty acid constituents with an added internal standard, as discussed earlier for diacylglycerol acetates. Approximately 10 mg of triacylglycerols can be separated on a 20 × 20 cm plate (0.5 mm thick layer), and excellent separations of large numbers of components have been obtained with 20 × 40 cm plates.

In addition, some separation of isomeric compounds is possible. For example, triacylglycerols of the type S_2M, in which the monoenoic component is in position 2, are separable with care from the related compounds in which the monoenoic component is in position 1 or 3. Presumably, the presence of long-chain fatty acids on either side of the monoenoic acid weaken the π-complex with the silver ion, permitting the component with the monoenoic fatty acid in position 2 to migrate ahead of isomers in which this residue is in position 1, as separations of this kind cannot be obtained with diacylglycerol acetates where one side of the molecule is comparatively open. When the fatty acid constituents of triacylglycerols contain *trans*-double bonds, the elution pattern is much more complicated as components with *trans*-acids migrate ahead of analogous compounds with fatty acids containing

cis-double bonds, producing a complex elution pattern. Molecular species of triacylglycerols with polar fatty acid constituents, separated by adsorption chromatography as described above, may be further fractionated by silver nitrate TLC.[149] Although column chromatography with silver nitrate-impregnated adsorbents has been used to separate triacylglycerols, it lacks the resolving power of thin-layer chromatography.

High-temperature GLC can be used to obtain efficient separations of triacylglycerols on a molecular weight basis, although the conditions necessary to elute them from the columns approach the limits of thermal stability both of the stationary phases and of the compounds themselves. Although triacylglycerols with carbon numbers up to 68 have been successfully resolved,[280] there is little margin for error in the preparation of columns and, in most laboratories, it is considered a sufficient achievement to separate components with carbon numbers up to 56 or 58. Short packed columns (50 to 100 cm × 3 mm), high carrier gas flow rates and low loads of stationary phases (1–2 per cent SE-30 or JXR) are normally used with temperature programming at 2–5°/min from 180° to 350°C, depending on the range of molecular weights of the compounds to be analysed. Individual components are recognised by their carbon numbers relative to authentic standards. They are separable into homologues, differing in carbon number by two units, with relative ease, but more care is necessary to separate components differing by one unit when odd-chain fatty acids are present in the sample.[451] Figure 8.5 illustrates the kind of separation obtainable with two natural triacylglycerol samples, A—coconut oil, B—pig adipose tissue (lard), although the operating conditions in this instance were not those generally considered optimal. Efficient separations of the lower molecular weight components of the former sample are readily achieved, but it is less easy to separate effectively the higher molecular weight components in the latter. Again, if the compounds are hydrogenated prior to the analysis, resolutions are improved and losses are reduced, as saturated compounds are more stable to the operating conditions. Poor resolutions are inevitable when packed columns are used with natural triacylglycerols, such as those of ruminant milk fats, that contain odd-chain and branched-chain fatty acids as

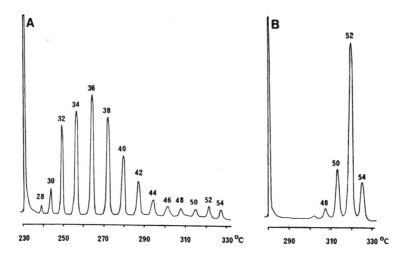

FIG. 8.5. GLC separation of intact triacylglycerols A. Coconut oil B. Pig adipose tissue. Column as in Fig. 8.2; temperature programmed from 230–330°C at 2°/min (A) and 280–330°C at 2°/min (B). The numbers above each peak refer to the carbon number of the component.

well as fatty acids with a very wide range of chain-lengths.

Quantification must be carefully checked with standard mixtures to ensure that losses of high molecular weight components relative to those of lower molecular weight are as low as possible; for example, if the recovery of a C_{54} standard is less than 90 per cent of that of a C_{30} standard, the column is not satisfactory. Greater losses of higher molecular weight triacylglycerols such as trierucin (C_{66}) are inevitable, however, and must at present be accepted, but all such losses can be compensated for by determining calibration factors with standard mixtures, provided that they are checked regularly. Results should again be converted to molar proportions. These aspects of high-temperature packed-column GLC are also discussed above (Section B.2) and have been reviewed elsewhere.[268, 410, 416, 449, 528]

Some separation of triacylglycerols according to degree of unsaturation in addition to chain-length has been achieved on packed GLC columns of Poly S179[36] and of Silar 10C[761]. Whether the resolution obtained is sufficient to be of value to lipid analysts is doubtful, however, especially as such columns will probably have a short life-time. On the other hand, gas chromatography using glass-walled capillary columns offers promise of greatly increased resolution of triacylglycerols into molecular species, although some technical problems remain in

obtaining quantitative recoveries.[679, 240–242] With capillary columns, mass spectrometry is of great assistance in identifying and quantifying components emerging in a single peak.[524]

GLC of triacylglycerols is used almost exclusively as an analytical technique although some preparative applications have been described, as discussed above (Section B.2).[419]

A number of liquid–liquid partition systems (TLC and column, including HPLC) have been devised for the separation of intact triacylglycerol molecules into simpler molecular species. The resultant fractions, separated both by molecular weight and by the number of double bonds, are sometimes defined by a "partition number" where

$$\text{partition number} = \text{carbon number} - 2\text{x (number of double bonds)}^{448}$$

Triacylglycerols having the same partition number, such as distearoolein and distearopalmitin, co-chromatograph in liquid–liquid systems. Reverse-phase TLC has been used more frequently than column chromatography because of the greater resolution attainable with the former. For example, Litchfield[448] separated triacylglycerols, differing in partition number over a range of 38 to 48, on 20 × 40 cm plates with silanised silicic acid impregnated with 8 per cent (w/w) hexadecane as stationary phase and nitroethane saturated with hexadecane as the mobile phase. 4–10 mg of Triacylglycerols were separated preparatively in this way, and bands were detected by exposing the edge of the plate to iodine vapour; adjacent areas, not contaminated with iodine, were scraped from the plate for analysis (there is a report that 2',7'-dichlorofluorescein sprays can be used in related circumstances).[827] A separation of this kind is illustrated in Fig. 8.6. Kaufmann and Das[384] described similar reverse-phase systems, for example with tetradecane on silica gel G as the stationary phase and acetone-acetonitrile (80:20, v/v), 80 per cent saturated with tetradecane, as the mobile phase. Litchfield's procedure may have particular advantages with highly unsaturated fats, such as those of fish oils, that are less easily analysed by silver nitrate TLC. Considerable practice is necessary with the technique, however, before separations of the kind illustrated (Fig. 8.6) can be obtained routinely.

A reverse-phase column procedure with heptane on silanised celite as stationary phase and acetonitrile–methanol (85:15, v/v) saturated with heptane as mobile phase has been used to effect remarkable preparative-scale separations of molecular species of triacylglycerols, including species containing trans-fatty acids not separable by silver nitrate chromatography.[544, 556] Most workers, however, would now prefer to utilise alkylated-Sephadex[557] as stationary phase, since it is much more convenient and stable in use. For example, Lindqvist et al.[446] separated butter fat into molecular species (0.5–2.5 g scale) on a 100 cm × 2.5 cm column

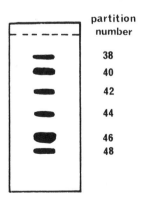

FIG. 8.6. Reverse-phase TLC separation of *Ephedra nevadensis* triacylglycerols (fatty acids from 14:0 to 20:4 are present). Stationary phase — silanised silicic acid impregnated with 8 per cent hexadecane; mobile phase — nitroethane saturated with hexadecane.[448]

packed with hydroxyalkoxypropyl–Sephadex LH-20 as illustrated in Fig. 8.7. In this instance, the detector was of the moving-wire flame-ionisation type so that gradient elution was possible; the temperature of the column was maintained at 40°C to ensure solubility of the higher molecular weight components.

Some excellent reverse-phase column separations of triacylglycerols have been achieved by means of HPLC.[65, 198, 302, 573, 574, 581] The stationary phases adopted are generally the commercially-available C_{18}-alkylated silicas; various mobile phases have been used, but they usually contain acetonitrile and/or acetone. For example, Bezard et al.[65] obtained clean separations of molecular species of triacylglycerols of seed oils on the basis of partition number on "μBondapack C18" columns, with acetonitrile–acetone (42:58) as mobile phase and differential refractive index detection. With infra-red detection (5.75 μm due to the carbonyl function), it is possible to use gradient elution techniques.[573] Lie Ken Jie[441] has demonstrated a need for careful calibration of the procedure if it is to be used quantitatively. On the other hand, as separation is achieved at ambient temperature, little harm can come to the samples.

A mass spectrometric system of determining molecular species of triacylglycerols has been devised.[307, 308] Mass spectra, obtained from a sample placed directly in the ion source, are fed into a computer programmed to recognise characteristic ions and calculate the proportions of various molecular species (separated by the chain length and degree of unsaturation of the constituent fatty acids) from the ion intensities. Large numbers of samples can be handled routinely, but few lipid analysts have such equipment available. Chemical ionisation-mass spectrometry can be used in an analogous manner.[525]

As discussed earlier for diacylglycerol acetates, greatly improved separations of molecular species of triacylglycerols are obtained if techniques are used in combination. Silver nitrate chromatography to separate by degree of unsaturation, followed by high-temperature GLC to separate by carbon number, is the most widely-used approach, but replacement of the GLC step with reverse-phase HPLC will give exactly the same fractions.[65] Again, the main limitation of such procedures is that

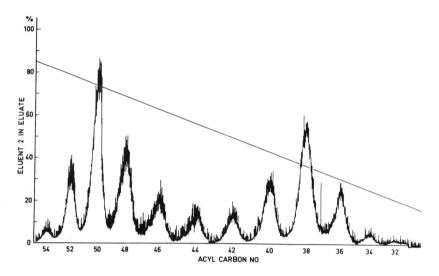

Fig. 8.7. Separation of triacylglycerols from 0.5 g of butter fat by reverse-phase column chromatography.[446] Stationary phase, hydroxy (C_{15}–C_{18}) alkoxypropyl–Sephadex LH20; mobile phase, gradient of heptane–acetone–water (4:15:1 by vol.; eluent 2) in isopropanol–chloroform–heptane–water (115:15:2:35, by vol.; eluate); Flow rate 75 ml/hr. Temperature, 40°; Detector, moving wire-flame ionisation. (Reproduced by kind permission of Dr. B. Lindqvist & the *Journal of Lipid Research*.)

positional isomers of triacylglycerols with specific fatty acid combinations are not in general separable, and it is necessary to apply stereospecific hydrolysis methods (see Chapter 9 for practical details) to each fraction to determine all possible components. Although this is now theoretically feasible,[274] the difficulties, not least of which is obtaining each fraction in sufficient quantity for analysis in a pure state, are enormous.

Molecular species of alkyldiacylglycerols and neutral plasmalogens have been isolated by procedures related to those described above.[840, 850]

D. MOLECULAR SPECIES OF COMPLEX LIPIDS

1. General approaches to the problem

Molecular species of complex lipids such as glycerophosphatides are most easily separated after the phosphorus moiety has been removed by enzymatic hydrolysis with phospholipase C (see Chapter 9 for practical details); the resulting 1,2-diacyl-*sn*-glycerols are normally purified and acetylated *immediately* to prevent acyl migration during subsequent separation procedures. Indeed,

even the diacylglycerol acetates should be purified by TLC on silica gel G (hexane–diethyl ether, 70:30, v/v is a suitable solvent system) before being analysed further. Diacylglycerol acetates from all phospholipids may be separated into molecular species by the procedures discussed earlier in this chapter. Alkyl- and alkenyl- forms should be isolated (see Chapter 7) before more detailed studies are commenced.

No satisfactory chemical method for removing the phosphorus moiety of glycerophosphatides has been developed, although acetylation with equal mixtures of acetic anhydride and acetic acid at 140°C in a sealed tube has been used for the purpose.[617] Unfortunately, some acyl migration with formation of a small proportion of 1,3-diacylglycerol acetates takes place, but it is argued that this occurs intra-molecularly not inter-molecularly. Therefore, the same fatty acids remain in molecular combination and the diacylglycerol acetates so produced are still suited to molecular species analysis (but not for analysis of fatty acid positional distribution).[282, 283] Trimethylsilyl-ether derivatives of diacylglycerols have been prepared from glycerophosphatides by subjecting the latter to very high temperatures (approx. 250°C) for brief periods and then silylating.[326, 327] Here also, 1,3-derivatives accompany the 1,2-compounds, and the latter do not accurately represent the original composition of the native phospholipid so that their use in structural analyses is limited.

In some studies it may be necessary to retain the phosphorus moiety in the molecule (e.g. in metabolic studies of compounds labelled with ^{32}P), and methods have been developed for isolating molecular species of glycerophosphatides in the native state or by rendering the polar head group less polar by conversion to an appropriate derivative. The preferred approach may differ for each lipid class. Silver nitrate TLC and reverse-phase TLC or HPLC, preferably used to complement each other, are the methods of choice for isolating simpler species; high-temperature GLC cannot be used as glycerophosphatides and their derivatives apparently pyrolyse at the temperatures necessary to elute them from the columns. In general, resolutions obtained with compounds that still contain the phosphorus moiety are somewhat inferior to those that can be achieved with diacylglycerol acetates. As was the case with simple lipids, it may be necessary to subject the fractions to hydrolysis with stereospecific lipolytic enzymes in order to obtain a complete picture of the lipid structure.

Glycosyl diacylglycerols and sphingolipids can sometimes be separated into simpler molecular species by methods related to those described above.

2. Phosphatidylcholine

Molecular species of phosphatidylcholine have been analysed as the diacylglycerols or diacylglycerol acetate derivatives in a number of laboratories.[130, 420, 555, 619, 793] The phospholipase C of *Clostridium welchii* (see Chapter 9) is generally used in their preparation although direct acetylation, in spite of the disadvantages, is occasionally used also. In addition, alkyl- and alkenyl- forms of phosphatidylcholines have been separated in this way.[157, 619, 621]

There have been a number of attempts to separate species of native phosphatidylcholine on thin layers impregnated with silver nitrate, but most have met with little success probably because the highly polar phosphorus-containing group masked the comparatively small changes in polarity produced by the formation of π-complexes between silver ions and the double bonds of the unsaturated fatty acids.

Arvidson[44, 46], however, has achieved some valuable separations by using highly-active TLC plates; layers, 0.35 mm thick, were prepared with silica gel H and silver nitrate in the proportions 10:3 (w/w), and were air-dried at room temperature (in the dark) for 24 hr, then either at 175°C for 5 hr or at 180°C for 24 hr. Plates prepared under the latter conditions were considerably more active than those dried at the lower temperature. Using a solvent system of chloroform–methanol–water (60:35:4, by vol.) and with the more active plates, species with one to six double bonds per molecule were separable from each other. Only the saturated and monoenoic fractions were not completely resolved, but they could be separated on the less active plates with the same solvent system. Up to 20 mg of phospholipid could be separated into molecular species on a 20 × 20 cm plate. Typical separations are illustrated in Fig. 8.8.

Lipids were recovered from the adsorbent after detection with a 2′,7′-dichlorofluorescein spray, by elution with chloroform–methanol–acetic acid–water (50:39:1:10 by vol.), washed subsequently with one third of a volume of 4 M ammonia to remove dye and excess silver ions. Fractions were determined by phosphorus analysis.[45, 46] Removal of all the silver ions sometimes presents difficulty, but the above procedure is as effective as most, and ion-exchange chromatography with CM-cellulose (see section C3 above) could no doubt be adapted to the purpose.

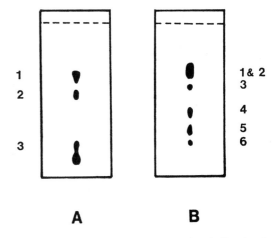

Fig. 8.8. Fractionation of native phosphatidylcholines by silver nitrate TLC[44, 46] (layers of silica gel H–silver nitrate, 10:3, w/w). Solvent system, chloroform–methanol–water (65:25:4 by vol.). Plate A activated at 180°C for 24 hr; plate B activated at 175°C for 5 hr. The numbers alongside the plate refer to the number of double bonds in each species.

Arvidson[45] has successfully separated native phosphatidylcholines by reverse-phase TLC into components of the same partition number. Undecane on silanised Kieselguhr G was the stationary phase and methanol–water (9:1, v/v), 70 per cent saturated with undecane, was the mobile phase. Each of the fractions from silver nitrate plates was separable into two components, largely those species that contained 16:0 and those that contained 18:0 (approximately 2 mg of material per 20 × 20 cm plate). Bands were visible, without spraying, as grey areas on a white background and could be recovered from the plate by elution with chloroform–methanol (2:1, v/v). "Green fingers" and practice are necessary before reproducible results can be obtained with either of the above two procedures.

Reverse-phase column chromatography has also been used with some success for separating molecular species of phosphatidylcholines. Arvidson[47], for example, used alkylated Sephadex as stationary phase and methanol–water as mobile phase for the purpose; others have used HPLC with alkylated silica ("μBondapak C18") as stationary phase.[155, 590]

The alternative approach to the problem is to render the head group less polar, i.e. phosphatidylcholine is converted enzymatically to phosphatidic acid, by reaction with phospholipase D, and this is in turn converted to the dimethyl ether by reaction with diazomethane (the preparation of these compounds is discussed in greater detail in Chapter 9). Although the choline group is lost, the phosphorus atom is retained. (The use of such derivatives to isolate alkyl- and alkenyl- forms of phosphatidylcholine is discussed in Chapter 7). Dimethyl phosphatidates are separable by silver nitrate and reverse-phase TLC (but not by high-temperature GLC) with almost the same facility as are diacylglycerols. For example, Renkonen[624] used silver nitrate TLC to obtain distinct fractions with zero to six double bonds per molecule. Chloroform–methanol (98:2, v/v) separated the less unsaturated species, and chloroform–methanol–water (90:10:1 by vol.) the more unsaturated species. Reverse-phase TLC has also been used to separate dimethylphosphatidates;[854] tetradecane on Kieselguhr G was the stationary phase and acetonitrile–acetone–water (8:1:1 by vol.) saturated with tetradecane was the mobile phase.

3. Phosphatidylethanolamine

Phosphatidylethanolamine is readily separated into molecular species in the form of diacylglycerol acetates, prepared following reaction with the phospholipase C of *Bacillus cereus* (see Chapter 9 for practical details). This approach has been followed in a number of laboratories.[318, 794, 850]

Species of native phosphatidylethanolamine analogous to those of phosphatidylcholine can be isolated by the silver nitrate and reverse-phase procedures of Arvidson[46], described above, under the same conditions. A similar approach was used by others.[325] However, most workers have preferred conversion to non-polar derivatives (retaining both the phosphorus and ethanolamine moieties), e.g. N-dinitrophenyl-O-methyl-,[624] N-acetyl- or N-benzoyl-O-methyl-,[744] and trifluoroacetamide[863] derivatives. All of these are readily separated by silver nitrate chromatography and some have been separated by reverse-phase TLC.

4. Other glycerophosphatides

Molecular species of a number of other glycerophosphatides have been isolated by methods involving the formation of diacylglycerols, including phosphatidylinositol,[22, 320, 841] polyphosphoinositides,[320] phosphatidylserine,[841] phosphatidylglycerol,[287] cardiolipin,[389, 841] phosphatidic acid,[304] and cytidine diphosphate diacylglycerol[770]. Acidic or highly ionic glycerophosphatides are normally hydrolysed to diacylglycerols more readily by the phospholipase C of *B. cereus*, but phosphatidylinositol and polyphosphoinositides have also been dephosphorylated by means of a specific phosphodiesterase from brain,[320] and phosphatidic acid has been dephosphorylated with a phosphatidic acid phosphatase[304]. The phospholipase C of *B. cereus* releases both diacylglycerol moieties from cardiolipin. The enzymes used in these structural studies are discussed in greater detail in Chapter 9.

In addition glycerophosphatides may be separated into molecular species in the form of the dimethylphosphatidates. For example, phosphatidic acid[591] itself has been analysed in this form, and phosphatidylserine and phosphatidylglycerol are both hydrolysed rapidly by phospholipase D so could undoubtedly be analysed in this way also. Although phosphatidylinositol is not readily hydrolysed by phospholipase D, it can be converted to phosphatidic acid by reaction with periodate.[463]

Native phosphatidylinositol has been fractionated by silver nitrate TLC on highly active plates with chloroform–methanol–water (65:35:5. by vol.) as the solvent system,[319] but mono- and dienoic species were not completely resolved. Luthra and Sheltawy[463], however, were able to retain the phosphorus and inositol moieties and improve the resolutions attainable by another approach. The inositol residue was acetylated with acetic anhydride and pyridine, yielding a triacetyl derivative as the main product. After purification, the phosphoric acid moiety was rendered non-polar by reaction with diazomethane, when the resulting compound was separable into distinct species with zero to six double bonds per mole of phospholipid by silver nitrate TLC with chloroform–acetone mixtures (75:25 or 50:50, v/v) as developing solvent. Two novel procedures were used to estimate the fractions separated. In the first, each fraction was transesterified in the presence of [³H]-methanol and in the second, the lipid was acetylated with [1–¹⁴C]-acetic anhydride; the [³H] methyl esters or [1–¹⁴C]-acetyl phosphatidylinositol derivatives were then determined by liquid scintillation counting.

Attempts have been made to fractionate native phosphatidylglycerol but with only limited success. The procedures used for separations of species of phosphatidylinositol could conceivably be of value with phosphatidylglycerol also.

Although it has proved possible to fractionate phosphatidylserine in unmodified form by silver nitrate TLC,[653] much better results are obtained if it is first converted to the trifluoroacetamide or azlactone (internal anhydride) derivative.[863]

5. Glycosyldiacylglycerols

Procedures have been developed for the separation of intact glycosyldiacylglycerols into simpler molecular species. For example, monogalactosyldiacylglycerols were separated into distinct fractions containing one to six double bonds per molecule, on thin layers of silica gel H containing 10 per cent (w/w) silver nitrate with chloroform–methanol–water (60:21:4, by vol.) as solvent system.[542] Somewhat better resolutions of both mono- and digalactosyldiacylglycerols are obtained if they are first acetylated with acetic anhydride and pyridine; the resulting compounds are similar in their chromatographic behaviour to diacylglycerol acetates, and have been separated into seven clear fractions, containing zero to six double bonds per mole, on silica gel G plates impregnated with 5 per cent (w/w) silver nitrate with a developing solvent consisting of chloroform–methanol (197:3, v/v).[43] High-temperature GLC of the acetate derivatives does not appear to have been attempted, but trimethylsilyl ether derivatives of mono- and digalactosyl-mono- and diacylglycerols (preferably after hydrogenation) have been separated into species according to carbon number, by high-temperature GLC under conditions similar to those used for triacylglycerols.[49, 829]

Diacylsulphoquinovosylglycerol has been successfully resolved into simpler molecular species in unmodified form by silver nitrate TLC.[543]

6. Sphingomyelin and glycosphingolipids

In Chapter 7, it was noted that sphingomyelin is separable into two fractions — one containing very long-chain fatty acids and one containing fatty acids of medium chain-length. Glycosphingolipids can sometimes be partially separated into two classes — one containing normal fatty acids and one containing 2-hydroxy fatty acids (see Chapter 7 also), but molecular species of intact sphingolipids are less easily isolated otherwise. Considerable simplification of sphingolipid species can, however, be achieved, if they are first converted to the less polar ceramides. By using the enzyme, phospholipase C from Clostridium welchii, sphingomyelin is hydrolysed to ceramide (see Chapter 9 for practical details). Unfortunately, ceramides are less easily prepared from glycosphingolipids. Compounds containing dihydroxy bases can be converted to ceramides by an elegant procedure devised by Carter et al.[108], in which the glycosidic ring is opened with periodate and the resulting product reduced with sodium borohydride, before being hydrolysed under mild acidic conditions. Derivatives of trihydroxy bases cannot be converted to ceramides by this procedure as the bases would be cleaved across the vicinal diol group by the periodate oxidation step. A number of enzymes have been described that will hydrolyse ceramide-carbohydrate bonds.[604] In particular, preparations of β-glucosidase (EC.3.2.1.21), that release ceramide from glucosylceramide, are readily obtained,[227] but they do not appear to have been used in studies of molecular species distributions, possibly because the reactions do not go to completion.

Karlsson and Pascher[378] have published a valuable paper on thin-layer chromatography of ceramides. There may be from two to four hydroxyl groups in the molecule (two or three in the base and zero or one in the fatty acid), and TLC on silica gel or better on silica gel containing diol-complexing agents, such as sodium tetraborate ($Na_2B_4O_7.10H_2O$) or sodium *meta*-arsenite ($NaAsO_2$), can be used to effect separations that depend on the number and configuration of the hydroxyl groups present. Layers are prepared by incorporating 1 per cent (w/v) of the salt into the water used to prepare the slurry of adsorbent. In addition, ceramides having a *trans*-double bond in position 4 of the long-chain base are separable on silica gel impregnated with sodium borate (saturated compounds migrate ahead of unsaturated) although the reason for the effect is not understood.[516] The long-chain base and fatty acid constituents may each contain up to two double bonds, so TLC with silica gel impregnated with silver nitrate (5 per cent, w/w) can be used to good effect.

From these studies, Karlsson and Pascher[378] were able to recommend a programme for the analysis of natural mixtures by TLC although they recognise that, as the components of natural ceramides vary greatly in chain length, bands may tend to spread more than is the case with pure model compounds. Four groups of ceramide can be separated on the arsenite layers with chloroform–methanol (95:5, v/v) as the solvent system; dihydroxy base-normal acid, dihydroxy base-hydroxy acid and similar species containing trihydroxy bases, as illustrated in Fig. 8.9 (plate A). Derivatives of dihydroxy bases isolated in this manner can then be separated on borate-impregnated layers using the same solvent mixture for development, according to whether they contain long-chain bases with *trans*-double bonds in position 4 or not (Fig. 8.9, plate B). Finally all the components isolated thus far can be further

separated into groups, according to the combined numbers of *cis*-double bonds (*trans*-double bonds have comparatively little effect) in the component fatty acids and in the long-chain bases on silver nitrate-impregnated layers. Better results are obtained in this instance if the free hydroxyl groups are first acetylated with acetic anhydride and pyridine (see Chapter 4), when separations of the kind illustrated in Fig. 8.10 are obtained; chloroform–benzene–acetone (80:20:10, by vol.) is a suitable solvent system for the development.

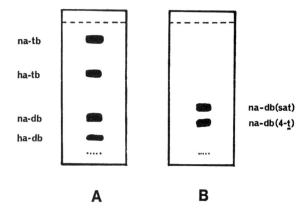

FIG. 8.9. Thin-layer chromatography of ceramides on layers of silica gel G, impregnated with 2 per cent sodium arsenite (plate A) and 2 per cent sodium borate (plate B), using a solvent system of chloroform–methanol (95:5, v/v).[378] Abbreviations: na, normal fatty acid; ha, hydroxy fatty acid; db, dihydroxy base; tb, trihydroxy base; db (sat), saturated dihydroxy base; db (4-t), $\triangle^4$-*trans*-unsaturated-dihydroxy base.

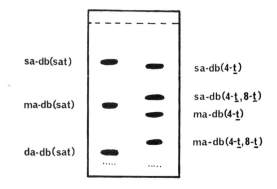

FIG. 8.10. TLC of ceramide acetates on layers of silica gel G impregnated with 5 per cent silver nitrate with a developing solvent of chloroform–benzene–acetone (80:20:10, by vol.).[788] Abbreviations: sa, saturated fatty acids; ma, monoenoic fatty acids; da, dienoic fatty acids; db (sat), saturated dihydroxy-bases; db (4-t), $\triangle^4$-*trans*-unsaturated dihydroxy-bases; db (4-t, 8-t) $\triangle^4$-*trans*, $\triangle^8$-*trans*-diunsaturated dihydroxy-bases.

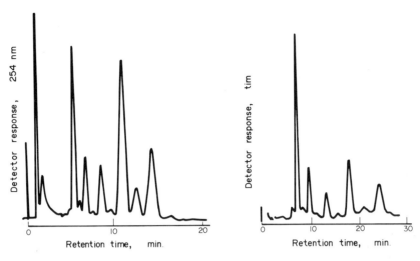

FIG. 8.11. A. HPLC of benzoylated glucosylceramide; μBondapack C18 column with methanol as eluting solvent (3 ml/min), detection at 254 nm. B. GLC of TMS derivatives of glucosylceramide; 1% OV-1 column (1.5 m long) at 310°. (Reproduced by kind permission of Professor T. Yamakawa and the *Journal of Biochemistry* (Tokyo)).[747]

Ceramides separated in this way can ultimately be resolved in the form of the acetate or trimethylsilyl ether derivatives according to the chain-lengths of the aliphatic components by high-temperature GLC, under conditions similar to those used in the separation of diacylglycerol acetates. As standards are not readily available, identification of peaks may present some difficulty unless a gas chromatograph-mass spectrometer combination can be used.[112, 271, 562, 654, 655] Ceramides that have not been simplified by TLC in the above manner are almost impossible to identify without such assistance.

Ceramide benzoates (prepared by reaction with benzoic anhydride) have been separated by HPLC with UV detection on columns of microparticulate silica gel into molecular species containing hydroxy-fatty acids, non-hydroxy-fatty acids and phyto-sphingosine.[343] Unmodified ceramides have been resolved by reverse phase HPLC (on "Nucleosil-5-C_{18}" columns), into species differing in the chain-length and degree of unsaturation of the alkyl moieties; the effect of a *cis*-double bond on the separation achieved differed somewhat in magnitude from that of a *trans*-double bond.[366] As UV absorption at 205 nm was the detection system used in this instance, the technique is probably not of value for quantitative purposes although it may be useful preparatively.

Intact monoglycosylceramides (cerebrosides)

have been subjected to high-temperature GLC after conversion of the free hydroxyl groups of the hexose and aliphatic residues to the trimethylsilyl ether derivatives. Compounds are separated according to the combined chain-lengths and number of hydroxyl groups in the long-chain base and fatty acid components, but the GLC recorder traces obtained are complex and components are not readily identified unless the column effluent is passed into a mass spectrometer.[273, 523, 562, 747] Column conditions similar to those used for triacylglycerol analysis are employed. Similarly, molecular species of intact ceramide aminoethylphosphonate (TMS derivative) have been resolved by high temperature GLC.[489]

Benzoylated cerebrosides have been successfully separated into molecular species by reverse-phase HPLC on C_{18}-alkylated stationary phases and with UV-detection.[747] In this instance, the separation is on the basis of the partition number of the alkyl moieties but, as cerebrosides tend to contain a relatively small proportion of unsaturated fatty acids, the separation is similar to that obtained by high-temperature GLC. Indeed, there was a remarkable similarity in the shapes of the recorder traces obtained by the two procedures with glucosylceramide as shown in Fig. 8.11.[747] The two procedures offer similar sensitivity and reproducibility. HPLC has been applied successfully to more complex glycosphingo-lipids.[748, 860]

CHAPTER 9

Structural Analysis of Lipids by means of Enzymatic Hydrolysis

A. INTRODUCTION

ESTER bonds to all three positions of the glycerol molecule are so similar in chemical reactivity that there is no simple chemical procedure for hydrolysing one or other of them selectively. Most living organisms, however, have developed lipolytic enzyme systems that are able to distinguish between bonds to the various positions of glycerol or between certain types of bonds in specific lipids and, in many cases, these enzymes can be isolated and used in simple incubations *in vitro* as an aid in structural analyses of lipids. A lipase (EC.3.1.1.3), prepared from pig pancreas extracts, for example, is used to hydrolyse quite specifically the fatty acids esterified to the primary positions of triacylglycerols (see Fig. 9.1) yielding 2-mono-acylglycerols, the fatty acid composition of which accurately reflects that of position 2 in the original triacylglycerols. While the same enzyme can be used in an analogous manner with more complex glycerolipids, it is usually simpler technically to hydrolyse glycerophosphatides with the enzyme, phospholipase A_2 (EC.3.1.1.4), available from snake venoms, which specifically hydrolyses the ester bond in position 2 of natural glycerophosphatides (Fig. 9.1). Such enzymes are useful not only for determining the manner in which fatty acids are distributed among the positions of intact lipids, but also for calculating the proportions of various positional isomers of molecular species of lipids, separated by the procedures described in Chapter 8.

Other enzymes exist which cleave the glycerol–phosphorus or phosphorus–nitrogenous base bonds in glycerophosphatides, i.e. phospholipase C (EC.3.1.4.3) and phospholipase D (EC.3.1.4.4)

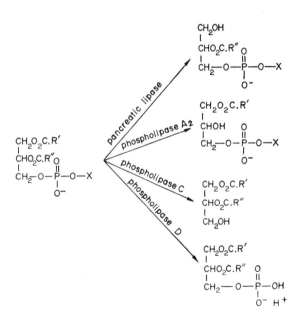

FIG. 9.1. Enzymatic hydrolysis of glycerolipids (X = choline, ethanolamine, etc.).

respectively (Fig. 9.1), and are useful in preparing derivatives of complex lipids, that are technically easier to separate into simpler molecular species. These enzymes are also active with sphingomyelin. Although enzymes have been isolated that hydrolyse the ceramide–carbohydrate bond in glycosphingolipids, they have not been widely used in lipid structural studies (see Chapter 8). The properties of phospholipases[166] in particular and lipolytic enzymes[92, 358] in general have been comprehensively reviewed.

In enzymatic structural studies of lipids, precautions must always be taken to minimise the effects of autoxidation (see Chapter 3).

B. POSITIONAL DISTRIBUTION OF FATTY ACIDS IN TRIACYLGLYCEROLS

1. Pancreatic lipase hydrolysis

A lipase (EC.3.1.1.3) that is completely specific for the primary ester bonds of glycerides is present in crude pig (and other animal) pancreas extracts. Although many other enzymes, including other lipolytic enzymes, and significant amounts of complex lipids are present, such preparations are available commercially at low cost and the impurities do not interfere with structural studies of triacylglycerols. The properties of the enzyme have been reviewed by Brockerhoff and Jensen[92] and by Litchfield[449].

All straight-chain saturated fatty acids of whatever chain-length are apparently hydrolysed from the primary positions at approximately the same rate, but the ester bonds of long-chain polyunsaturated fatty acids such as 22:6(n-3), trans-3-hexadecenoic acid and phytanic acid to glycerol are hydrolysed more slowly, possibly as a result of steric hindrance, caused by the proximity of functional groups to the ester bonds. Unfortunately, the enzyme hydrolyses triacylglycerol molecules that contain short-chain fatty acids more rapidly than molecules that contain only longer-chain fatty acids, e.g. with a triacylglycerol such as 1-butyro-2,3-dipalmitin, both the fatty acids on the primary positions are hydrolysed at the same rate, but faster than from a triacylglycerol such as tripalmitin[358]. However, when the enzyme is used on triacylglycerols that contain a more normal range of fatty acid components, little fatty acid specificity is evident.

Calcium ions are essential and bile salts are helpful for the reaction and it is necessary that the triacylglycerols be well dispersed, as they must be in a micellar form for hydrolysis to occur. For this reason, methyl oleate[54] or hexane[88] are sometimes added as carriers to solid highly saturated fats. In triacylglycerol structural studies, the concentrations of the various cations, bile salts and the enzyme, the pH of the reaction medium and the temperature are adjusted so that the optimum degree of hydrolysis is obtained (50–60 per cent) in the shortest possible time. Only the 2-monoacylglycerol formed need be isolated and transesterified, for analysis by gas chromatography (any 1-monoacylglycerols found are products of acyl migration from 2-monoacylglycerols). Although diacylglycerols produced as intermediates in the enzymatic hydrolysis may differ slightly from those expected of a random reaction, the fatty acid composition of the monoacylglycerols rarely appears to deviate far from the true result in practice.

A semimicro method developed by Luddy et al.[460] has been widely used, and was found to give the best results of a number of published procedures[27]. As the activities of enzyme preparations from different sources may vary, it is advisable to test the method with some natural triacylglycerols, e.g. lard or corn oil, as substrate to ensure that hydrolysis proceeds at a satisfactory rate.

"Tris(hydroxymethyl)methylamine (tris) buffer (1 M; pH 8.0; 1 ml), calcium chloride solution (2.2 per cent, 0.1 ml) and a solution of bile salts (0.05 per cent; 0.25 ml) are added to the triacylglycerols (up to 5 mg) in a stoppered test-tube, and the whole is allowed to equilibrate at 40°C in a water-bath for 1 min before the pancreatic lipase preparation (1 mg) is added. The mixture is shaken vigorously at this temperature by means of a mechanical shaker for 2–4 min until the desired degree of hydrolysis is attained, when the reaction is stopped by the addition of ethanol (1 ml) followed by 6M hydrochloric acid (1 ml). The solution is extracted three times with diethyl ether (10 ml portions), with centrifugation if necessary to break any emulsions, the solvent layer is washed twice with distilled water (5 ml portions) and dried over anhydrous sodium sulphate. On removal of the solvent, the products are separated by TLC on silica gel G layers (0.5 mm thick on 20 × 10 cm glass plates) with hexane–diethyl ether–formic acid (80:20:2 by vol.) as solvent system. After spraying with 2′,7′-dichlorofluorescein solution, the mono-acylglycerol band is identified (it is normally the bottom band just above the origin) and recovered from the adsorbent by elution with chloroform–methanol (95:5, v/v) for transesterification. No more than the minimum delay between stages is advisable."

The fatty acid composition of the monoacylglycerol fraction, which represents that of position 2 of the original triacylglycerols, is ultimately determined by gas chromatography, and it is customary to present results, as in most lipid structural studies, in mol per cent. On the other hand, the free fatty acids released may not be similar to the mean composition of fatty acids originally present in positions 1 and 3 of the triacylglycerols, as hydrolysis may not be completely random and there may be contamination by fatty acids liberated from lipids present in the enzyme preparation, or by fatty acids released from position 2 after they have migrated to the primary positions. The mean composition of each fatty acid in positions 1 and 3 can, however, be calculated from its concentrations in the intact triacylglycerol and in position 2 by means of the relationship

$$[\text{positions 1 and 3}] = \frac{3 \times [\text{triacylglycerol}] - [\text{position 2}]}{2}$$

although it is, of course, subject to the cumulative errors of the analyses.

Pancreatic lipase hydrolysis can also be used to determine the fatty acid composition of each position in diacylglycerols and diacylglycerol acetates, but it is necessary to saturate the incubation medium with diethyl ether for accurate and reproducible results[420]. 1-Alkyl-2,3-diacylglycerols can be subjected to hydrolysis with the enzyme, the product, a 1-alkyl-2-acyl-glycerol, isolated and the fatty acid composition of position 2 determined.[724] With both the above, as with triacylglycerols, the free fatty acids released do not contain only components from the primary positions, the compositions of which must again be calculated from those of the intact lipid and of the 2-acyl hydrolysis products. Certain other lipases related to pancreatic lipase (see section B2) may be more suitable for these compounds.

Pancreatic lipase hydrolysis can be used to determine the fatty acid composition of position 2 of most triacylglycerols containing a normal range of fatty acids or fatty acids with polar or non-polar substituents remote from the ester linkage. Estolides can also be hydrolysed by the enzyme; only the glycerol-fatty acid bond is broken while estolide linkages remain intact[126, 509]. The method is not suitable for triacylglycerols containing long-chain polyunsaturated fatty acids in the primary positions, such as fish oils[79], and is of limited value for triacylglycerols containing significant quantities of short-chain fatty acids, such as ruminant milk fats[359]. Fortunately, a chemical method has been developed that may be appropriate to these circumstances.[866] In this, triacylglycerols are reacted with a Grignard reagent that liberates fatty acids (converting them to tertiary alcohols) with formation of partial glycerides, provided that the reaction is not allowed to go to completion. The various products are isolated by TLC on layers of silica gel G impregnated with boric acid, and although some acyl migration inevitably occurs, the amount of each fatty acid in position 2 can be calculated with acceptable accuracy from the concentration of each component in the original triacylglycerols and in either the α,β-diacylglycerols[866] or the α,α^1-diacylglycerols (Author, unpublished), formed in the reaction, by means of the relationships

$$[\text{position 2}] = 4 \times [\alpha,\beta\text{-diacylglycerols}] - 3 \times [\text{triacylglycerols}]$$
$$\text{or,} \quad = 3 \times [\text{triacylglycerols}] - 2 \times [\alpha,\alpha^1\text{-diacylglycerols}]$$

The result is of course subject to the cumulative error of the analyses. The Grignard reaction is discussed in greater detail in the next section.

2. Lipase of *Rhizopus arrhizus*

The mould *Rhizopus arrhizus* secretes an extracellular lipase which has a near absolute specificity for the primary bonds of glycerolipids (reviewed by Benzonana[61]). It, therefore, resembles pancreatic lipase in many respects but is able to hydrolyse micelles as well as droplets of triacylglycerols containing short-chain fatty acids (the latter at a slower rate than pancreatic lipase); it is not activated by bile salts and does not have an absolute requirement for Ca^{2+}[683]. In general, it appears to offer no advantage over pancreatic lipase for the analysis of triacylglycerols, but as it is readily isolated in a relatively pure state it is of value for the analysis of phospholipids and glycosyldiacylglycerols (see below), especially as the unesterified fatty acids released are much less liable to contamination from fatty acids in the lipase preparation itself.

3. Stereospecific analysis of triacylglycerols

No lipolytic enzyme has yet been discovered that distinguishes between positions 1 and 3 of a triacyl-*sn*-glycerol. Nonetheless, a number of ingenious stereo specific analysis procedures for determining the composition of fatty acids esterified to positions 1, 2 and 3 of L-glycerol in triacylglycerols have been devised, principally by Brockerhoff, who has reviewed the available methods[91] (also reviewed by Breckenridge[83]).

Most methods require the preparation of partial glycerides that can be chemically phosphorylated and reacted with a stereospecific lipase or are phosphorylated by stereospecific enzymes. In the most widely used approach (Fig. 9.2), α,β-diacylglycerols (an equimolar mixture of the 1,2- and 2,3- *sn*-isomers) are prepared by hydrolysing the triacylglycerols partially with a Grignard reagent, and are converted synthetically to phospholipids which are in turn hydrolysed by the phospholipase A of snake venom; this enzyme reacts only with the "natural" 1,2-diacyl-*sn*-glycerophosphatide (see below)[87]. The products are a lysophosphatide that contains the fatty acids originally present in position

sn-1, free fatty acids released from position *sn*-2 and the unchanged "unnatural" 2,3-diacyl-*sn*-phosphatide. After transesterification, the fatty acid composition (in mol per cent) of each product is determined by gas chromatography. In addition, the composition of position *sn*-2 can be determined independently by means of pancreatic lipase hydrolysis, or from the 1,3-diacylglycerols generated by Grignard hydrolysis (see section B1).

Only the fatty acid composition of position *sn*-3 is not determined directly, but the amount of each fatty acid in this position can be calculated from the analysis of the original triacylglycerol and those of positions *sn*-1 and *sn*-2, or from the analysis of the unchanged 2,3-diacyl-*sn*-phosphatide and that of position *sn*-2, i.e. for each component

$$[\text{position } sn\text{-}3] = 3 \times [\text{triacylglycerol}] - [\text{position } sn\text{-}1] - [\text{position } sn\text{-}2]$$
$$[\text{position } sn\text{-}3] = 2 \times [2,3\text{-diacyl-}sn\text{-phosphatide}] - [\text{position } sn\text{-}2]$$

The key to success with the procedure lies in the preparation of the intermediate diacylglycerols, which must be generated in a random manner so that the fatty acid compositions of the various positions are the same as in the original triacylglycerols, i.e. there must be no selectivity for specific fatty acids or fatty acid combinations, and a minimum of acyl migration during their formation. Hydrolysis with pancreatic lipase was originally used for the purpose, but is not always suitable as it exhibits specificity for

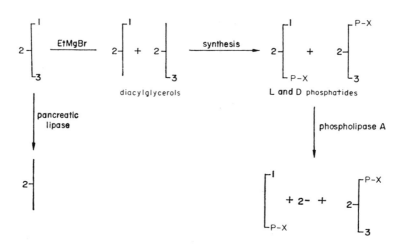

FIG. 9.2. Stereospecific analysis of triacylglycerols. (-P-X = phosphorylphenol or phosphorylcholine).

molecules containing certain fatty acid combinations. Therefore, chemical hydrolysis is now preferred and Grignard reagents are used for the purpose, as they have no fatty acid specificities and cause less acyl migration than other chemical methods[866]. Methyl magnesium bromide was originally used, but has largely been replaced by ethyl magnesium bromide, which reacts rapidly and randomly with the ester bonds of triacylglycerols converting the fatty acids to tertiary alcohols and, if the reaction is stopped when it is between 30 and 70 per cent complete, a complex mixture of partial glycerides including α,β- and α,α^1- (or 1,3-sn)-diacylglycerols. Unfortunately, some acyl migration always occurs during the reaction, and the 1,3-diacylglycerols contain 6 to 10 per cent of fatty acids that have migrated from position 2, but although position 2 of the α,β-diacylglycerols is contaminated by up to 4 per cent of fatty acids that have migrated from positions 1 and 3, the fatty acid compositions of the primary positions appear to be identical to those of the original triacylglycerols[139]. The α,β-diacylglycerols required for the procedure must not come in contact with polar solvents or be heated, and are isolated *immediately* by TLC on layers of silica gel G impregnated with boric acid for conversion without delay to phosphatides, which can also be purified by TLC or other means.

In the original Brockerhoff procedure, phosphatidylphenols[87, 88] were prepared from the diacylglycerols but it has more recently been shown that it is relatively easy to prepare phosphatidylcholines[138, 531]. The chromatographic properties of the latter are better understood, and they can be hydrolysed with some stereoselectivity with phospholipase C, offering additional analytical possibilities.

The following modification of Brockerhoff's original procedure for determining the composition of each of positions 1, 2 and 3 of a triacyl-sn-glycerol incorporates features developed in other laboratories[138, 139, 531].

"Triacylglycerols (40 mg) are dissolved in dry diethyl ether (2 ml), and freshly prepared ethyl magnesium bromide in dry diethyl ether (1 ml, 0.5 M) is added. The mixture is shaken for 1 min and then glacial acetic acid (0.05 ml) followed by water (2 ml) are added to stop the reaction, and the products are extracted with diethyl ether (three

10 ml portions). The ether extract is washed first with aqueous potassium bicarbonate (2 per cent, 5 ml), then with water (5 ml) and is dried over anhydrous sodium sulphate. After removal of the solvent at ambient temperature, the required α,β-diacylglycerols are obtained by preparative TLC on silica gel G layers containing 5 per cent (w/w) boric acid (0.5 mm thick on a 20 × 20 cm plate), developed in hexane–diethyl ether (1:1, v/v). Bands are visualised under UV light after spraying with aqueous Rhodamine 6G (0.01 per cent, w/v) (Fig. 9.3) and that containing the α,β-diacylglycerols is scraped into a small chromatographic column and eluted with diethyl ether (100 ml). Small amounts of boric acid are also eluted but do not interfere with subsequent steps. The 1,3-diacylglycerols can be recovered for analysis if required.

A mixture of chloroform–pyridine–phosphorus oxychloride (47.5:47.5:5 by vol.; 0.65 ml) at 4°C is added to the diacylglycerols (yield 6–7 mg) and the whole is allowed to stand for 1 hr at 4°C, then for 1 hr at room temperature. At the end of this time, dry powdered choline chloride (200 mg) is added and the mixture stirred overnight at 30°C, then for a further 30 min following addition of water (20 μl). Much of the solvent is removed by evaporation under nitrogen, the products are extracted with chloroform–methanol–water–acetic acid (50:39:10:1 by vol.; 12 ml) and partitioned with 4M ammonia (4 ml). The aqueous phase is re-

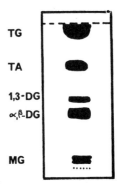

FIG. 9.3. Stereospecific analysis of triacylglycerols. TLC separation of Grignard deacylation products on layers of silica gel G impregnated with 5% (w/w) boric acid; solvent system, hexane–diethyl ether (1:1, v/v). Abbreviations: TG, triacylglycerols; TA, tertiary alcohols; DG, diacylglycerols; MG, monoacylglycerols.

extracted with fresh lower phase (2 ml). The organic layer is dried and evaporated, and the required phosphatidylcholine is obtained by chromatography on CM-cellulose (Whatman CM-52; Na$^+$ form). The adsorbent is packed into a Pasteur pipette in chloroform and the lipids are added to the column in this solvent; chloroform (10 ml) elutes any unchanged diacylglycerols, and chloroform–methanol (9.25:0.75, v/v; 10 ml) elutes pure phosphatidylcholines.

The phosphatidylcholines are hydrolysed by means of the phospholipase A$_2$ of snake venom, and the products are isolated by preparative TLC for analysis. A suitable procedure is described below (Section C1)."

Although the procedure has two synthetic steps, two enzymatic hydrolyses (including a separate pancreatic lipase hydrolysis), two thin-layer and one CM-cellulose separations, followed by five trans-esterifications and GLC analyses, consistent and reliable results can be obtained with practice.

Analysis of the lysophosphatide in the final stage of the procedure gives trustworthy results for the fatty acid composition of position sn-1, but that of position sn-2 can be determined with greater accuracy by means of an independent hydrolysis with pancreatic lipase (with the exception of fish oils or milk fats), rather than by analysis of the free fatty acids released by phospholipase A. The composition of position sn-2 can also be determined from an analysis of the 1,3-diacylglycerols, produced as by-products of the Grignard hydrolysis step, as a check (section B1). The result for position sn-3 can be calculated by the two procedures described above (the first method is generally favoured), and is subject to the cumulative experimental error of the other analyses, but is probably accurate to 2–3 per cent (absolute) for major components. Analyses should not be accepted unless the results for major components in positions sn-2 and sn-3, determined by both methods, agree within 4 per cent (absolute).

Ethyl magnesium bromide cannot be used to generate diacylglycerols when the triacylglycerols contain fatty acids with polar functional groups, such as estolide bonds, that may also react with the reagent.[126] Pancreatic lipase must then be used for the purpose and the errors potentially associated with this method must be accepted. Similarly, when

the triacylglycerols contain isotopically-labelled lipids, some modification of the method may be necessary.[131, 136]

The main drawback of this approach to stereospecific analysis of triacylglycerols is that the fatty acid composition of position sn-3 is not determined directly, so that errors in the results for minor fatty acid components in this position may be considerable. Indeed, small negative values are sometimes obtained and small positive values may arise in calculations for a component that is not present in the position at all. This could no doubt be overcome by reacting the unchanged 2,3-diacyl-phosphatide with the lipase of *Rhizopus arrhizus* to release the fatty acids from position sn-3. Brockerhoff devised an alternative approach (Fig. 9.4, scheme A)[90] in the hope of resolving the difficulty. 1,3-Diacylglycerols, prepared by reaction with the Grignard reagent, are isolated in this procedure, and are phosphorylated and reacted with phospholipase A as before. The fatty acids in position sn-1 are released, leaving a lysophosphatide that contains the fatty acids originally present in position sn-3. Unfortunately, 1,3-diacylglycerols are contaminated by significant amounts of isomerised α,β-diacylglycerols, as discussed above, so that positions sn-1 and sn-3 may contain 6–10 per cent of fatty acids originally present in position sn-2. As a result, the method is much less precise than the first one described.

Another important analytical method has been described by Lands and his colleagues (Fig. 9.4, scheme B)[432]. α,β-Diacylglycerols are again prepared (pancreatic lipase hydrolysis was used for the purpose in the original procedure, but reaction with ethyl magnesium bromide would probably now be considered more suitable), and are phosphorylated enzymatically by a diacylglycerol kinase from *E. coli*, that is completely stereospecific for 1,2-diacyl-sn-glycerols. The fatty acid composition of the phosphatidic acid formed is determined and that of position sn-2 is obtained separately by pancreatic lipase hydrolysis so that the results for the concentrations of each acid in positions sn-1 and sn-3 can be calculated, i.e.

[position sn-1] = 2 × [phosphatidic acid] − [position sn-2]
[position sn-3] = 3 × [triacylglycerol] − 2 × [phosphatidic acid]

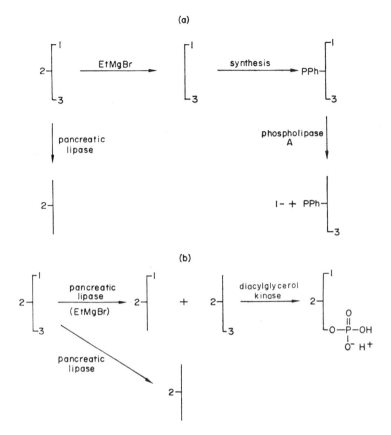

FIG. 9.4. Stereospecific analysis of triacylglycerols. Scheme A, Brockerhoff's second procedure[90] (PPh = phosphorylphenol). Scheme B, Land's procedure[432].

The procedure has been little used, probably because the relevant enzyme is not available commercially, but merits greater notice as the 1,2-diacyl-*sn*-phosphatidic acid formed is structurally identical to an important intermediate in triacylglycerol biosynthesis. 1,2-diacyl-*sn*-glycerols of importance to biochemical studies can also be generated by a controlled reaction of phospholipase C with racemic phosphatidylcholines, prepared as described above, and this can serve as the basis of an alternative procedure for stereospecific analysis of triacylglycerols[531].

It should be recognised that an analysis of a natural triacylglycerol that indicates that positions *sn*-1 and *sn*-3 have identical fatty acid compositions does not necessarily imply that there is an absence of specific stereoisomers; simpler molecular species must be subjected to stereo-specific analysis before this conclusion can be justified unequivocally.

C. ENZYMATIC HYDROLYSIS OF COMPLEX LIPIDS

1. Phospholipase A_2

Phospholipase A_2 (EC.3.1.1.4) hydrolyses the ester bond in position 2 of glycerophosphatides (see Fig. 9.1), releasing the fatty acids in this position. The products, free fatty acids and a lysophosphatide containing the fatty acids of position 1, are isolated for analysis so that the distribution of fatty acids in both positions of the glycerol moiety is determined. Although the enzyme is found in many animal tissues, the most important sources for use in structural studies are snake venoms. That of *Crotalus adamanteus* or *C. atrox* has been used most often, but is now being supplanted by that of *Ophiophagus hannah*[554]. The venom of the honey bee,

Apis mellifera, is also being used increasingly for specific purposes. The molecular properties and specificities of these enzymes have been reviewed.[92] The enzymes from snake venoms are completely stereospecific for L-glycerophosphatides and will react with most natural compounds of this type, including phosphatidylcholine, phosphatidylethanolamine, phosphatidylinositol, phosphatidylglycerol, phosphatidylserine, phosphatidic acid and cardiolipin,[790] but not with polyphosphoinositides. They also react with the phosphatidylphenols which can be intermediates in stereospecific analysis of triacylglycerols, and with synthetic phospholipids containing the phosphorus group in position 2, in which case the fatty acids in position 1 of L-glycerol are released. Although the phospholipase A in some venoms exhibits specificity for certain fatty acid combinations, this need not be troublesome in structural studies provided that the hydrolysis is taken to completion. Fortunately, the enzyme in *O. hannah* does not exhibit such specificities and is especially suited to structural studies[554]. Calcium ions are essential for the reaction, which is also stimulated by diethyl ether.

It is necessary to modify the reaction conditions according to the nature of the phospholipid to be hydrolysed; phosphatidylcholines have been most often studied in this manner and the following procedure, essentially that of Robertson and Lands[630], is widely used for the purpose.

"*Ophiophagus hannah* snake venom (2.5 mg) is dissolved in tris buffer solution (0.5 ml; pH 7.5) containing calcium chloride (4 mM). Phosphatidylcholine (5 to 50 mg) is dissolved in diethyl ether (2 ml), 100 μl of the snake venom solution is added and the whole shaken vigorously for 1 hr. The mixture is washed into a conical flask with methanol (10 ml) followed by chloroform (20 ml) and the solution dried over anhydrous sodium sulphate. After filtering and removing the solvent *in vacuo*, the products are applied to a silica gel G TLC plate (20 × 20 cm; 0.5 mm thick layer), which is developed first in hexane–diethyl ether–formic acid (60:40:2 by vol.). The top half of the plate is sprayed with 2′,7′-dichlorofluorescein solution, and the free fatty acid band scraped off and esterified. When the plate is dry, it is re-developed in chloroform–methanol–acetic acid–

water (25:15:4:2 by vol.), and the lysophosphatidylcholine (Rf approx. 0.2) is detected and transesterified for GLC analysis."

The accuracy of the procedure should be checked by summing the results for the concentrations (mol per cent) of each fatty acid in positions 1 and 2, dividing by two, and comparing this quantity with the analysis for each component in the original phosphatidylcholine; the two results should agree within 1 per cent (absolute) for each component.

Poor hydrolysis is sometimes found, when the procedure is used with other glycerophosphatides, particularly with phosphatidylethanolamine, but the use of borate buffers appears to alleviate the problem[559, 794]. The author has found the following procedure satisfactory for phosphatidylethanolamine and phosphatidylinositol[794].

"Phosphatidylethanolamine (8 mg) is dissolved in diethyl ether (2 ml), and 0.1 M borate buffer (0.25 ml; pH 7), 0.01 M calcium chloride (0.125 ml), water (0.125 ml) and *O. hannah* snake venom (1 mg) are added. The mixture is shaken thoroughly for 1 hr then worked up as above."

It may be necessary to prolong the reaction with more acidic lipids and to vary the polarity of the second solvent system in the TLC separation according to the polarity of the lipid hydrolysed. With phosphatidylserine, it is necessary to render the buffer solution alkaline with 0.1 M potassium hydroxide solution to enable the reaction to go to completion.[857] The accuracy of the procedure should always be checked in the manner described above for phosphatidylcholine.

Phospholipase A$_2$ from *Crotalus* species hydrolyses diacyl phosphatides much more rapidly than the analogous ether compounds[812], and has been used to obtain fractions enriched in the latter (see Chapter 7). When ether-containing phospholipids are present in a sample, it is therefore necessary to prolong the hydrolysis time to ensure complete reaction if the fatty acid composition of position 2 of the mixture is required. The enzyme in *Naja naja* venom hydrolyses alkyl phosphatides at the same rate as the diacyl compounds but twice as rapidly as it hydrolyses alkenyl phosphatides[812]. Bee venom is apparently able to hydrolyse alkenyl and acyl forms of phosphatidylcholine and phosphatidylethanolamine with comparable rapidity.[858] Although the phospholipase A of snake venom does not hydrolyse the secondary ester bond in di- and triphospho-inositides, the related enzyme in bee venom does and has been used in structural studies[769].

The existence of a phospholipase A₁ that hydrolyses the primary ester bond of glycerophosphatides has been demonstrated in rat liver and pancreas, but it has yet to be isolated in a sufficiently pure state for use in structural studies.

2. Pancreatic lipase and related enzymes

(I) GLYCEROPHOSPHATIDES

Pancreatic lipase will hydrolyse the ester bond to position 1 of glycerophosphatides (Fig. 9.1), leaving a lysophosphatide containing the fatty acids of position 2. The following method is generally satisfactory.[709, 710]

"The phospholipids (10 mg), bile salts (3 mg) and bovine serum albumin (4.5 mg) are dispersed by ultrasonification in 0.1 M borate buffer (pH 8.0; 1 ml) made 5×10^{-3} M with calcium chloride, and then pancreatic lipase (0.5 mg) is added. The mixture is shaken vigorously for 0.5 to 6 hr (determined by experiment) until hydrolysis is complete. The mixture is extracted three times with chloroform–methanol (7:3, v/v; 4 ml portions), and the lysophosphatide is obtained for analysis, after removing the solvent, by TLC on silica gel G layers as described above for phospholipase A hydrolyses."

Unfortunately, unless highly purified enzyme preparations are used (and they are not available commercially), the free fatty acids released will be contaminated by fatty acids endogenous to the enzyme preparation and possibly by some that have migrated from position 2 to position 1 before being hydrolysed. Only the result for position 2 is then meaningful, but that for position 1 can of course be obtained by difference as the composition of the intact lipid is known. It is possible to circumvent the problem by preparing purified pancreatic lipase, normally a tedious task, although a simplified method has recently been described[505].

The lipase of similar specificity to pancreatic lipase from *Rhizopus arrhizus* may also be used for the hydrolysis of phospholipids.[709, 710] It is utilised *in vitro* in a similar manner to the above except that it has a pH optimum of 6.5.

Both enzymes have been used in the preparation of concentrates of alkyl- and alkenyl-phosphatides (see Chapter 7). They have not been much used in structural studies of lipids to date, as they only offer significant advantages over phospholipase A₂ in a few circumstances. On the other hand, they may be useful with semi-synthetic non-polar phosphatides, such as dimethylphosphatidic acid, which are sometimes prepared from the more common phospholipids for separation into simpler molecular species (see Chapter 8), or for the analysis of diacylglycerol acetates prepared from phospholipids for similar reasons (section B1).

(II) GLYCOSYLDIACYLGLYCEROLS

Pancreatic lipase hydrolyses the ester bond to position 1 of mono- and digalactosyldiacylglycerols, with production of free fatty acids and glycosyl-monoacylglycerols containing the fatty acids of position 2. Both products can be isolated by TLC and esterified for GLC analysis. For the reasons given above, only the fatty acid composition of position 2 is in general determined directly, while that of position 1 is calculated by difference from the known composition of the intact lipid. The following procedure is based on that of Noda and Fujiwara[551].

"Pancreatic lipase (10 mg), 45 per cent calcium chloride solution (0.03 ml), 1 per cent sodium deoxycholate solution (0.01 ml) and the glycosyl-diacylglycerol (15 mg) are dispersed by vigorous shaking in 0.2 M tris buffer (pH 7.6; 1 ml). The mixture is incubated at 40°C for 1 hr with constant shaking, before it is acidified to pH 3 with 1 M sulphuric acid. The products are extracted with chloroform–methanol, as in the phospholipase A procedure above, and are separated by TLC on silica gel G layers in a similar two-step system. Free acids are obtained in the top half of the plate with hexane–diethyl ether–formic acid (60:40:2, by vol.), then the monoacyl products are separated by means of a further development with a solvent system of chloroform–methanol–acetone (80:16:4, by vol.; monogalactosylmonoacylglycerols) or chloroform–methanol–acetic acid–water (80:40:2:3, by vol.; digalactosylmonoacylglycerols)."

Native mono- and digalactosyldiacylglycerols have been subjected to hydrolysis in this manner in a number of laboratories,[49, 551, 649] but Arunga and Morrison[43] obtained better results when the carbohydrate moieties were acetylated prior to the reaction.

Other workers have preferred to use the related enzyme from *Rhizopus arrhizus*, which has proved suitable for the hydrolysis of mono-, di- and trigalactosyldiacylglycerols, sulphoquinovosyldi-acylglycerol and glucosyldiacylglycerols.[212, 650, 652] In this instance, both products of hydrolysis are sufficiently pure for analysis. The following method is suitable.[212]

"The glycosyldiacylglycerol (3–6 μmol) is dissolved in chloroform–methanol (2:1, v/v), Triton X-100^R (2–4 mg) in the same solvent is added and the mixture taken to dryness. 0.04 M Tris/HCl buffer (1.0 ml; pH 7.2) is added, and the mixture sonicated until a homogenous dispersion is obtained, when the lipase preparation (1–3 mg) is added. The mixture is incubated with vigorous shaking for 30 min at room temperature, then the reaction is stopped by adjusting to pH 4 with acetic acid. The various products are isolated as described above."

3. Phospholipase C

A variety of enzymes capable of releasing 1,2-diacylglycerols from phosphoglycerides or of ceramide from sphingomyelin, and termed 'phospholipase C (EC.3.1.4.3), have been isolated from various microorganisms, but especially from *Clostridium welchii* and *Bacillus cereus*. Enzymes from both sources are commercially available, and highly active preparations of that from *B. cereus* can be obtained by ammonium sulphate precipitation from the supernatant fluid used as a growth medium for the microorganisms[565]; if the enzyme is to be used in structural studies, little further purification is necessary. The properties and substrate specificities of corresponding enzyme preparations from different sources can vary greatly (reviewed by Brockerhoff and Jensen[92]) and in structural studies, it is necessary to choose the enzyme from the most appropriate source for a particular lipid class. The

enzymes do not possess an absolute specificity for a phosphate bond in position 3 of L-glycerol, and that of *B. cereus* will react with synthetic phospholipids with the phosphate bonds in positions *sn*-1, *sn*-2 and *sn*-3[173]; that of *C. welchii* hydrolyses D-phosphatidyl-choline but much more slowly than the L-isomer[531]. Although there is some evidence that molecular species containing shorter-chain fatty acids are hydrolysed more rapidly than those with longer-chain components, the effect need not be troublesome as the reaction can usually be taken to completion with care.

Phospholipase C preparations from *C. welchii* are used most often for the preparation of 1,2-diacyl-*sn*-glycerols from phosphatidylcholine, or of ceramides from sphingomyelin. It can also be used to prepare monoacylglycerols from lysophosphatidylcholine[226], and ceramides from ceramide aminoethylphos-phonate[489]. It is used as follows.[617]

"The phospholipase C of *C. welchii* (1 mg) in 0.5 M tris buffer (pH 7.5; 2 ml), 2×10^{-3}M in calcium chloride, is added to phosphatidylcholine (10 mg) in diethyl ether (2 ml). The mixture is shaken at room temperature for 3 hr when it is extracted (3 times) with diethyl ether. The ether layer is dried over anhydrous sodium sulphate before the solvent is removed at ambient temperature. Pure, 1,2-diacyl-*sn*-glycerols are obtained by preparative TLC on layers of silica gel G impregnated with boric acid (10 per cent, w/w), with hexane-diethyl ether (50:50, v/v) as solvent system. The appropriate band is located with the 2',7'-dichlorofluorescein spray, and is eluted from the adsorbent with diethyl ether (boric acid, which is also eluted, does not interfere with subsequent stages)."

No unnecessary delays are allowable at any stage and the compounds should not be heated or permitted to come in contact with polar solvents, otherwise acyl migration may occur and, if molecular species of the diacylglycerols themselves are required, they should be fractionated immediately. Alternatively, the diacylglycerols may be acetylated at once with acetic anhydride and pyridine (see Chapter 4 for practical details), as the diacylglycerol acetates can be stored indefinitely in an inert atmosphere at low temperatures without coming to harm (although they should be purified by preparative TLC before being analysed further). A small portion of the diacylglycerols or diacylglycerol acetates should be transesterified in order that its fatty acid composition may be compared with that of the original lipid to ensure that random hydrolysis of molecular species has occurred.

Ceramides, prepared by this method, are comparatively stable but should be purified on thin layers of silica gel G with chloroform–methanol (90:10, v/v) as solvent system.

Although *C. welchii* preparations can be used to hydrolyse phosphatidylethanolamine, provided that some lysophosphatidylcholine[763] or sphingomyelin[617] are present to activate the enzyme, better results are obtained with the phospholipase C of *B. cereus*, which can also be used to prepare diacylglycerols from phosphatidylserine, phosphatidylinositol and cardiolipin. The author has found the following method[841, 850] satisfactory for the purpose.

"Phosphatidylethanolamine (5 mg) and sphingomyelin (3 mg) are mixed with 0.2 M phosphate buffer (pH 7.0; 0.5 ml) containing 0.001 M 2-mercaptoethanol and 0.0004 M zinc chloride, and phospholipase C of *B. cereus* (1 mg) in the above buffer (0.5 ml) is added. The mixture is shaken vigorously for 2 hr at 37°C, when the diacylglycerols produced are extracted and purified as in the procedure immediately above."

The phospholipase C of *Clostridium perfringens* is specific for the diacyl forms of phospholipids, yet that of *B. cereus* hydrolyses the 1-0-alkyl forms of phospholipids three times as quickly as the diacyl or alkenyl forms[812]. The latter enzyme will only hydrolyse sphingomyelin under exceptional circumstances[563]. Phospholipase C preparations that are specific for phosphatidylinositol have been obtained from *Staphylococcus aureus*[456] and *B. thuringiensis*[759]. A related enzyme, found in brain, has been used to prepare diacylglycerols from mono-, di- and triphosphoinositides[320, 769]. While hydrolysis does not go to completion, representative diacylglycerols appear to be obtained, although some acyl migration occurs with formation of 1,3-diacylglycerols, probably because of the low pH optimum of the enzyme. Phosphatidic acid has been dephosphorylated by a phosphatase found in liver mitochondria and microsomes[304], but the preparation of the enzyme is fraught with difficulty. The acidic phosphatase (EC.3.1.3.2) from wheat germ will perform the same feat,[71] and as it is readily available commercially, it will probably be favoured in future.

4. Phospholipase D

Phosphatidic acid is formed from phosphatidylcholine, phosphatidylethanolamine, phosphatidylserine and phosphatidylglycerol by reaction with phospholipase D (EC.3.1.4.4) (see Fig. 9.1), but phosphatidylinositol is attacked very slowly and cardiolipin not at all. The corresponding bond in sphingomyelin is also hydrolysed. Although the enzyme from vegetable sources is available commercially, highly active preparations can be made with very little effort simply by homogenising cabbage (or brussel sprouts) with an equal volume of water, filtering and centrifuging at 13,000 g for 30 min; the supernatant fluid is used directly[854]. Hydrolyses with the enzyme are carried out in the following manner.[854]

"0.1 M sodium acetate (1.25 ml; pH 5.6), 1 M calcium chloride (0.25 ml) and fresh enzyme solution (2 ml) are added to the glycerophosphatide (10 mg) in diethyl ether (1 ml). The mixture is shaken vigorously for 3 hr, then brought to pH 2.5 by addition of glacial acetic acid, before it is extracted three times with chloroform–methanol (2:1, v/v; 4 ml portions). The lower layer, containing the phosphatidic acid, is taken to dryness and can be purified if necessary by TLC on layers of silica gel G with chloroform–methanol-water (25:10:1, by vol.) as developing solvent (the product is found just below the solvent front)."

Alcohols must be excluded from the reaction medium as the enzyme also catalyses esterification of the phosphate group. Phosphatidic acid formed in this manner is converted to the dimethyl derivative, for separation into simpler molecular species (see Chapters 7 and 8), by reaction with diazomethane in diethyl ether[854].

"Crude phosphatidic acid in diethyl ether solution (1 ml) is allowed to react with excess diazomethane in diethyl ether for 16–18 hr at room temperature. The solvent is removed in a stream of nitrogen, and the dimethylphosphatidate purified by preparative TLC on layers of silica gel G, with carbon tetrachloride–chloroform–glacial acetic acid–ethanol (120:80:1:6, by vol.) as developing solvent. The required product has an R_f value of approximately 0.4 to 0.5."

Phospholipase D has also been used to obtain concentrates of alkyl, acyl- and alkenyl, acyl-phospholipids as it reacts much less rapidly with them than with the diacyl forms of a given lipid (see Chapter 7). For example, phospholipase D from carrot and cabbage hydrolyses the 1-0-alkyl and 1-0-alkenyl forms of phosphatidylcholine at one third and one-tenth the rate respectively of the diacyl components[812].

D. THE USE OF ENZYMATIC HYDROLYSIS IN THE DETERMINATION OF MOLECULAR SPECIES OF LIPIDS

None of the methods described in Chapter 8 can be used to separate molecular species of glycerolipids differing in the positional distribution of specific fatty acids; for example, 1-palmityl-2-oleyl- and 1-oleyl-2-palmityl-phosphatidylcholine or their derivatives cannot be resolved physically by any method yet developed. If the molecular species containing the mixture of these components can be isolated by chromatographic techniques, however, the proportions of each can be calculated from the results of phospholipase A (phospholipid) or pancreatic lipase (diacylglycerol acetate) hydrolysis, simply by obtaining the ratio of palmitic acid in each position. Similar considerations apply to the analyses of most other lipids.

One major difficulty arises with molecular species of triacylglycerols that contain three different types of fatty acid[89] (the so-called "three acid problem"), such as saturated, monoenoic and dienoic acids for example. If these are denoted by S, M and D respectively, six different stereoisomers exist — SMD, SDM, MSD, MDS, DSM and DMS, but stereospecific analysis can only be used to ascertain the combined amounts of the pairs SMD and SDM, MSD and MDS, and DSM and DMS, and not the proportions of each isomer in the pair. The problem can be resolved when the stereospecific analysis procedure of Lands and Slakey[433, 705] is used, if the 1,2-diacylphosphatides formed in the final stage of the procedure (see section B2 above) are separated by silver nitrate TLC and hydrolysed with phospholipase A. Hammond, indeed, has shown theoretically[274] how all the species present in a complex natural triacylglycerol mixture, such as coconut oil, can be calculated by applying various stereospecific analysis procedures to the fractions obtained by combinations of silver nitrate TLC and high-temperature GLC. The technical difficulties are enormous, however.

CHAPTER 10

The Analysis and Radioassay of Isotopically-labelled Lipids

A. INTRODUCTION

THE SENSITIVITY of methods for detecting radio-tracers is much greater than that of most other chemical and physical procedures, and the use of lipids labelled with β-emitters such as ^{14}C, ^3H or ^{32}P has revolutionised the study of dynamic biological processes. Indeed, the accuracy and ease of quantifying radioactive lipids is such that it is often convenient to use isotopically-labelled lipids as model compounds, when testing separatory procedures or in evaluating hydrolysis or derivitisation procedures. It is also possible to convert lipids to radioactive derivatives as a means of quantifying them. For example, diacylglycerols, phosphatidylinositol and glycolipids have been estimated following acetylation with ^{14}C-acetic anhydride (see appropriate chapters above); free fatty acids have been assayed as the ^{63}Ni salts.[172] On the other hand, an extra dimension can be added to the problem of the analysis of labelled lipids, if both the mass and radioactivity (and thence the specific activity) of individual components of a mixture are required.

The analyst must be aware of and take precautions against the hazards associated with radiation; this information is available in a number of textbooks. In most countries, legislation exists to control the handling of isotopes. With low levels of ^{14}C- or ^3H-labelled lipids, such as are likely to be encountered in most biological experiments, the physical dangers to the operator are negligible, but it is none the less a wise precaution to wear protective clothing, especially rubber gloves, whenever these compounds are handled, and it is inadvisable to consume food, smoke or apply cosmetics in the laboratory.

Contaminated glassware should be segregated from that used for other purposes, all accidental spillages should be noted and scrupulous cleanliness should be observed, so that radioactivity is not carelessly transmitted to other parts of the laboratory or carried over to future experiments. Compounds labelled with higher energy β-emitters such as ^{32}P are potentially more hazardous to the operator and should be handled with particular care. Thin-layer chromatography of isotopically-labelled lipids should always be performed in fume cupboards to prevent radioactive dust spreading throughout the laboratory. Similarly, the outlets from gas chromatographs used in the analysis of radioactive tracers should be led into a fume cupboard or to the outside of the building.

It is not possible to present here a detailed account of the physics of the processes for detection of radioactivity or to discuss the merits of the various systems, as it is assumed that the reader has or can readily acquire such knowledge from one of the many textbooks currently available on the subject. Nor is it possible to discuss the availability commercially or by synthetic and biosynthetic routes, of labelled lipids. If they are to be used in biological experiments, the chemical and radiological purity of labelled lipids should not be taken for granted and may occasionally have to be checked by appropriate methods (see below). Fatty acids labelled with ^{14}C have been stored for over a decade with no radiation-induced decomposition occurring.[485] On the other hand, [^3H]-labelled lipids lose activity at a rate that can be significant, if excessive storage times are involved. ^{32}P has a half-life of only 14 days. It is probably a wise precaution to store isotopically-labelled lipids at low temperatures in inert solvents, so that the energy of radioactive decay is dissipated quickly.

The methods used to separate and estimate isotopically-labelled lipids do not differ markedly from those described above for unlabelled materials, and the amounts of the various radioactive compounds recovered from chromatographic processes can be determined by the methods already described, provided that the compounds are not altered in such a way that the radioactivity cannot also be measured by appropriate means. For example, radioassay should in general be performed on a separate aliquot of a sample, if the remainder is to be estimated by a destructive procedure, such as photodensitometry after charring. In addition, radiodetection systems are extremely sensitive and metabolites may be formed in biological experiments that have, to all intents and purposes, negligible mass, but are significantly radioactive; they might go undetected if only those compounds that could be seen by visual techniques were separated for counting. It is therefore important to ensure that the yield of radioactivity at the end of a separatory process is close to that present initially and to check, by alternative chromatographic procedures, that radioactivity apparently associated with a given compound is due to that compound and not to some other substance that co-chromatographs with it.

With those liquid–solid or liquid–liquid column chromatographic procedures that involve collection of discrete fractions, monitoring of radioactivity is not a problem as, after removal of solvents, the lipids can be dissolved in toluene containing suitable fluors and counted in a liquid scintillation counter. A small amount of methanol can be added to effect solution of highly polar lipids such as lysophosphatidylcholine, or if the label is present in the fatty acid moiety, transmethylation will ensure efficient counting. Measurement of radioactivity on thin-layer adsorbents or in gas chromatography effluents does present some special difficulties, as discussed below. Although radioactivity detection systems have been designed specifically for TLC and GLC, liquid scintillation counting can generally be adapted to the purpose, and as a result of its great versatility and high sensitivity, this is undoubtedly the most important single counting method. The

counting efficiency should always be determined so that counts per sec (cps) can be converted to disintegrations per sec (dps).

In all the following procedures, it is assumed that the analyst will observe all the relevant safety precautions.

B. CHROMATOGRAPHY AND RADIOASSAY

1. Thin-layer chromatography

Radioactive lipids on thin-layer plates can be detected for counting by spraying the edge of the plate with 2′,7′-dichlorofluorescein solution, or with water (less sensitive), or by exposing the plate to iodine vapour, marking the bands and allowing the iodine to evaporate off. Alternatively, "cold" standards can be run alongside the labelled compounds and only the former detected by these methods, permitting the position of the latter to be fixed. Bands may then be scraped into small chromatographic columns, simple lipid components eluted from the absorbent with chloroform–methanol (95:5, v/v), and complex lipids with chloroform–methanol–water (5:5:1, by vol.). The lipids are transferred to scintillation vials and, after removal of solvents, dissolved in toluene containing an appropriate mixture of fluors (e.g. 4 g 2,5-diphenyloxazole (PPO) and 0.2 g 1,4-di-(2-(5-phenyloxazolyl))-benzene (POPOP) per litre of toluene) for liquid scintillation counting. It is especially important that all traces of chloroform be removed, as this is a powerful quenching agent. With lipids of very high specific activity, it may be necessary to add an unlabelled ("cold") carrier at an early stage in the analysis to ensure quantitative recovery of radioactivity. For example, if it is not intended that the masses of particular components be determined, it is advantageous to add the carrier to the sample before separating by TLC. Unfortunately, the dyes used to detect lipids on thin-layer adsorbents may produce strong chemiluminescence effects, enhancing the apparent amount of radioactivity detected, but this phenomenon can be minimised by using the least possible amount of dye and storing the samples in the dark in a warm room for several hours before counting. There are also commercial scintillant "cocktails" formulated to minimise the problem. If precautions are taken to circumvent chemiluminescence, the whole plate can be sprayed and any dubiety about the position of bands is removed. Coloured materials in lipid extracts can cause problems if they bring about excessive quenching; it proved possible to decolourise faecal extracts effectively prior to counting, for example, by bleaching with 5.25% sodium hypochlorite solution, before destroying excess chlorine with nitrite.[161] Commercial equipment is available for oxidation of such materials to $^{14}CO_2$ and $^{3}H_2O$, which are trapped separately and can then be counted with high efficiency.

After having been subjected to liquid scintillation counting, non-polar compounds such as methyl esters, cholesterol esters and triacylglycerols can be recovered from the scintillant solution for further analyses, if this is necessary, by evaporating the solvent and subjecting the fluor-lipid mixture to preparative TLC on silica gel G layers, with hexane–diethyl ether (80:20, v/v) as developing solvent; the fluors remain in the bottom half of the plate. Alternatively, the labelled fatty acid moieties can be recovered from scintillant solutions by hydrolysing the mixture with alkali and removing the fluors with the non-saponifiable components. The procedure described in Chapter 4 is suitable, although a method has been devised specifically for this purpose.[331]

Another approach to the problem of radioassay of compounds separated by thin-layer chromatography consists in suspending them, together with the TLC adsorbent, in a liquid scintillation "cocktail" containing toluene, a detergent, water and a fluor, for counting. Many proprietory brands of emulsifier–fluors are available, but it is also possible to make up suitable preparations from basic ingredients.[837] The main difficulty is that the sample may continue to adhere to the silica gel to a variable extent so that the efficiency of counting is erratic. Ideally, the sample should be entirely in solution, and the scintillation medium should be formulated so as to solubilise as wide a spectrum of different lipid classes as possible. There has been a great deal of debate in the literature concerning the accuracy of such procedures, especially when weak β-emitters such as ^{3}H, are present (c.f. References 632, 689, 818). It has been the author's experience that such methods can give entirely satisfactory results with

^{14}C, provided that quench-correction curves are drawn up with a set of standards in a scintillation medium identical to that of the samples. If the counter is not cooled the suspended particles of silica gel can precipitate on standing with some commercial scintillant preparations.[550] Such procedures are of limited value only for ^{3}H-labelled lipids as self-absorption effects are more marked with this isotope. It is particularly important that recoveries of label from the TLC plates be monitored, as poor recoveries can be a good indication that counting problems are developing.

Procedures that involve eluting lipids from TLC adsorbents are tedious, time-consuming and subject to a number of errors, so some effort has been devoted to devising means of measuring the radioactivity of compounds while they are still on the plate. Unfortunately, many of these techniques (reviewed by Mukherjee and Mangold[522], require costly dedicated equipment that is not often available to the analyst. For example, the developed thin-layer plate can be laid on an X-ray photographic film so that the radiation produces an image, which can be developed and is proportional in intensity to the amount of radioactivity present. The relative intensities of the various bands may be determined by photodensitometry.[70] The TLC plate is unaltered, and can then be sprayed with reagents to detect or identify the various components or to determine the masses of each. In an alternative procedure, the developed TLC plate is passed through a radiochromatogram scanner in which radioactivity is measured with two windowless geiger detectors on each side of the plate. The signals from these are transmitted to a recorder, so that a continuous trace is obtained illustrating peaks of radioactivity on the plate. Specific spray reagents are then used to detect and identify components, so that the radioactivity peaks can be assigned to specific compounds. Again, the masses of the various components can be separately determined by any of the methods described in earlier chapters, in order that their specific activities can be calculated. Figure 10.1 illustrates a TLC separation with the radioactivity trace alongside the developed plate. A number of manufacturers supply equipment for the purpose, but such procedures are less accurate than those involving liquid scintillation counting.

A procedure for estimating lipids after charring on

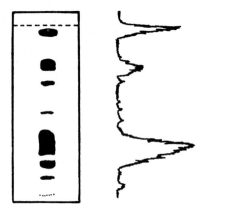

FIG. 10.1. TLC plate of labelled lipids together with radioactivity scan (bands visualised after radioactivity measurement).

TLC plates by suspending the charred bands in a scintillant and measuring the degree of quenching via the external standard channels ratio was described earlier (see Chapters 3, 6 and 7).[685] Noble and Shand[549] have shown that following mass determination by this means, a reproducible and efficient ^{14}C-count can be obtained from labelled lipids. Although there was some loss of radioactivity, accurate dpm measurements could be made and the specific activity of individual lipids obtained.

It should again be stressed that all steps in the thin-layer chromatography of radioactive lipids should be performed in an efficient fume cupboard, to prevent contamination of laboratories with radioactive dust.

2. Gas–liquid chromatography

Isotopically-labelled lipids may be subjected to gas chromatography under exactly the same conditions as unlabelled compounds (see Chapter 5), except that it is doubly important to use silanised supports and catalyst-free liquid phases to minimise retention of components on the columns that would cause the background count to rise and memory effects to become apparent. Facilities for on-column injection are essential and all parts of the instrument that may be exposed to radioactive vapours should be readily replaceable. It is also advisable to cover the whole instrument with a ventilated hood to remove radioactive combustion products.

Combined gas chromatography-radioassay of labelled lipids (especially methyl esters of fatty acids) has been comprehensively reviewed.[380, 489a] There are two main approaches to the problem. In the first, conventional preparative GLC is adapted to the collection of samples for liquid scintillation counting while, in the second, the GLC effluent is passed through a sensitive radiation detector so that the mass and radiactivity of components are monitored at the same time in a continuous manner.

I) RADIOASSAY AFTER PREPARATIVE GLC

Preparative gas chromatography of methyl esters of fatty acids is described in Chapter 5, and of higher molecular weight compounds in Chapter 8. These procedures, in which a stream splitter diverts part of the column effluent to a flame ionisation detector so that a mass trace is obtained while the bulk of the sample is vented into the collection system, can be used without modification to collect isotopically-labelled components for subsequent assay by conventional liquid scintillation techniques. For example, Watson and Williams[815] have described a conventional system in which components in a GLC effluent are condensed in a trap containing silicone liquid coated on to glass beads (40–60 mesh; 20 per cent, w/w), from which they can be recovered by elution with toluene–fluor for counting. Better than 90 per cent recoveries are claimed. The counting procedure can be simplified, however, if crystals of primary scintillators such as anthracene or p-terphenyl coated with 5 per cent (w/v) silicone oil (a slurry of the crystals and the silicone oil in acetone is prepared, then the solvent is evaporated) are used to trap components.[380, 381] Traps approximately 4 cm long are filled with this material, and will collect with great efficiency the labelled lipid in a given peak issuing from a GLC column. There is no need to recover the compounds which can be counted simply by placing the trap in a conventional liquid scintillation counter. In this system, ^{14}C is counted with an efficiency comparable to that in solution in a liquid scintillator, but the efficiency of ^{3}H counting is poor and not reproducible. With compounds labelled with ^{3}H, therefore, better results are obtained if all the material in the trap (p-terphenyl only, *not* anthracene) is dissolved in a conventional toluene–fluor for counting. Indeed the contents of the trap can be placed in the vial with the toluene–fluor, as the organic material will dissolve, while the glass in the trap does not appear to interfere with the scintillation process. The author[131] has obtained satisfactory results by trapping the GLC effluent on glass beads moistened with toluene, and subsequently washing this with toluene–fluor directly into a scintillation vial.

A number of procedures have also been described in which components in the GLC effluent are condensed by bubbling the exit gas into toluene–fluor in a scintillation vial, which can later be placed in a liquid scintillation counter for assay. Possibly the most effective of these systems is one devised by Dutton[192], that has been used in a number of laboratories. The outlet of the gas chromatograph is connected via narrow bore stainless-steel tubing, surrounded by a heated aluminium jacket that serves as a heat sink, to a three-way syringe valve. A syringe containing toluene–fluor is attached to one of the remaining outlets, while the third outlet is connected again to a length of narrow bore stainless-steel tubing directed into a scintillation vial containing 15 ml toluene–fluor. When samples are to be taken, the GLC effluent is bubbled directly into the scintillation solution and, at the end of the sampling period, the sampling device is flushed with 1 ml of toluene–fluor from the syringe reservoir, to minimise tailing of radioactivity, before the scintillation vial is changed. Fractions are collected by reference to particular peaks on the mass trace, or preferably at regular predetermined time intervals for counting. The system was automated[768] by arranging to pass toluene–fluor continuously past the heated column exit into a mechanical fraction collector loaded with scintillation vials. In addition, the resolution and sensitivity of the analysis were improved by combining and averaging the results of several replicate analyses by means of a computer.

Inaccuracies associated with uneven recovery of components were overcome by an elegant procedure described by Biship et al.[67] Free or combined ^{3}H-labelled fatty acids are converted to their methyl-^{14}C esters, and ^{14}C-labelled fatty acids to their methyl-^{3}H esters. If the specific radioactivity of the methyl group is known, the ^{14}C to ^{3}H ratio in the esters, separated by GLC, gives a direct measure of the specific activity of the fatty acid and quantitative collection is unnecessary. The fatty acids or lipids must be methylated with the minimum amount of labelled methanol for reasons of economy, but suitable methods are available.

Techniques of collection followed by counting have several advantages over continuous radioassay systems, of which the most significant is that comparatively low levels of radioactivity can be measured by extending the counting time in the liquid scintillation counter. Also, ^{14}C- and ^{3}H-labelled lipids can be handled with equal facility, and labelled fatty acids, in particular, can be recovered from the toluene–fluor if necessary using the methods discussed in the previous section. The main disadvantage of non-automated systems is that the collection of large numbers of fractions is tedious and requires the constant attention of a skilled operator. If the fractions collected are too large, it may not be easy to assign the measured radioactivity to specific peaks in the mass trace, and if the entire column effluent is not collected, some compounds that unexpectedly contain radioactivity may be missed. Some lipids, such as cholesterol[380], may not condense with sufficient rapidity in the toluene–fluor and so could be lost.

II) CONTINUOUS MONITORING OF THE COLUMN EFFLUENT

The principal advantages of monitoring the effluent of a gas chromatographic column by flow counting techniques are that selective losses are unlikely to occur and the mass and radioactivity profiles of the components of a mixture can be displayed simultaneously on the same recorder trace. Thus, routine analyses of numbers of samples can be performed. Although many radioactivity detectors are operated at the temperatures of the gas chromatographic columns, better results are obtained if the effluent is combusted first so that ^{14}C is counted as $^{14}CO_2$, and ^{3}H as $^3H_2$, at ambient temperature. If this is not done, radioactive compounds may be adsorbed on the walls of the detector, reducing the amount of radioactivity detected (as such compounds are counted in 2π geometry) and causing "memory" effects with other compounds eluting from the column, while generally raising the background radioactivity.

In most commercial radio-gas chromatography systems, the effluent is split as it leaves the column and a portion passes into a flame ionisation detector so that the masses of components are recorded. At the same time, the bulk of the effluent is diluted with cold carbon dioxide and passed into a quartz tube packed with cupric oxide wire, maintained at approximately 700°C, where organic compounds are converted to carbon dioxide and water. If ^3H is to be assayed also, the effluent is then passed into a column of steel wool, also at 700°C and flushed continuously with hydrogen, where the ^3H_2O is reduced to elemental hydrogen. Finally, the gases are passed through a water trap containing anhydrous magnesium perchlorate into the radioactivity detector.

Although many types of detector have been tried, proportional counters have proved most successful and are used in most commercial instruments. The sensitivity of such detectors is dependent on the time the substances remain in the detector, and comparatively large cells (50–100 ml volume) are often used to obtain the maximum possible radioactivity count. On the other hand if the cell is too large, mixing of adjacent components in the GLC effluent may occur and there will be a loss of resolution (with peak broadening in the radio-activity trace relative to the mass trace). For this reason, some manufacturers use smaller detector cells (approximately 10–20 ml volume) and accept lower counting efficiencies for the sake of the higher resolution that can be achieved. With the large cells, a counting efficiency of 80 per cent is possible with ^{14}C, but with smaller cells the apparent efficiency may be as low as 20 per cent, depending on the flow rate used. With proper shielding of the detector, the background count is as low as 2 cpm.[697] Argon is the most effective carrier gas for use with proportional counters, and is generally diluted with 5 per cent (v/v) carbon dioxide immediately before the combustion train, to prevent adsorption effects and to serve as a quencher in the proportional counter; it is also possible to use argon–carbon dioxide (95:5, v/v) as the carrier gas for the entire chromatographic process.

For the visual presentation of results, the detector signal is sent via a ratemeter to one channel of a dual pen recorder, the other channel of which is used to record the mass response from the flame ionisation detector. The radioactivity associated with each component in the sample is then seen directly as in the specimen trace in Fig. 10.2. Some loss of resolution of radioactivity peaks in comparison to the mass peaks is inevitable, even in the most efficient system, because of the dilution of the sample and the

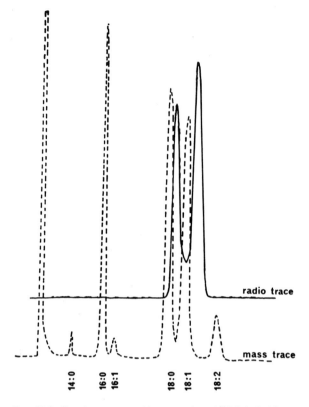

FIG. 10.2. Gas chromatographic separation of ^{14}C-labelled fatty acids with radioassay of components (trace obtained with Panax Radiogas Detector System).

large volume of the apparatus through which it must pass after leaving the column. As well as the graphic record, the signal from the ratemeter can be passed into a scaler unit with a digital print-out, so that the numbers of counts in predetermined time intervals are obtained directly.

When the background radioactivity is low, the number of counts recorded from a sample is proportional to the time the sample is in the detector cell, which is in turn dependent on the gas flow rate and the volume of the detector cell. Higher counting efficiencies can, therefore, be obtained by reducing the flow of carrier gas; in instruments with a small detector cell, this improvement can be appreciable. For example, if the detector cell volume is 10 ml, a flow rate of 10 ml/min will permit an overall apparent efficiency of counting of up to 80%, whereas a more usual carrier-gas flow rate of 50 ml/min will reduce the apparent efficiency of counting to one fifth of this value. In instruments with small cells, it is then advisable to use the lowest flow rate commensurate with the required degree of resolution in the gas chromatographic column, compensating for the increased retention time of components by raising the column temperature. Although it is always advisable to calibrate the instrument with precise amounts of $^{14}CO_2$ or $^3H_2$, it is possible to remove doubts about the efficiency of the counting process by adding an internal standard of known specific activity to the sample. [1-^{14}C]-Pentadecanoic acid and [1-^{14}C]-heptadecanoic acid have been used in this way in the analysis of [^{14}C]-labelled fatty acids in the author's laboratory.

The main disadvantage of continuous flow systems is that the compounds to be assayed must have comparatively high specific activities, as the overall efficiency of the combined chromatographic and counting process may be low. There should be sufficient activity in a sample of less than 1 mg, as chromatographic columns are unlikely to take larger amounts without loss of resolution. Radioactive decay is a random process and for a given number of counts (C) in a radioactive peak, the standard deviation (σ) is given by

$$\sigma = \sqrt{C}$$

It is therefore necessary to have 10,000 counts in a peak to obtain a 1 per cent standard deviation, while 100 counts would give a 10 per cent standard deviation. It is not possible to extend the counting time and so improve the accuracy of the measurement as in liquid scintillation counting.

If need be, it is possible to by-pass the proportional counter and collect the $^{14}CO_2$ formed in the combustion unit in a scintillation vial containing 10 ml toluene–fluor and hyamine hydroxide in the ratio 15:8 (v/v).[775] Fractions collected in this way are counted in a liquid scintillation counter with greater efficiency and improved statistical accuracy. It is claimed that better recoveries are obtained than are possible with conventional preparative GLC techniques.

3. Liquid column chromatography

Equipment has been devised for monitoring the eluate from liquid chromatography columns by liquid scintillation counting (reviewed by Mukherjee and Mangold[522]). Both homogeneous and heterogeneous systems have been devised. In the former, the column eluate is mixed with the scintillant, permitting the counting of isotopes such as ^{3}H or ^{14}C with very high efficiency; fractions are not easily recovered from the scintillant for further analysis, however. In heterogeneous systems, the column eluate is passed in a narrow-bore tube through a vial containing the scintillant. Uncontaminated fractions are recovered but ^{3}H and ^{14}C are counted with rather low efficiency (0.2 and 17% respectively). Phospholipids labelled with ^{32}P are counted with acceptable efficiency (35%), and excellent HPLC separations and analyses of such compounds have been recorded.[74]

C. LOCATION OF ^{14}C IN ALIPHATIC CHAINS

The portion of a radioactive lipid that contains the label can be ascertained simply by hydrolysing the lipid and isolating the various component parts, using the procedures described in earlier chapters, assaying the radioactivity of each in a liquid scintillation counter. It may then be important to discover which carbon atoms in the aliphatic chains carry the ^{14}C-label. The reactions used to solve the problem are most easily performed with fatty acids, and it is usually necessary to oxidise fatty alcohols or aldehydes to acids first. Before proceeding to the analysis, the individual fatty acids concerned must be isolated in the greatest degree of purity possible, using the methods described in Chapter 5, so that traces of contaminants of high specific activity do not influence the results.

Fatty acids are more easily labelled with ^{14}C in the carboxyl group than in any other position by the available synthetic methods and, as a result, most ^{14}C-labelled fatty acids that are used in biological experiments contain the label in this position. Methods of determining the activity of C-1 fatty acids therefore have considerable importance. Of these, the Schmidt reaction (reaction a, Fig. 10.3) is simplest and most suitable for the purpose. The fatty acid is reacted with hydrazoic acid (sodium azide and sulphuric acid) causing the carboxyl group to be released as carbon dioxide which can be recovered for liquid scintillation counting. A number of semi-

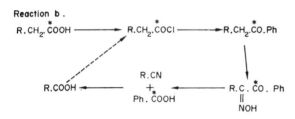

Reaction a.

$$R.\overset{*}{C}OOH + HN_3 \longrightarrow R.NH_2 + \overset{*}{C}O_2 + N_2$$

Reaction b.

FIG. 10.3. Reaction a. Schmidt reaction for degradation of fatty acids. Reaction b. Stepwise degradation of radioactive fatty acids.

micro modifications of the procedure have been described for use with ^{14}C-labelled fatty acids and the following is based on that of Aronsson and Gurtler[41]. It is necessary to hydrogenate unsaturated fatty acids before carrying out the reaction.

"A micro-glass apparatus is required, consisting of a flask and condenser on which is placed a pressure-equalising dropping funnel connected to an argon line led through the condenser to the bottom of the flask; there is a side-arm outlet from the top of the condenser, below the funnel, so that the gases pass into a suitable trap. The whole is set on top of a magnetic stirrer-heater. All the reagents should be of the highest possible purity. The ^{14}C-labelled fatty acid and a fatty acid carrier (40 mg in total) in dry benzene (2 ml) are added to the flask together with sodium azide (75 mg). 2-Methoxy-ethanol (2 ml) and ethanolamine (1 ml) are placed in the CO_2 collection vessel. The argon flow is set at 2 ml/min, and concentrated sulphuric acid (0.5 ml) is introduced into the flask from the dropping funnel. The reaction is allowed to proceed at 55°C with stirring for 1 hr, when the contents of the CO_2 trap are transferred to a scintillation vial with 10 ml of a toluene-based fluor (toluene, 268 ml; 2-methoxy-ethanol, 134 ml; 2,5-diphenyloxazole, 2 g) for counting. $^{14}CO_2$ recoveries of 99 per cent are possible, but this should be checked with standards."

Although it has apparently not been applied to fatty acids yet, a new reagent, diphenylphosphoryl-azide, can be used in a mild Curtius reaction to remove C-1.[693] No simple procedure exists for the stepwise degradation of fatty acids, but the procedure of Dauben et al.[160], outlined in Fig. 10.3 (reaction b), though complex, avoids the use of strong oxidising agents so that over-reaction is unlikely to occur. The acid is converted to the acid chloride and thence to a phenylketone in a Friedel–Crafts condensation; the ketone is reacted with nitrous acid forming an α-oximinoketone, which is cleaved by reaction with p-toluene-sulphonyl chloride and alkali to give benzoic acid and an aliphatic nitrile, one carbon atom shorter than the original fatty acid. Both products can be recovered for radioassay, and the nitrile hydrolysed to a fatty acid and put through the cycle again. Steinberg et al.[733] were able to obtain an overall yield of 65 per cent of the next lower homologue when the products were isolated by chromatographic procedures.

Unfortunately, methods of locating ^{14}C in other parts of a fatty acid chain tend to be less elegant and precise, and often much more tedious than those described above for C-1. Sometimes information can be obtained by making use of structural features of the fatty acids; for example, unsaturated fatty acids can be oxidised at double bonds (by one of the methods described in Chapter 5), and the mono- and dibasic fragments formed can be separated and counted by radio-gas chromatography so that the distribution of radioactivity between the proximal and distal portions of the molecule is obtained. Polyunsaturated fatty acids are better subjected to partial reduction to obtain monoenoic acids (also discussed in Chapter 5) for structural analysis in this manner. The vigorous permanganate oxidation reaction, used to locate methyl branches in aliphatic chains (see Chapter 5), can also be adapted for determining the radioactivity of specific carbon atoms in the chain of saturated fatty acids; the homologous series of fatty acids produced in the reaction must be assayed by radio-gas chromatography, preferably in the form of the butyl ester derivatives.[537] Although simple in concept, methods that produce large numbers of fragments suffer from the disadvantages that the fatty acids must initially have a very high activity so that sufficient remains in each of the fragments for the radio-gas chromatographic analysis to be meaningful, and that the calculation of the activities of particular carbon atoms may be fraught with errors if more than one carbon contains the label.

Determinations of the precise point of labelling of fatty acid metabolites isolated from biological experiments should undoubtedly be performed much more frequently than is the case at present, especially when the label was originally present in C-1, as it is well established that partial β-oxidation of fatty acids occurs in mammalian tissues; labelled acetate is then released that can be assimilated into other fatty acids. Similarly, labelled acetate fed to animals or incubated with animal tissues *in vitro* may be used to resynthesise fatty acids that have been partially degraded. It is not always realised how prevalent these reactions are.

CHAPTER 11

The Separation of Plasma Lipoproteins

A. THE NATURE OF PLASMA LIPOPROTEINS

A HIGH proportion of the lipids in plasma exist in the form of complexes with proteins known as *lipoproteins*, which have much of the solubility characteristics of proteins in that they are soluble in water or in aqueous salt solutions. Lipoproteins are thus able to transport essentially hydrophobic lipids from their sites of synthesis in the liver or intestines to peripheral tissues. The nature of the association between lipids and proteins is only poorly understood; it does not involve covalent binding, but rather interactions (London–Van der Waals forces) between the hydrophobic hydrocarbon "tails" of the fatty acid moieties and hydrophobic amino acid side-chains of proteins, although lipid–lipid interactions are also important. In consequence, the lipids are readily released from this binding by extraction with organic solvents.

Plasma lipoproteins are very heterogeneous, differing in the nature and composition of both the lipid and protein ("apoproteins") components, but some subdivision into classes or families of compounds is possible using various physical methods. It is increasingly being realised that an analysis of these subclasses can be of great value in diagnosing the presence of certain disease states in a patient or in assessing the risk of disease occurring. Methods of isolating lipoproteins are, therefore, of some importance. Separatory procedures[284, 443, 444, 535] and the chemistry and metabolism of lipoproteins[367, 719] have been thoroughly reviewed.

The techniques most often used to fractionate lipoprotein classes are electrophoresis and flotation by ultracentrifugation, and the nomenclature used to delineate fractions tends to reflect the isolation procedure used.

High density lipoproteins (HDL) comprise the fraction which, upon ultracentrifugation, distributes in the solution density band of 1.063–1.21 g/ml. On zonal electrophoresis, they migrate in the alpha region so are also termed "α-lipoproteins".

Low density lipoproteins (LDL) are found, on ultracentrifugation, in the density region 1.006–1.063 g/ml. Because of their electrophoretic mobility, they are also known as "β-lipoproteins".

Very low density lipoproteins (VLDL) are found, on ultracentrifugation, in the density region 0.93–1.006 g/ml, and are also known as pre-β-lipoproteins", migrating ahead of β-lipoproteins in electrophoresis.

Chylomicrons are large particles of very low density (<0.95) of enteric origin. They remain at the start in electrophoresis.

The distinctive flotation properties of each class of lipoproteins are a consequence, in large part, of the nature of the lipid components in each: triacylglycerols are the main lipid class of chylomicrons and VLDL, whereas cholesterol esters, cholesterol and phospholipids are more abundant in LDL and HDL. With care, it is possible to further separate many of the above fractions into subclasses. It is important to recognise that the lipoprotein class definitions do not in general define their metabolic origin or physiological function. Also, the above definitions were applied initially to human plasma lipoproteins, and the properties of those of other species can be rather different.[119]

The techniques used for lipoprotein fractionation are generally considered to be the province of the protein rather than the lipid chemist. However, most lipid analysts should be able to overcome any innate hydrophobia and tackle the simplified separatory procedures described below. These procedures require fresh, *not* frozen, plasma samples.

B. SEPARATION BY ULTRACENTRIFUGATION

The most widely-used technique for the isolation of plasma lipoproteins on a preparative scale consists of floating different fractions sequentially in solutions of increasing density. It was pioneered by de Lalla and Gofman[174] and with modifications, mainly by Lindgren and colleagues,[443, 444] has become the method against which others are judged. The following somewhat simplified procedure can be used for the isolation of VLDL, LDL and HDL fractions, and involves the addition of solutes and solutions of controlled density to the plasma prior to ultracentrifugation. It does not give quantitative yields of individual lipoprotein classes, but it does give fractions of high purity. Initially and subsequently at intervals, the density of the solutions should be checked by precision refractometry.[443, 444]

Swinging-bucket rotors are more suitable than anglehead rotors for the purpose, as there is less danger of VLDL sticking to the sides of the tube, but anglehead rotors tend to be more readily available and can also be used if need be.

"Three solutions are necessary (a) NaCl (0.85 g) and Na_2EDTA (0.037 g) in water (final volume 100 ml). (b) as solution a but also containing KBr (8.81 g per 100 ml final volume) (c) as solution a but also containing KBr (29.3 g per 100 ml final volume).

Fresh plasma (25 ml) is placed in a 90 ml capacity centrifuge tube, and solution a (30 ml) is layered very carefully on top of this by means of a peristaltic pump. The tube is centrifuged at 60,000 g for $1\frac{1}{2}$ hr at 12°C. The top 10 ml, which contains the VLDL and chylomicrons, is sucked off carefully from the surface by means of a syringe.

The bottom 20 ml of the remaining liquid is removed by syringe and transferred to another centrifuge tube. Solid KBr (1.85 g) is added to make the density of the solution up to 1.063 g/ml. Solution b (30 ml) is layered on top of this, and the whole is centrifuged as before, but for 6 hr. The top 10 ml, which contains the LDL, is removed carefully by means of a syringe.

The bottom 20 ml of the solution is transferred to a fresh centrifuge tube, KBr (4.5 g) is added to increase the density to 1.21 g/ml and solution c (30 ml) is layered on top. The whole is centrifuged for 20 hr as before, when the HDL is found in the top 10 ml."

These conditions are suitable for separating the three main plasma lipoprotein fractions from the human, rat or ruminants. If further subdivision of fractions is required or if problems arise, the major review articles cited above should be consulted.

Density-gradient procedures using discontinuous gradients have recently been described that permit the preparative-scale isolation of the major lipoprotein classes and some sub-fractionation of these in a single run.[120, 616] A related method, in which a vertical rotor is used has been described.[144]

C. SEPARATION BY ELECTROPHORESIS

The use of electrophoresis for the analytical scale separation and characterization of serum lipoprotein is very well established (reviewed by Narayan[535]). Various support media have been used but that currently most widespread in agarose gel.[309] A number of commercial biochemical suppliers provide kits designed for the purpose. Lindgren *et al.*[445] critically compared a procedure of this kind with the standard flotation method, and found that the former offered acceptable precision and was particularly suited to the routine screening of large numbers of samples. The procedure that follows suggests the use of Tris–Tricine buffer,[506] in place of the traditional barbital buffer systems, as the former appears to give better resolution in routine use, and does not suffer from the disadvantages of toxicity and limited solubility inherent in the use of the latter.

Electrophoresis is performed in thin slabs of 0.5% agarose mounted on glass or Gel-Bond® film supported horizontally (ideally on a cooled platten). Gel-Bond is particularly easy to use, having excellent adhesion to the agarose and providing a ready means of storing the stained and dried gels. The detailed procedure for electrophoresis will depend upon the equipment being used; the reader is referred to the instructions of individual manufacturers and to the paper of Axelson *et al.*[50] All reagents should be of Analar grade. A generalised procedure is as follows:

"0.08 M Tris–0.024 M Tricine buffer (pH 8.6) is made by dissolving Tris base (9.69 g) and Tricine (4.3 g) in a final volume of 1 litre. No adjustment of pH should be necessary. Sodium azide (0.02%) may be added for longterm storage. Agarose (0.5 g per 100 ml) is dissolved in the buffer by heating on a boiling water-bath with occasional swirling for 15 min. This solution is dispensed into tubes in a water bath at 50°C in volumes appropriate to one plate per tube. (As a rough guide, 0.14 ml 0.5% agarose solution covers 1 cm^2 of glass plate or a slightly greater area of Gel-Bond.) When the agarose has cooled to 50°C, it is poured onto plates on a flat level surface. The plates should be left for 24 hr in a water-saturated atmosphere before use.

The precise method of loading samples on the plates will depend on the manufacturer's instructions, but 1–2 μl of plasma diluted with an equal volume of Tris-Tricine buffer will generally give good staining and resolution. With a voltage gradient of 15 v.cm^{-1} across the gel, a run-time of 30 min is satisfactory; during this period, the most rapidly-migrating lipoprotein (HDL) should move about 2 cm towards the anode.

Immediately after removal of the gels from the electrophoresis apparatus, they are soaked in 7% acetic acid for fixing. They are dried thoroughly in an oven at up to 70°C or using a hair dryer. Drying times may be shortened by first "pressing" the wet gel.[50]

In order to reveal the presence of lipoproteins, dried gels can be stained for protein using Coomassie blue. To stain for lipid, a 12:100 mixture of water:fat red in methanol (225 mg/ml) is poured onto the dried gel (the water and methanolic fat red should only be mixed immediately before use), which is agitated for precisely 3 min in the stain. The gel is transferred rapidly to a clearing solution (water–methanol, 1:1, v/v) and rinsed in this until the pink lipoprotein bands stand out clearly against a colourless background (<1 min). Finally, the gel is transferred to a rinse-bath of 2% aqueous glycerol for approximately 1 min, before it is dried as before."

Separated lipoproteins can be identified by their rates of migration relative to authentic standards and quantified by scanning the stained gels spectrophotometrically at 540 nm.

D. PRECIPITATION OF LIPOPROTEINS BY POLYANIONS

Lipoprotein fractions can be selectively precipitated by adding polyanionic reagents such as heparin, dextran sulphate or phosphotungstate to plasma solutions in which the pH and ionic strength of the solution (divalent cations are required) are controlled. The available methods have been reviewed.[438] The advantages of such procedures are that large quantities of lipoproteins can be isolated in a relatively short period of time without the need to make use of ultracentrifugation. Disadvantages are that VLDL and LDL tend to coprecipitate and even under optimum conditions are not fully resolved. Burstein *et al.*[99] have provided four detailed procedures for various circumstances, while Friedewald *et al.*[218] have described a simplified procedure, which has been critically evaluated by others.[445]

E. ANALYSIS AND QUANTIFICATION OF LIPOPROTEINS

Lipids can be extracted from lipoproteins by the standard chloroform–methanol procedure described in Chapter 2 and analysed by the methods described in subsequent chapters. Lipoproteins may have to be first dispersed in Triton X-100® solution (final concentration, 1% w/v) to minimise possible lipid "masking" of the protein in the intact lipoprotein particle; the detergent tends to interfere with the standard "Lowry" procedure for protein determination,[457] unless modified appropriately.[813] If the protein content of the fraction is low, as in chylomicrons for example, a more sensitive fluorescence method may have to be used, such as that using fluorescamine (Triton X-100 does not interfere in this instance).[784]

APPENDIX A

Commercial Sources of Lipids

Analabs Inc., P.O. Box 501, North Haven, Con. (06473), U.S.A. U.K. Distributor — Auriema Ltd., 442 Bath Road, Slough, Bucks.

Applied Science Laboratories Inc., P.O. Box 440, State College, Pa. (16801), U.S.A. U.K. Distributor — Field Instrument Co., Queens House, Holly Rd., Twickenham, Middlesex.

B.D.H. Chemicals Ltd., Poole, Dorset, U.K.

Calbiochem-Behring Corp., U.K. Distributor — C.P. Laboratories Ltd., P.O. Box 22, Bishop Stortford, Herts CM22 7RQ.

Nu-Chek-Prep, Inc., P.O. Box 172, Elysian, Minn. (56028), U.S.A. European distributor — Bast of Copenhagen, 44 Ingerslevsgade DK 1705, Copenhagen V., Denmark.

P–L Biochemical Inc., 1037 West McKinley Ave., Milwaukee, Wisc., U.S.A.

Amersham International Ltd., Amersham, Bucks, U.K.

Sigma Chemical Co., P.O. Box 14508, Saint Louis, Missouri (63178), U.S.A. U.K. Distributor — Sigma London Chemical Co. Ltd., Fancy Rd., Poole, Dorset.

Supelco Inc., Bellefonte, Pa (16823), U.S.A.

APPENDIX B
Sources of Information

Journals

The following journals are devoted entirely to papers on the chemistry, biochemistry and analysis of lipids.

Biochimica Biophysica Acta — Lipids and Lipid Metabolism Section
Chemistry and Physics of Lipids
Journal of the American Oil Chemists' Society
Journal of Lipid Research
Lipids

In addition many other journals publish papers on the subject including — *Archives of Biochemistry and Biophysics, Biochemical Journal, Biochemistry, Canadian Journal of Biochemistry, Comparative Biochemistry and Physiology, Hoppe-Seyler's Zeitschrift für Physiologische Chemie, Journal of Biochemistry (Tokyo), Journal of Biological Chemistry, Journal of Chromatography.*

Reviews

The following review volumes appear annually and are devoted entirely to various aspects of the study of lipids:

Progress in the Chemistry of Fats and other Lipids (edited by R. T. Holman, Pergamon Press, Oxford). Sixteen volumes to date.
Progress in Lipid Research (edited by R. T. Holman, Pergamon Press, Oxford) continues the above series (3 further volumes).
Advances in Lipid Research (edited by R. Paoletti and D. Kritchevsky, Academic Press, New York). Eighteen volumes to date.

Text books

The following books have been published since the first edition of "Lipid Analysis".

Lipid Biochemistry: An Introduction (3rd Edition) by M. I. Gurr and A. T. James (Chapman and Hall Ltd., London, 1980).

Lipid Metabolism in Ruminant Animals (edited by W. W. Christie, Pergamon Press, Oxford, 1981).
Methods in Enzymology, Vol 35. *Lipids,* Part B (1975).
Methods in Enzymology, Vol 71. *Lipids,* Part C (1981).
Methods in Enzymology, Vol 72. *Lipids,* Part D (1981) (edited by J. M. Lowenstein, Academic Press, New York).
Lipids in Evolution, by W. R. Nes and W. D. Nes (Plenum Press, New York, 1980).
The Biochemistry of Plants, Vol. 4. *Lipids: Structure and Function* (edited by P. K. Stumpf and E. E. Conn Academic Press, Inc., New York. 1980).
Analysis of Lipids and Lipoproteins (edited by E. G. Perkins, American Oil Chemists' Society, Champaign, Illinois, 1975).
Glycolipid Methodology (edited by L. A. Witting, American Oil Chemists' Society, Champaign, Illinois, 1976).
Geometrical and Positional Fatty Acid Isomers (edited by E. A. Emken and H. J. Dutton, American Oil Chemists' Society, Champaign, Illinois, 1976).
Fatty Acids (edited by E. H. Pryde, American Oil Chemists' Society, Champaign, Illinois, 1979).
Cholesterol, by J. R. Sabine (Marcel Dekker, Inc., New York, 1977).
Low Density Lipoproteins (edited by C. E. Day and R. S. Levy, Plenum Press, New York, 1976).
Chemistry and Biochemistry of Natural Waxes (edited by P. E. Kolattukudy, Elsevier Publishing Co., Amsterdam, 1976).
Lipids and Biomembranes of Eukaryotic Microorganisms (edited by J. A. Erwin, Academic Press, New York, 1973).
The Role of Fats in Human Nutrition (edited by A. J. Vergroesen, Academic Press, London, 1975).
Lipid Metabolism in Mammals (2 volumes) (edited by F. Snyder, Plenum Press, New York, 1977).
Handbook of Lipid Research, Vol 1 (edited by D. J. Hanahan and A. Kuksis, Plenum Press, New York, 1978).
Form and Function of Phospholipids (edited by G. B. Ansell, J. N. Hawthorne and R. M. C. Dawson, Elsevier Publishing Co., Amsterdam, 1973).
Biochemistry of Lipids, II (edited by T. W. Goodwin, University Park Press, Baltimore, 1977).
Lipolytic Enzymes, by H. Brockerhoff and R. G. Jensen (Academic Press, New York, 1974).
Lipid Chromatographic Analysis (Second Edition — 3 volumes) (edited by G. V. Marinetti, Marcel Dekker, Inc., New York, 1976).

References

1. AASEN, A. J., HOFSTETTER, H. H., IYENGAR, B. T. R. and HOLMAN, R. T., *Lipids,* **6,** 502 (1971).
2. ABRAHAMSSON, S., STÄLLBERG-STENHAGEN, S. and STENHAGEN, E., *Prog. Chem. Fats,* **7,** 1 (1964).
3. ACKMAN, R. G., *J. Am. Oil Chem. Soc.,* **40,** 558 (1963).
4. ACKMAN, R. G., *J. Gas Chromatogr.,* **2,** 173 (1964).
5. ACKMAN, R. G., *Lipids,* **2,** 502 (1967).
6. ACKMAN, R. G., *J. Chromat.,* **28,** 225 (1967).
7. ACKMAN, R. G., *J. Chromat.,* **34,** 165 (1968).
8. ACKMAN, R. G., *Prog. Chem. Fats,* **12,** 165 (1972).
9. ACKMAN, R. G., *Lipids,* **12,** 293 (1977).
10. ACKMAN, R. G. and BURGHER, R. D., *J. Am. Oil Chem. Soc.,* **42,** 38 (1965).
11. ACKMAN, R. G. and EATON, C. A., *Fette Seifen Anstrichm.,* **80,** 21 (1978).
12. ACKMAN, R. G. and HANSEN, R. P., *Lipids,* **2,** 357 (1967).
13. ACKMAN, R. G., MANZER, A. and JOSEPH, J., *Chromatographia,* **7,** 107 (1974).
14. ACKMAN, R. G. and SIPOS, J. C., *J. Chromat.,* **16,** 298 (1964).
15. ACKMAN, R. G. and SIPOS, J. C., *J. Chrom. Sci.,* **14,** 568 (1976).
16. ACKMAN, R. G., SIPOS, J. C., EATON, C. A., HILAMAN, B. L. and LITCHFIELD, C., *Lipids,* **8,** 661 (1973).
17. ACKMAN, R. G., SIPOS, J. C. and JANGAARD, P. M., *Lipids,* **2,** 251 (1967).
18. AGRANOFF, B. W., BRADLEY, R. M. and BRADY, R. O., *J. biol. Chem.,* **233,** 1077 (1958).
19. AITZETMULLER, K., *J. Chromat.,* **113,** 231 (1975).
20. AITZETMULLER, K. and KOCH, J., *J. Chromat.,* **145,** 195 (1978).
21. AKESSON, B., *Eur. J. Biochem.,* **9,** 463 (1969).
22. AKINO, T. and SHIMOJO, T., *Biochim. biophys. Acta,* **210,** 343 (1970).
23. ALBRO, P. W. and DITTMER, J. C., *J. Chromat.,* **38,** 230 (1968).
24. ALLEN, R. R., *J. Am. Oil Chem. Soc.,* **46,** 552 (1969).
25. AMENTA, J. S., *J. Lipid Res.,* **5,** 270 (1964).
26. AMES, B. N., *Methods in Enzymology,* **8,** 115 (1966).
27. ANDERSON, R. E., BOTTINO, N. R. and REISER, R., *Lipids,* **2,** 440 (1967).
28. ANDERSON, R. E., GARRETT, R. D., BLANK, M. L. and SNYDER, F., *Lipids,* **4,** 327 (1969).
29. ANDERSON, R. L. and HOLLENBACH, E. J., *J. Lipid Res.,* **6,** 577 (1965).
30. ANDERSSON, B. A., *Prog. Chem. Fats,* **16,** 279 (1978).
31. ANDERSSON, B. A., CHRISTIE, W. W. and HOLMAN, R. T., *Lipids,* **10,** 215 (1975).
32. ANDERSSON, B. A. and HOLMAN, R. T., *Lipids,* **9,** 185 (1974).
33. ANDERSSON, B. A. and HOLMAN, R. T., *Lipids,* **10,** 716 (1975).
34. ANDO, N., ANDO, S. and YAMAKAWA, T., *J. Biochem. (Tokyo),* **70,** 341 (1971).
35. ANDO, S., CHANG, N-N. and YU, R. K., *Analyt. Biochem.,* **89,** 437 (1978).
36. ANEJA, R., BHATI, A., HAMILTON, R. J., PADLEY, F. B. and STEVEN, D. A., *J. Chromat.,* **173,** 392 (1979).
37. ANTHONY, G. M., BROOKS, C. J. W., MACLEAN, I. and SANGSTER, I., *J. Chromat. Sci.,* **7,** 623 (1969).
38. APON, J. M. B. and NICOLAIDES, N., *J. Chromat. Sci.,* **13,** 467 (1975).
39. Applied Science Laboratories Inc., *Gas-Chrom Newsletter,* Vol. 11, No. 4 (1970).
40. ARGOUDELIS, C. J. and PERKINS, E. G., *Lipids,* **3,** 379 (1968).
41. ARONSSON, P. and GURTLER, J., *Biochim. biophys. Acta,* **248,** 21 (1971).
42. ARTISS, J. D., DRAISEY, T. F., THIBERT, R. J. and TAYLOR, K. E., *Microchem. J.,* **24,** 239 (1979).
43. ARUNGA, R. O. and MORRISON, W. R., *Lipids,* **6,** 768 (1971).
44. ARVIDSON, G. A. E., *J. Lipid Res.,* **6,** 574 (1965).
45. ARVIDSON, G. A. E., *J. Lipid Res.,* **8,** 155 (1967).
46. ARVIDSON, G. A. E., *Eur. J. Biochem.,* **4,** 478 (1968).
47. ARVIDSON, G. A. E., *J. Chromat.,* **103,** 201 (1975).
48. AUDIER, H., BORY, S., FETIZON, M., LONGEVIALLE, P. and TOUBIANA, R., *Bull. Soc. chim. Fr.,* 3034 (1964).
49. AULING, G., MEINZ, E. and TULLOCH, A. P., *Hoppe-Seyler's Z. physiol. Chem.,* **352,** 905 (1971).
50. AXELSEN, N. H., KROLL, J. and WEEKE, B., *Scand. J. Immunol.,* **2,** supplement no. 1 (1973).
51. BAER, E., *Lipids,* **3,** 384 (1968).
52. BAGBY, M. O., SMITH, C. R. and WOLFF, I. A., *J. org. Chem.,* **30,** 4227 (1965).
53. BANSCHBACH, M. W., GEISON, R. L. and O'BRIEN, J. F., *Analyt. Biochem.,* **59,** 617 (1974).
54. BARFORD, R. A., LUDDY, F. E. and MAGIDMAN, P., *Lipids,* **1,** 287 (1966).
55. BARTLETT, J. C. and IVERSON, J. L., *J. Assoc. off. Analyt. Chem.,* **49,** 21 (1966).
56. BARTLETT, G. R., *J. biol. Chem.,* **234,** 466 (1959).
57. BARTON, F. E., HIMMELSBACH, D. S. and WALTERS, D. B., *J. Am. Oil Chem. Soc.,* **55,** 574 (1978).
58. BATLEY, M., PACKER, N. H. and REDMOND, J. W., *J. Chromat.,* **198,** 520 (1980).
59. BAUMANN, W. J. and ULSHÖFER, H. W., *Chem. Phys. Lipids,* **2,** 114 (1968).
60. BEIJER, K. and NYSTROM, E., *Analyt. Biochem.,* **48,** 1 (1972).
61. BENZONANA, G., *Lipids,* **9,** 166 (1974).
62. BERGELSON, L. D., *Prog. Chem. Fats,* **10,** 239 (1970).
63. BERGER, H. and HANAHAN, D. J., *Biochim. biophys. Acta,* **231,** 584 (1971).

64. BERNARD, D. M. and VERCAUTEREN, R. E., *J. Chromat.* **120,** 211 (1976).

65. BEZARD, J. A. and OUEDRAOGO, M. A., *J. Chromat.,* **196,** 279 (1980).

66. BIERL, B. A., BEROZA, M. and ALDRIDGE, M. H., *Analyt. Chem.,* **43,** 636 (1971).

67. BISHOP, C., GLASCOCK, R. F., NEWELL, E. M. and WELCH, V. A., *J. Lipid Res.,* **12,** 777 (1971).

68. BJERVE, K. S., DAAE, L. N. W. and BREMER, J., *Analyt. Biochem.,* **58,** 238 (1974).

69. BLANK, M. L., CRESS, E. A., PIANTADOSI, C. and SNYDER, F., *Biochim. biophys. Acta,* **380,** 208 (1975).

70. BLANK, M. L., SCHMIT, J. A. and PRIVETT, O. S., *J. Am. Oil Chem. Soc.,* **41,** 371 (1964).

71. BLANK, M. L. and SNYDER, F., *Biochemistry,* **9,** 5034 (1970).

72. BLAU, K. and KING, G. S. (Editors) *Handbook of Derivatives for Chromatography* (1978) (Heyden & Son, London).

73. BLIGH, E. G. and DYER, W. J., *Can. J. Biochem. Physiol.,* **37,** 911 (1959).

74. BLOM, C. P., DEIERKAUF, F. A. and RIEMERSMA, J. C., *J. Chromat.,* **171,** 331 (1979).

75. BLOMBERG, J., *Analyt. Biochem.,* **88,** 302 (1978).

76. BORCH, R. F., *Analyt. Chem.,* **47,** 2437 (1975).

77. BORKA, L. and PRIVETT, O. S., *Lipids,* **1,** 104 (1966).

78. BOTTINO, N. R., *J. Lipid Res.,* **12,** 24 (1971).

79. BOTTINO, N. R., VANDENBURG, G. A. and REISER, R., *Lipids,* **2,** 489 (1967).

80. BOUHOURS, J-F., *J. Chromat.,* **169,** 462 (1979).

81. BOYNE, A. W. and DUNCAN, W. R. H., *J. Lipid Res.,* **11,** 293 (1970).

82. BRANDT, A. E. and LANDS, W. E. M., *Lipids,* **3,** 178 (1968).

83. BRECKENRIDGE, W. C., in *Handbook of Lipid Research,* Vol. 1, pp 197–232 (1978) (edited by D. J. Hanahan and A. Kuksis, Plenum Press, New York).

84. BREMER, E. G., GROSS, S. K. and McCLUER, R. H., *J. Lipid Res.,* **20,** 1028 (1979).

85. BRIAN, B. L. and GARDNER, E. W., *Appl. Microbiol.,* **16,** 549 (1968).

86. BRIAN, B. L., GRACY, R. W. and SCHOLES, V. E., *J. Chromat.,* **66,** 138 (1972).

87. BROCKERHOFF, H., *J. Lipid Res.,* **6,** 10 (1965).

88. BROCKERHOFF, H., *Arch. Biochem. Biophys.,* **110,** 586 (1965).

89. BROCKERHOFF, H., *Lipids,* **1,** 162 (1966).

90. BROCKERHOFF, H., *J. Lipid Res.,* **8,** 167 (1967).

91. BROCKERHOFF, H., *Lipids,* **6,** 942 (1971).

92. BROCKERHOFF, H. and JENSEN, R. G., *Lipolytic Enzymes* (1974) (Academic Press, New York).

93. BROOKS, C. J. W. and MACLEAN, I., *J. Chromat. Sci.,* **9,** 18 (1971).

94. BROWN, H. C., SIVASANKARAN, K. and BROWN, C. A., *J. org. Chem.,* **28,** 214 (1963).

95. BROWN, J. B. and KOLB, D. K., *Prog. Chem. Fats,* **3,** 57 (1955).

96. BRUNDISH, D. E., SHAW, D. N. and BADDILEY, J., *Biochem. J.,* **97,** 158 (1965).

97. BRUNNGRABER, E. G., TETTAMANTI, G. and BERRA, B., in *Glycolipid Methodology,* pp 159–186(1976)(edited by L. A. Witting, American Oil Chemists' Soc., Champaign, Illinois).

98. BU'LOCK, J. D. and SMITH, G. N., *J. Chem. Soc.,* (C) 332 (1967).

99. BURSTEIN, M.; SCHOLNICK, H. R. and MORFIN, R., *J. Lipid Res.,* **11,** 583 (1970).

100. BUS, J., SIES, I. and LIE KEN JIE, M. S. F., *Chem. Phys. Lipids,* **17,** 501 (1976).

101. BUS, J., SIES, I. and LIE KEN JIE, M. S. F., *Chem. Phys. Lipids,* **18,** 130 (1977).

102. BUSSELL, N. E. and MILLER, R. A., *J. Liquid Chromat.,* **2,** 697 (1979).

103. CAPELLA, P. and ZORZUT, C. M., *Analyt. Chem.,* **40,** 1458 (1968).

104. CARREAU, J. P. and DUBACQ, J. P., *J. Chromat.,* **151,** 384 (1978).

105. CARROLL, K. K., *J. Lipid Res.,* **2,** 135 (1961).

106. CARROLL, K. K., *J. Am. Oil Chem. Soc.,* **40,** 413 (1963).

107. CARROLL, K. K., in *Lipid Chromatographic Analysis Second Edition,* Vol. 1, pp 173–214 (1976) (edited by G. V. Marinetti, Marcel Dekker Inc., N.Y.).

108. CARTER, H. E., ROTHFUS, J. A. and GIGG, R., *J. Lipid Res.,* **2,** 227 (1961).

109. CARTER, H. E., SMITH, D. B. and JONES, D. N., *J. biol. Chem.,* **232,** 681 (1958).

110. CARTER, H. E. and WEBER, E. J., *Lipids,* **1,** 16 (1966).

111. CARTER, T. P. and KANFER, J. N., *Lipids,* **8,** 537 (1973).

112. CASPARRINI, G., HORNING, E. C. and HORNING, M. G., *Chem. Phys. Lipids,* **3,** 1 (1969).

113. CASTELLANI, T., KJELGAARD-NIELSEN, P. and WOLFF-JENSEN, J., *J. Chromat.,* **104,** 123 (1975).

114. CASTER, W. O., AHN, P. and POGUE, R., *Chem. Phys. Lipids,* **1,** 393 (1967).

115. CHAN, H. W. S. and LEVETT, G., *Lipids,* **12,** 99 (1977).

116. CHAN, H. W. S. and LEVETT, G., *Chem. Ind.* (London), 578 (1978).

117. CHAPMAN, D., *The Structure of Lipids by Spectroscopic and X-ray Techniques,* (1965) (Methuen, London).

118. CHAPMAN, D., *J. Am. Oil Chem. Soc.,* **42,** 353 (1965).

119. CHAPMAN, M. J., *J. Lipid Res.,* **21,** 789 (1980).

120. CHAPMAN, M. J., GOLDSTEIN, S., LAGRANGE, D. and LAPLAUD, P. M., *J. Lipid Res.,* **22,** 339 (1981).

121. CHEN, S. L., STEIN, R. A. and MEAD, J. F., *Chem. Phys. Lipids,* **16,** 161 (1976).

122. CHERNICK, S. S., *Methods in Enzymology,* **14,** 627 (1969).

123. CHRISTIANSEN, K., MAHADEVAN, V., VISWANATHAN, C. V. and HOLMAN, R. T., *Lipids,* **4,** 421 (1969).

124. CHRISTIE, W. W., *J. Chromat.,* **34,** 405 (1968).

125. CHRISTIE, W. W., *J. Chromat.,* **37,** 27 (1968).

126. CHRISTIE, W. W., *Biochim. biophys. Acta,* **187,** 1 (1969).

127. CHRISTIE, W. W., in *Topics in Lipid Chemistry,* Vol. 1, pp 1–49 (1970) (edited by F. D. Gunstone, Logos Press, London).

128. CHRISTIE, W. W., *Analyst* (London), **97,** 221 (1972).

129. CHRISTIE, W. W., *Topics in Lipid Chemistry,* Vol. 3, pp 171–197 (1972) (edited by F. D. Gunstone, Logos Press, London).

130. CHRISTIE, W. W., *Biochim. biophys. Acta,* **316,** 204 (1973).

131. CHRISTIE, W. W., *J. Chromat. Sci.,* **13,** 411 (1975).

132. CHRISTIE, W. W., GUNSTONE, F. D., ISMAIL, I. A. and WADE, L., *Chem. Phys. Lipids,* **2,** 196 (1968).

133. CHRISTIE, W. W., GUNSTONE, F. D. and PRENTICE, H. G., *J. Chem. Soc.,* 5768 (1963).

134. CHRISTIE, W. W., GUNSTONE, F. D., PRENTICE, H. G. and SEN GUPTA, S. C., *J. Chem. Soc.,* 5833 (1964).

135. CHRISTIE, W. W. and HOLMAN, R. T., *Chem. Phys. Lipids,* **1,** 407 (1967).

136. CHRISTIE, W. W. and HUNTER, M. L., *Biochim. biophys. Acta,* **316,** 282 (1973).

137. CHRISTIE, W. W. and HUNTER, M. L., *J. Chromat.,* **171,** 517 (1979).

138. CHRISTIE, W. W. and HUNTER, M. L., *Biochem. J., 191,* 637 (1980).
139. CHRISTIE, W. W. and MOORE, J. H., *Biochim. biophys. Acta,* **176,** 445 (1969).
140. CHRISTIE, W. W. and MOORE, J. H., *J. Sci. Fd Agric.,* **22,** 120 (1971).
141. CHRISTIE, W. W., NOBLE, R. C. and MOORE, J. H., *Analyst* (London), **95,** 940 (1970).
142. CHRISTIE, W. W., REBELLO, D. and HOLMAN, R. T., *Lipids,* **4,** 229 (1969).
143. CHRISTOPHERSON, S. W. and GLASS, R. L., *J. Dairy Sci.,* **52,** 1289 (1969).
144. CHUNG, B. H., WILKINSON, T., GEER, J. C. and SEGREST, J. P., *J. Lipid Res.,* **21,** 284 (1980).
145. COLBORNE, A. J. and LAIDMAN, D. L., *Phytochemistry,* **14,** 2639 (1975).
146. COMFURIUS, P. and ZWAAL, R. F. A., *Biochim. biophys. Acta,* **488,** 36 (1977).
147. CONACHER, H. B. S., *J. Assoc. off. Analyt. Chem.,* **58,** 488 (1975).
148. CONACHER, H. B. S., *J. Chromat. Sci.,* **14,** 405 (1976).
149. CONACHER, H. B. S., GUNSTONE, F. D., HORNBY, G. M. and PADLEY, F. B., *Lipids,* **5,** 434 (1970).
150. CONACHER, H. B. S., IYENGAR, J. R. and BEARE-ROGERS, J. L., *J. Assoc. off. Analyt. Chem.,* **60,** 899 (1977).
151. COOPER, M. J. and ANDERS, M. W., *Analyt. Chem.,* **46,** 1849 (1974).
152. COOPER, P. H. and HAWTHORNE, J. N., *J. Chromat.,* **87,** 267 (1973).
153. COREY, E. J. and VENKATESWARLU, A., *J. Am. Chem. Soc.,* **94,** 6190 (1972).
154. CRASKE, J. D. and EDWARDS, R. A., *J. Chromat.,* **53,** 253 (1970).
155. CRAWFORD, C. G., PLATTNER, R. D., SESSA, D. J. and RACKIS, J. J., *Lipids,* **15,** 91 (1980).
156. CROCKEN, B. J. and NYC, J. F., *J. biol. Chem.,* **239,** 1727 (1964).
157. CURSTEDT, T., *Biochim. biophys. Acta,* **489,** 79 (1977).
158. CURSTEDT, T. and SJÖVALL, J., *Biochim. biophys. Acta,* **360,** 24 (1974).
159. DARBRE, A., *Handbook of Derivatives for Chromatography* pp 36–103 (1978) (edited by K. Blau and G. S. King, Heyden & Son, London).
160. DAUBEN, W. G., HOERGER, E. and PETERSEN, J. W., *J. Am. Chem. Soc.,* **75,** 2347 (1953).
161. DAVIDSON, L. M. and KRITCHEVSKY, D., *Analyt. Biochem.,* **66,** 287 (1975).
162. DAVIES, J. E. D., HODGE, P., BARVE, J. A., GUNSTONE, F. D. and ISMAIL, I. A., *J. Chem. Soc. Perkin,* **II,** 1557 (1972).
163. DAVIES, J. E. D., HODGE, P., GUNSTONE, F. D. and LIE KEN JIE, M. S. F., *Chem. Phys. Lipids,* **15,** 48 (1975).
164. DAVIES, J. E. D., LIE KEN JIE, M. S. F. and LAM, C. H., *Chem. Phys. Lipids,* **15,** 157 (1975).
165. DAWSON, G., in *lipid Chromatographic Analysis,* 2nd Edition, Vol. 2, pp 663–700 (1976) (edited by G. C. Marinetti, Marcel Dekker Inc., N.Y.).
166. DAWSON, R. M. C., in *Form and Function of Phospholipids,* pp 97–116 (1973) (edited by G. B. Ansell, J. N. Hawthorne, and R. M. C. Dawson, Elsevier Publishing Co., Amsterdam).
167. DAWSON, R. M. C., in *Lipid Chromatographic Analysis,* 2nd Edition, Vol. 1, pp 149–172 (1976) (edited by G. V. Marinetti, Marcel Dekker Inc., N.Y.).
168. DAWSON, R. M. C. and EICHBERG, J., *Biochem. J., 96,* 634 (1965).
169. DAWSON, R. M. C., HEMINGTON, N. and DAVENPORT, J. B., *Biochem. J., 84,* 497 (1962).
170. DAWSON, R. M. C. and KEMP, P., *Biochem. J., 105,* 837 (1967).
171. DE BOER, T. J. and BACKER, H. J., *Organic Syntheses,* Coll. Vol. 4, pp 250–255 (1963) (edited by N. Rabjohn, John Wiley & Sons, London).
172. DE BRABANDER, H. F. and VERBEKE, R., *Analyt. Biochem.,* **110,** 240 (1981).
173. DE HAAS, G. H. and VAN DEENEN, L. L. M., *Biochim. biophys. Acta,* **106,** 315 (1965).
174. DE LALLA, O. and GOFMAN, J., in *Methods of Biochemical Analysis,* Vol. 1, pp 459–478 (1954) (edited by E. Glick, Interscience, New York).
175. DHOPESHWARKAR, G. A. and MEAD, J. F., *Proc. Soc. exp. Biol. Med.,* **109,** 425 (1962).
176. DINH-NGUYEN, N., RYHAGE, R. and STALLBERG-STENHAGEN, S., *Arkiv. Kemi,* **15,** 433 (1960).
177. DIRINGER, H., *Hoppe-Seyler's Z. physiol. Chem.,* **353,** 39 (1972).
178. DITTMAR, K. E. J., HECKERS, H. and MELCHER, F. W., *Fette Seifen Anstrichm.,* **80,** 297 (1978).
179. DITTMER, J. C., FEMINELLA, J. L. and HANAHAN, D. J., *J. biol. Chem.,* **233,** 862 (1958).
180. DITTMER, J. C. and LESTER, R. L., *J. Lipid Res.,* **5,** 126 (1964).
181. DITTMER, J. C. and WELLS, M. A., *Methods in Enzymology,* **14,** 482 (1969).
182. DOLEV, A., ROHWEDDER, W. K. and DUTTON, H. J., *Lipids,* **1,** 231 (1966).
183. DOMMES, V., WIRTZ-PEITZ, F. and KUNAU, W. H., *J. Chromat. Sci.,* **14,** 360 (1976).
184. DOWNING, D. T. and GREENE, R. S., *Lipids,* **3,** 96 (1968).
185. DROZD, J., *J. Chromat.,* **113,** 303 (1975).
186. DUCK-CHONG, C. G., *Lipids,* **14,** 492 (1979).
187. DUDLEY, P. A. and ANDERSON, R. E., *Lipids,* **10,** 113 (1975).
188. DUDZINSKI, A. E., *J. Chromat.,* **31,** 560 (1967).
189. DUNCOMBE, W. G., *Biochem. J., 88,* 7 (1963).
190. DUNCOMBE, W. G. and RISING, T. J., *J. Lipid Res.,* **14,** 258 (1973).
191. DURST, H. D., MILANO, M., KIKTA, E. J., CONNELLY, S. A. and GRUSHKA, E., *Analyt. Chem.,* **47,** 1797 (1975).
192. DUTTON, H. J., *J. Am. Oil Chem. Soc.,* **38,** 631 (1961).
193. DYATLOVITSKAYA, E. V., VOLKOVA, V. I. and BERGELSON, L. D., *Bull. Acad. Sci. USSR Div. chem. Sci.,* 946 (1966).
194. EBERLEIN, K. and GERCKEN, G., *J. Chromat.,* **61,** 285 (1971).
195. EBERLEIN, K. and GERCKEN, G., *J. Chromat.,* **106,** 425 (1975).
196. EISELE, T. A., LIBBEY, L. M., PAWLOSKI, N. E., NIXON, J. E. and SINNHUBER, R. O., *Chem. Phys. Lipids,* **12,** 316 (1974).
197. ELBEIN, A. D. and FORSEE, W. T., in *Glycolipid Methodology,* pp 345–368 (1976) (edited by L. A. Witting, American Oil Chemists' Society, Champaign, Illinois).
198. EL-HAMDY, A. H. and PERKINS, E. G., *J. Am. Oil Chem. Soc.,* **58,** 49 (1981).
199. ELLINGBOE, J., NYSTROM, E. and SJOVALL, J., *Biochim. biophys. Acta,* **152,** 803 (1968).
200. ELLINGBOE, J., NYSTROM, E. and SJOVALL, J., *J. Lipid Res.,* **11,** 266 (1970).
201. EMKEN, E. A., *Lipids,* **6,** 686 (1971).
202. EMKEN, E. A., *Lipids,* **7,** 459 (1972).
203. EMKEN, E. A. and DUTTON, H. J., *Lipids,* **9,** 272 (1974).
204. EMKEN, E. A., HARTMAN, J. C. and TURNER, C. R., *J. Am. Oil Chem. Soc.,* **55,** 561 (1978).

205. ETTRE, L. S., *Open Tubular Columns in Gas Chromatography* (1965) (Plenum Press, N.Y.).

206. ETTRE, L. S., PURCELL, J. E. and NOREM, S. D., *J. Gas Chromat.*, **3**, 181 (1965).

207. FANG, M. and MARINETTI, G. V., *Methods in Enzymology*, **14**, 598 (1969).

208. FERRELL, W. J., RADLOFF, J. F. and JACKIW, A. B., *Lipids*, **4**, 278 (1969).

209. FEWSTER, M. E., BURNS, B. J. and MEAD, J. F., *J. Chromat.*, **43**, 120 (1969).

210. FIRESTONE, D. and HORWITZ, W., *J. Assoc. off. Analyt. Chem.*, **62**, 709 (1979).

211. FISCHER, W., *Biochim. biophys. Acta*, **487**, 74 (1977).

212. FISCHER, W., HEINZ, E. and ZEUS, M., *Hoppe-Seyler's Z. physiol. Chem.*, **354**, 1115 (1973).

213. FOLCH, J., LEES, M. and STANLEY, G. H. S., *J. biol. Chem.*, **226**, 497 (1957).

214. FREDMAN, P., NILSSON, O., TAYOT, J. L. and SVENNERHOLM, L., *Biochim. biophys. Acta*, **618**, 42 (1980).

215. FREEMAN, C. P. and WEST, D., *J. Lipid Res.*, **7**, 324 (1966).

216. FREEMAN, N. K., *J. Am. Oil Chem. Soc.*, **45**, 798 (1968).

217. FREEMAN, N. K., LINDGREN, F. T., NG, Y. C. and NICHOLS, A. V., *J. biol. Chem.*, **227**, 449 (1957).

218. FRIEDEWALD, W. T., LEVY, R. I. and FREDRICKSON, D. S., *Clin. Chem.*, **18**, 499 (1972).

219. FROST, D. J. and BARZILAY, J., *Analyt. Chem.*, **43**, 1316 (1971).

220. FROST, D. J. and GUNSTONE, F. D., *Chem. Phys. Lipids*, **15**, 53 (1975).

221. FROST, D. J. and SIES, I., *Chem. Phys. Lipids*, **13**, 173 (1974).

222. FULK, W. K. and SHORB, M. S., *J. Lipid Res.*, **11**, 276 (1970).

223. GALANOS, D. S. and KAPOULAS, V. M., *J. Lipid Res.*, **3**, 134 (1962).

224. GALLAI-HATCHARD, J. J. and GRAY, G. M., *Biochim. biophys. Acta*, **116**, 532 (1966).

225. GARDNER, H. W., *J. Lipid Res.*, **9**, 139 (1968).

226. GASKELL, S. J. and BROOKS, C. J. W., *J. Chromat.*, **142**, 469 (1977).

227. GATT, S., *Methods in Enzymology*, **14**, 152 (1969).

228. GAVER, R. C. and SWEELEY, C. C., *J. Am. Oil. Chem. Soc.*, **42**, 294 (1965).

229. GAVER, R. C. and SWEELEY, C. C., *J. Am. Chem. Soc.*, **88**, 3643 (1966).

230. GERBER, J. G., BARNES, J. S. and NIES, A. S., *J. Lipid Res.*, **20**, 912 (1979).

231. GEURTS VAN KESSEL, W. S. M., HAX, W. M. A., DEMEL, R. A. and DE GIER, J., *Biochim. biophys. Acta*, **486**, 524 (1977).

232. GEYER, R. G. and GOODMAN, H. M., *Proc. Soc. Exp. Biol. Med.*, **133**, 404 (1970).

233. GILLILAND, K. M. and MOSCATELLI, E. A., *Biochim. biophys. Acta*, **187**, 221 (1969).

234. GLASS, C. A. and DUTTON, H. J., *Analyt. Chem.*, **36**, 2401 (1964).

235. GONZALEZ-SASTRE, F. and FOLCH-PI, J., *J. Lipid Res.*. **9**, 532 (1968).

236. GRAY, G. M., *Nature*, **207**, 505 (1965).

237. GRAY, G. M., *Biochim. biophys. Acta*, **144**, 511 (1967).

238. GRAY, G. M., *Biochim. biophys. Acta*, **144**, 519 (1967).

239. GRAY, G. M., *Lipid Chromatographic Analysis*, Second Edition, Vol. 3, pp 897–923 (1976) (edited by G. V. Marinetti, Marcel Dekker Inc., N.Y.).

240. GROB, K., *J. Chromat.*, **178**, 387 (1979).

241. GROB, K., *J. Chromat.*, **205**, 289 (1981).

242. GROB, K., NEUKOM, H. P. and BATTAGLIA, R., *J. Am. Oil Chem. Soc.*, **57**, 282 (1980).

243. GROSS, R. W. and SOBEL, B. E., *J. Chromat.*, **197**, 79 (1980).

244. GROSS, S. K. and MCCLUER, R. H., *Analyt. Biochem.*, **102**, 429 (1980).

245. GUNSTONE, F. D. and INGLIS, R. P., in *Topics in Lipid Chemistry*, Vol. 2, pp 287–307 (1971) (edited by F. D. Gunstone, Logos Press, London).

246. GUNSTONE, F. D. and INGLIS, R. P., *Chem. Phys. Lipids*, **10**, 73 (1973).

247. GUNSTONE, F. D. and ISMAIL, I. A., *Chem. Phys. lipids*, **1**, 337 (1967).

248. GUNSTONE, F. D., ISMAIL, I. A. and LIE KEN JIE, M., *Chem. Phys. Lipids*, **1**, 376 (1967).

249. GUNSTONE, F. D. and JACOBSBERG, F. R., *Chem. Phys. Lipids*, **9**, 26 (1972).

250. GUNSTONE, F. D., KILCAST, D., POWELL, R. G. and TAYLOR, G. M., *Chem. Commun.*, 295 (1967).

251. GUNSTONE, F. D. and LIE KEN JIE, M., *Chem. Phys. Lipids*, **4**, 131 (1970).

252. GUNSTONE, F. D., LIE KEN JIE, M. S. F. and WALL, R. T., *Chem. Phys. Lipids*, **3**, 297 (1969).

253. GUNSTONE, F. D., MCLAUGHLAN, J., SCRIMGEOUR, C. M. and WATSON, A. P., *J. Sci. Fd Agric.*, **27**, 675 (1976).

254. GUNSTONE, F. D. and MORRIS, L. J., *J. Chem. Soc.*, 2127 (1959).

255. GUNSTONE, F. D. and PADLEY, F. B., *J. Am. Oil Chem. Soc.*, **42**, 957 (1965).

256. GUNSTONE, F. D., PADLEY, F. B. and QURESHI, M. I., *Chem. Ind. (London)*, **483** (1964).

257. GUNSTONE, F. D., POLLARD, M. R., SCRIMGEOUR, C. M., GILMAN, N. W. and HOLLAND, B. C., *Chem. Phys. Lipids*, **17**, 1 (1976).

258. GUNSTONE, F. D., POLLARD, M. R., SCRIMGEOUR, C. M. and VEDANAYAGAM, H. S., *Chem. Phys. Lipids*, **18**, 115 (1977).

259. GUNSTONE, F. D. and SUBBARAO, R., *Chem. Ind.* (London), 461 (1966).

260. HAAHTI, E. and NIKKARI, T., *Acta Chem. Scand.*, **17**, 536 (1963).

261. HAAHTI, E., NIKKARI, T. and JUVA, K., *Acta Chem. Scand.*, **17**, 538 (1963).

262. HADORN, H. and ZUERCHER, K., *Mitt. Lebensmittelunters. Hyg.*, **58**, 236 (1967).

263. HAINES, T. H., *Prog. Chem. Fats*, **11**, 297 (1971).

264. HAJRA, A. K. and RADIN, N. S., *J. Lipid Res.*, **3**, 131 (1962).

265. HAKOMORI, S. and STRYCHARZ, G. D., *Biochemistry*, **7**, 1279 (1968).

266. HALLGREN, B. and LARSSON, S., *J. Lipid Res.*, **3**, 31 (1962).

267. HALLGREN, B., RYHAGE, R. and STENHAGEN, E., *Acta Chem. Scand.*, **13**, 845 (1959).

268. HAMILTON, R. J. and ACKMAN, R. G., *J. Chromat. Sci.*, **13**, 474 (1975).

269. HAMILTON, R. J. and SEWELL, P. A., *Introduction to High Performance Liquid Chromatography* (1977) (Chapman and Hall, London).

270. HAMILTON, R. J. and YAQUB RAIE, M., *Chem. Ind.* (London), 1228 (1971).

271. HAMMARSTRÖM, S., *J. Lipid Res.*, **11**, 175 (1970).

272. HAMMARSTRÖM, S., *Methods in Enzymology*, **35**, 326 (1975).

273. HAMMARSTRÖM, S. and SAMUELSSON, B., *J. biol. Chem.*, **247**, 1001 (1972).

274. HAMMOND, E. G., *Lipids*, **4**, 246 (1969).

275. HANAHAN, D. J., EKHOLM, J. and JACKSON, C. M., *Biochemistry*, **2**, 630 (1963).

276. HANAHAN, D. J., TURNER, M. B. and JAYKO, M. E., *J. biol. Chem.*, **192**, 623 (1951).

277. HANSEN, R. P., *Chem. Ind.* (London), 1640 (1967).
278. HANSEN, R. P. and SMITH, J. F., *Lipids,* **1,** 316 (1966).
279. HARA, A. and RADIN, N. S., *Analyt. Biochem.,* **90,** 420 (1978).
280. HARLOW, R. D., LITCHFIELD, C. and REISER, R., *Lipids,* **1,** 216 (1966).
281. HARRINGTON, C. A., FENIMORE, D. C. and EICHBERG, J., *Analyt. Biochem.,* **106,** 307 (1980).
282. HASEGAWA, K. and SUZUKI, T., *Lipids,* **8,** 631 (1973).
283. HASEGAWA, K. and SUZUKI, T., *Lipids,* **10,** 667 (1975).
284. HATCH, F. T. and LEES, R. S., *Adv. Lipid Res.,* **6,** 1 (1968).
285. HAUSER, G. and EICHBERG, J., *Biochim. biophys. Acta,* **326,** 201 (1973).
286. HAUSER, G., EICHBERG, J. and GONZALEZ-SASTRE, F., *Biochim. biophys. Acta,* **248,** 87 (1971).
287. HAVERKATE, F. and VAN DEENEN, L. L. M., *Biochim. biophys. Acta,* **106,** 78 (1965).
288. HAWTHORNE, J. N. and KEMP, P., *Adv. Lipid Res.,* **2,** 127 (1964).
289. HAX, W. M. A. and GEURTS VAN KESSEL, W. S. M., *J. Chromat.,* **142,** 735 (1977).
290. HAY, J. D. and MORRISON, W. R., *Biochim. biophys. Acta,* **202,** 237 (1970).
291. HAYES, L., LOWRY, R. R. and TINSLEY, I. J., *Lipids,* **6,** 65 (1971).
292. HAZLEWOOD, G. P. and DAWSON, R. M. C., *Biochem. J.,* **153,** 49 (1976).
293. HEATH, R. R., TUMLINSON, J. H. and DOOLITTLE, R. E., *J. Chromat. Sci.,* **15,** 10 (1977).
294. HECKERS, H., DITTMAR, K., MELCHER, F. W. and KALINOWSKI, H. O., *J. Chromat.,* **135,** 93 (1977).
295. HECKERS, H., MELCHER, F. W. and SCHLOEDER, U., *J. Chromat.,* **136,** 311 (1977).
296. HEINZ, E., *Biochim. biophys. Acta,* **144,** 333 (1967).
297. HEINZ, E. and TULLOCH, A. P., *Hoppe-Seyler's Z. physiol. Chem.,* **350,** 493 (1969).
298. HELMY, F. M. and HACK, M. H., *Lipids,* **1,** 279 (1966).
299. HENDRICKSON, H. S. and BALLOU, C. E., *J. biol. Chem.,* **239,** 1369 (1964).
300. HENLY, R. S., *J. Am. Oil Chem. Soc.,* **42,** 673 (1965).
301. HENLY, R. S. and ROYER, D. J., *Methods in Enzymology,* **14,** 450 (1969).
302. HERSLOF, B., PODLAHA, O. and TOREGARD, B., *J. Am. Oil Chem. Soc.,* **56,** 864 (1979).
303. HEYNEMAN, R. A., BERNARD, D. M. and VERCAUTEREN, R. E., *J. Chromat.,* **68,** 285 (1972).
304. HILL, E. E., HUSBANDS, D. R. and LANDS, W. E. M., *J. biol. Chem.,* **243,** 4440 (1968).
305. HIRSCH, J. and AHRENS, E. H., *J. biol. Chem.,* **233,** 311 (1958).
306. HITCHCOCK, C. and NICHOLS, B. W., *Plant Lipid Biochemistry* (1971) (Academic Press, London).
307. HITES, R. A., *Analyt. Chem.,* **42,** 1736 (1970).
308. HITES, R. A., *Methods in Enzymology,* **35,** 348 (1975).
309. HJERTEN, S., *Biochim. biophys. Acta,* **53,** 514 (1961).
310. HOFSTETTER, H. H., SEN, N. and HOLMAN, R. T., *J. Am. Oil Chem. Soc.,* **42,** 537 (1965).
311. HOLLA, K. S. and CORNWELL, D. G., *J. Lipid Res.,* **6,** 322 (1965).
312. HOLLA, K. S., HORROCKS, L. A. and CORNWELL, D. G., *J. Lipid Res.,* **5,** 263 (1964).
313. HOLLOWAY, P. J. and CHALLEN, S. B., *J. Chromat.,* **25,** 336 (1966).
314. HOLMAN, R. T., *Prog. Chem. Fats,* **9,** 1 (1966).
315. HOLMAN, R. T., *Prog. Chem. Fats,* **9,** 275 (1966).
316. HOLMAN, R. T. and HOFSTETTER, H. H., *J. Am. Oil Chem. Soc.,* **42,** 540 (1965).
317. HOLMAN, R. T. and RAHM, J. J., *Prog. Chem. Fats,* **9,** 13 (1966).
318. HOLUB, B. J. and KUKSIS, A., *Lipids,* **4,** 466 (1969).
319. HOLUB, B. J. and KUKSIS, A., *J. Lipid Res.,* **12,** 510 (1971).
320. HOLUB, B. J., KUKSIS, A. and THOMPSON, W., *J. Lipid Res.,* **11,** 558 (1970).
321. HOOPER, N. K. and LAW, J. H., *J. Lipid Res.,* **9,** 270 (1968).
322. HOPKINS, C. Y., *Prog. Chem. Fats,* **8,** 213 (1966).
323. HOPKINS, C. Y., *J. Am. Oil Chem. Soc.,* **45,** 778 (1968).
324. HOPKINS, C. Y. and BERNSTEIN, H. J., *Can. J. Chem.,* **37,** 775 (1959).
325. HOPKINS, S. M., SHEEHAN, G. and LYMAN, R. L., *Biochim. biophys. Acta,* **164,** 272 (1968).
326. HORNING, M. G., CASPARRINI, G. and HORNING, E. C., *J. Chromat. Sci.,* **7,** 267 (1969).
327. HORNING, M. G., MURAKAMI, S. and HORNING, E. C., *Am. J. clin. Nutr.,* **24,** 1086 (1971).
328. HORNING, E. C., KARMEN, A. and SWEELEY, C. C., *Prog. Chem. Fats,* **7,** 167 (1964).
329. HORROCKS, L. A. and CORNWELL, D. G., *J. Lipid Res.,* **3,** 165 (1962).
330. HORVATH, C. and MELANDER, W., *J. Chromat. Sci.,* **15,** 393 (1977).
331. HOWARD, C. F. and KITTINGER, G. W., *Lipids,* **2,** 438 (1967).
332. HOWARD, G. A. and MARTIN, A. J. P., *Biochem. J.,* **46,** 532 (1950).
333. HUANG, A. and FIRESTONE, D., *J. Assoc. off. Analyt. Chem.,* **54,** 1288 (1971).
334. HUANG, T. C., CHEN, C. P., WEFLER, V. and RAFTERY, A., *Analyt. Chem.,* **33,** 1405 (1961).
335. IOANNOU, P. V. and GOLDING, B. T., *Prog. Lipid Res.,* **17,** 279 (1979).
336. ITABASHI, Y. and TAKAGI, T., *Lipids,* **15,** 205 (1980).
337. ITAYA, K., *J. Lipid Res.,* **18,** 663 (1977).
338. IUPAC–IUB Commission on Biochemical Nomenclature, *Eur. J. Biochem.,* **2,** 127 (1967); *Biochem. J.,* **105,** 897 (1967).
339. IUPAC–IUB Commission on Biochemical Nomenclature, *Hoppe-Seyler's Z. physiol. Chem.,* **358,** 599 (1977).
340. IVERSON, J. L. and SHEPPARD, A. J., *J. Chromat. Sci.,* **13,** 505 (1975).
341. IVERSON, J. L. and SHEPPARD, A. J., *J. Assoc. off. Analyt. Chem.,* **60,**, 284 (1977).
342. IVERSON, J. L. and WEIK, R. W., *J. Assoc. off. Analyt. Chem.,* **50,** 1111 (1967).
343. IWAMORI, M., COSTELLO, C. and MOSER, H. W., *J. Lipid Res.,* **20,** 86 (1979).
344. IWAMORI, M. and NAGAI, Y., *Biochim. biophys. Acta,* **528,** 257 (1978).
345. IYENGAR, B. T. R. and SCHLENK, H., *Biochemistry,* **6,** 396 (1967).
346. JACOB, J., *J. Chromat. Sci.,* **13,** 415 (1975).
347. JAEGER, H., KLOR, H. U. and DITSCHUNEIT, H., *J. Lipid Res.,* **17,** 185 (1976).
348. JAMIESON, G. R., *Topics in Lipid Chemistry,* Vol. 1, pp 107–159 (1970) (edited by F. D. Gunstone, Logos Press, London).
349. JAMIESON, G. R., *J. Chromat. Sci.,* **13,** 491 (1975).
350. JAMIESON, G. R., MCMINN, A. L. and REID, E. H., *J. Chromat.,* **178,** 555 (1979).
351. JAMIESON, G. R. and REID, E. H., *J. Chromat.,* **20,** 232 (1965).
352. JAMIESON, G. R. and REID, E. H., *J. Chromat.,* **40,** 160 (1969).

353. JAMIESON, G. R. and REID, E. H., *J. Chromat.*, **42**, 304 (1969).
354. JAMIESON, G. R. and REID, E. H., *J. Chromat.*, **128**, 193 (1976).
355. JANSSEN, G. and PARMENTIER, G., *Biomed. Mass Spectrom.*, **5**, 439 (1978).
356. JATZKEWITZ, H. and MEHL, E., *Hoppe-Seyler's Z. physiol. Chem.*, **320**, 251 (1960).
357. JENNINGS, W., *Gas chromatography with Glass Capillary Columns*, (1978) (Academic Press, New York).
358. JENSEN, R. G., *Prog. Chem. Fats*, **11**, 347 (1971).
359. JENSEN, R. G., SAMPUGNA, J., CARPENTER, D. L. and PITAS, D. E., *J. Dairy Sci.*, **50**, 231 (1967).
360. JOHNSON, A. R., MURRAY, K. E., FOGERTY, A. C., KENNETT, B. H., PEARSON, J. A. and SHENSTONE, F. S., *Lipids*, **2**, 316 (1967).
361. JOHNSON, C. B. and HOLMAN, R. T., *Lipids*, **1**, 371 (1966).
362. JOHNSON, C. B., PEARSON, A. M. and DUGAN, L. R., *Lipids*, **5**, 958 (1970).
363. JOHNSTON, A. E., DUTTON, H. J., SCHOLFIELD, C. R. and BUTTERFIELD, R. O., *J. Am. Oil Chem. Soc.*, **55**, 486 (1978).
364. JUNGALWALA, F. B., EVANS, J. E. and MCCLUER, R. H., *Biochem. J.*, **155**, 55 (1976).
365. JUNGALWALA, F. B., HAYES, L. and MCCLUER, R. H., *J. Lipid Res.*, **18**, 285 (1977).
366. JUNGALWALA, F. B., HAYSSEN, V., PASQUINI, J. M. and MCCLUER, R. H., *J. Lipid Res.*, **20**, 579 (1979).
367. KANE, J. P., in *Lipid Metabolism in Mammals* Vol. 1, pp 209–257 (1977) (edited by F. Snyder, Plenum Press, New York).
368. KANFER, J. N., *Methods in Enzymology*, **14**, 660 (1969).
369. KANNAN, R., SENG, P. N. and DEBUCH, H., *J. Chromat.*, **92**, 95 (1974).
370. KAPOULAS, V. M., *Biochim. biophys. Acta*, **176**, 324 (1969).
371. KARIM, S. M. M. (editor), *Prostaglandins: Chemical and Biochemical Aspects* (1976) (MTP Press, Lancaster, U.K.).
372. KARLSSON, K-A., *Acta Chem. Scand.*, **18**, 565 (1964).
373. KARLSSON, K-A., *Acta Chem. Scand.*, **21**, 2577 (1967).
374. KARLSSON, K-A., *Acta Chem. Scand.*, **22**, 3050 (1968).
375. KARLSSON, K-A., *Chem. Phys. Lipids*, **5**, 6 (1970).
376. KARLSSON, K-A., *Lipids*, **5**, 878 (1970).
377. KARLSSON, K-A. and MARTENSSON, E., *Biochim. biophys. Acta*, **152**, 230 (1968).
378. KARLSSON, K-A. and PASCHER, I., *J. Lipid Res.*, **12**, 466 (1971).
379. KARLSSON, K-A., SAMUELSSON, B. E. and STEEN, G. O., *Chem. Phys. Lipids*, **11**, 17 (1973).
380. KARMEN, A., *Methods in Enzymology*, **14**, 465 (1969).
381. KARMEN, A. and TRITCH, H. R., *Nature*, **186**, 150 (1960).
382. KATES, M., *Can. J. Biochem. Physiol.*, **34**, 967 (1956).
383. KATES, M., *Lipid Chromatographic Analysis* Vol. 1, pp 1–39 (1967) (edited by G. V. Marinetti, Edward Arnold Limited, London).
384. KAUFMANN, H. P. and DAS, B., *Fette Seifen Anstrich.*, **64**, 214 (1962).
385. KAUFMANN, H. P. and MAKUS, Z., *Fette Seifen Anstrich.*, **63**, 235 (1961).
386. KAUFMANN, H. P. and VISWANATHAN, C. V., *Fette Seifen Anstrich.*, **65**, 925 (1963).
387. KAUFMANN, H. P. and WESSELS, H., *Fette Seifen Anstrich.*, **66**, 81 (1964).
388. KAULEN, H. D., *Analyt. Biochem.*, **45**, 664 (1972).
389. KEENAN, T. W., AWASTHI, Y. C. and CRANE, F. L., *Biochem. Biophys. Res. Commun.*, **40**, 1102 (1970).

390. KENNER, G. W. and STENHAGEN, E., *Acta Chem. Scand.*, **18**, 1551 (1964).
391. KENNERLY, D. A., PARKER, C. W. and SULLIVAN, T. J., *Anal. Biochem.*, **98**, 123 (1979).
392. KHAN, G. R. and SCHEINMANN, F., *Prog. Chem. Fats other Lipids*, **15**, 343 (1978).
393. KISHIMOTO, Y. and RADIN, N. S., *J. Lipid Res.*, **4**, 130 (1963).
394. KISIC, A. and RAPPORT, M. M., *J. Lipid Res.*, **15**, 179 (1974).
395. KITTEREDGE, J. S. and ROBERTS, E., *Science*, **164**, 37 (1969).
396. KIUCHI, K., OHTA, T. and EBINE, H., *J. Chromatogr.*, **133**, 226 (1977).
397. KLEIMAN, R., BOHANNON, M. B., GUNSTONE, F. D. and BARVE, J. A., *Lipids*, **11**, 599 (1976).
398. KLEIMAN, R. and SPENCER, G. F., *J. Amer. Oil Chem. Soc.*, **50**, 31 (1973).
399. KLEIMAN, R., SPENCER, G. F. and EARLE, F. R., *Lipids*, **4**, 118 (1969).
400. KLEIN, R. A. and KEMP, P., in *Methods in Membrane Biology*, Vol. 8, pp. 51–217 (1977) (edited by E. D. Korn, Plenum Press, New York).
401. KLEM, H. P., HINTZE, U. and GERCKEN, G., *J. Chromat.*, **75**, 19 (1973).
402. KLOPFENSTEIN, W. E., *J. Lipid Res.*, **12**, 773 (1971).
403. KOLATTUKUDY, P. E. (editor) *Chemistry and biochemistry of natural waxes* (1976) (Elsevier, Amsterdam).
404. KOMAREK, R. J., JENSEN, R. G. and PICKETT, B. W., *J. Lipid Res.*, **5**, 268 (1964).
405. KORITALA, S. and ROHWEDDER, W. K., *Lipids*, **7**, 274 (1972).
406. KRAMER, J. K. G. and HULAN, H. W., *J. Lipid Res.*, **17**, 674 (1976).
407. KUHN, R. and WIEGANDT, H., *Chem. Ber.*, **96**, 866 (1963).
408. KUKSIS, A., *Can. J. Biochem.*, **42**, 407 (1964).
409. KUKSIS, A., *Fette Seifen Anstrich.*, **73**, 130 (1971).
410. KUKSIS, A., *Fette Seifen Anstrich.*, **73**, 332 (1971).
411. KUKSIS, A., *Can. J. Biochem.*, **49**, 1245 (1971).
412. KUKSIS, A., *J. Chromat. Sci.*, **10**, 53 (1972).
413. KUKSIS, A., *Prog. Chem. Fats*, **12**, 1 (1972).
414. KUKSIS, A., *Fette Seifen Anstrich.*, **75**, 420 (1973).
415. KUKSIS, A., *Fette Seifen Anstrich.*, **75**, 517 (1973).
416. KUKSIS, A., in *Analysis of Lipids and Lipoproteins*, pp 36–62 (1975) (edited by E. G. Perkins, American Oil Chemists' Society, Champaign, Illinois).
417. KUKSIS, A., *J. Chromat.*, **143**, 3 (1977).
418. KUKSIS, A., *Separation Purification Methods*, **6**, 353 (1977).
419. KUKSIS, A. and LUDWIG, J., *Lipids*, **1**, 202 (1966).
420. KUKSIS, A. and MARAI, L., *Lipids*, **2**, 217 (1967).
421. KUKSIS, A., MYHER, J. J., GEHER, K., HOFFMAN, A. G. D., BRECKENRIDGE, W. C., JONES, G. J. L. and LITTLE, J. A., *J. Chromat.*, **146**, 393 (1978).
422. KUKSIS, A., MYHER, J. J., MARAI, L. and GEHER, K., *J. Chromat. Sci.*, **13**, 423 (1975).
423. KUKSIS, A., MYHER, J. J., MARAI, L. and GEHER, K., *Analyt. Biochem.*, **70**, 302 (1976).
424. KUKSIS, A., STACHNYK, O. and HOLUB, B. J., *J. Lipid Res.*, **10**, 660 (1969).
425. KUNDU, S. K., CHAKRAVARTY, S. K., ROY, S. K. and ROY, A. K., *J. Chromat.*, **170**, 65 (1979).
426. KUNDU, S. K. and ROY, S. K., *J. Lipid Res.*, **19**, 390 (1978).
427. KUPKE, I. R. and ZEUGNER, S., *J. Chromat.*, **146**, 261 (1978).
428. KUSAMRAN, K. and POLGAR, N., *Lipids*, **6**, 961 (1971).
429. KYRIAKIDES, E. C. and BALINT, J. A., *J. Lipid Res.*, **9**, 142 (1968).
430. LAINE, R. A., ESSELMAN, W. J. and SWEELEY, C. C., *Methods in Enzymology*, **28**, 159 (1972).
431. LAINE, R. A., STELLNER, K. and HAKOMORI, S. I., in

Methods in Membrane Biology, Vol. 2, pp 205–244 (1974) (edited by E. D. Korn, Plenum Press, New York).

432. LANDS, W. E. M., PIERINGER, R. A., SLAKEY, P. M. and ZSCHOCKE, A., *Lipids,* **1,** 444 (1966).

433. LANDS, W. E. M. and SLAKEY, P. M., *Lipids,* **1,** 295 (1966).

434. LEDEEN, R., *J. Am. Oil Chem. Soc.,* **43,** 57 (1966).

435. LEDEEN, R., *Chem. Phys. Lipids,* **5,** 205 (1970).

436. LEDEEN, R., SALSMAN, K. and CABRERA, M., *J. Lipid Res.,* **9,** 129 (1968).

437. LEDEEN, R. W., YU, R. K. and ENG, L. F., *J. Neurochem.,* **21,** 829 (1973).

438. LEE, D. M., in *Low Density Lipoproteins,* pp 3–47 (1976) (edited by C. E. Day and R. S. Levy, Plenum Press, New York).

439. LEE, K. J., *J. Lipid Res.,* **12,** 635 (1971).

440. LEIKOLA, E., NIEMINEN, E. and SOLOMAA, E., *J. Lipid Res.,* **6,** 490 (1965).

441. LIE KEN JIE, M. S. F., *J. Chromat.,* **192,** 457 (1980).

442. LIN, S. N. and HORNING, E. C., *J. Chromat.,* **112,** 465 (1975).

443. LINDGREN, F. T., in *Analysis of Lipids and Lipoproteins,* pp 204–224 (1975) (edited by E. G. Perkins, American Oil Chemists' Society, Champaign, Illinois).

444. LINDGREN, F. T., JENSEN, L. C. and HATCH, F. T., in *Blood Lipids and Lipoproteins,* pp 181–274 (1972) (edited by G. J. Nelson, Wiley-Interscience, New York).

445. LINDGREN, F. T., SILVERS, A., JUTAGIR, R., LAYSHOT, L. and BRADLEY, D. D., *Lipids,* **12,** 278 (1977).

446. LINDQVIST, B., SJOGREN, I. and NORDIN, R., *J. Lipid Res.,* **15,** 65 (1974).

447. LINSTEAD, R. P. and WHALLEY, M., *J. Chem. Soc.,* 2987 (1950).

448. LITCHFIELD, C., *Lipids,* **3,** 170 (1968).

449. LITCHFIELD, C., *Analysis of Triglycerides (1972) (Academic Press, New York).*

450. LITCHFIELD, C., ACKMAN, R. G., SIPOS, J. C. and EATON, C. A., *Lipids,* **6,** 674 (1971).

451. LITCHFIELD, C., HARLOW, R. D. and REISER, R., *Lipids,* **2,** 363 (1967).

452. LONGONE, F. T. and MILLER, A. H., *Chem. Commun.,* 447 (1967).

453. LOUGH, A. K., *Biochem. J.,* **90,** 4C (1964).

454. LOUGH, A. K., *Prog. Chem. Fats,* **14,** 1 (1975).

455. LOUGH, A. K., FELINSKI, L. and GARTON, G. A., *J. Lipid Res.,* **3,** 478 (1962).

456. Low, M. G. and FINEAN, J. B., *Biochem. J.,* **167,** 281 (1977).

457. LOWRY, O. H., ROSENBROUGH, N. J., FARR, A. L. and RANDALL, R. J., *J. biol. Chem.,* **193,** 265 (1951).

458. LOWRY, R. R., *J. Lipid Res.,* **9,** 397 (1968).

459. LUCAS, C. C. and RIDOUT, J. H., *Prog. Chem. Fats,* **10,** 1 (1970).

460. LUDDY, F. E., BARFORD, R. A., HERB, S. F., MAGIDMAN, P. and RIEMENSCHNEIDER, R. W., *J. Am. Oil Chem. Soc.,* **41,** 693 (1964).

461. LUNDBERG, W. O. and JARVI, P., *Prog. Chem. Fats,* **9,** 377 (1971).

462. LUTHRA, M. G. and SHELTAWY, A., *Biochem. J.,* **126,** 251 (1972).

463. LUTHRA, M. G. and SHELTAWY, A., *Biochem. J.,* **126,** 1231 (1972).

464. McCLOSKEY, J. A., *Methods in Enzymology,* **14,** 382 (1969).

465. McCLOSKEY, J. A., in *Topics in Lipid Chemistry,* Vol. 1, pp 369–440 (1970) (edited by F. D. Gunstone, Logos Pres, London).

466. McCLOSKEY, J. A. and LAW, J. H., *Lipids,* **2,** 225 (1967).

467. McCLOSKEY, J. A. and McCLELLAND, M. J., *J. Amer. chem. Soc.,* **87,** 5090 (1965).

468. McCLUER, R. H., *Chem. Phys. Lipids,* **5,** 220 (1970).

469. McCLUER, R. H. and EVANS, J. E., *J. Lipid Res.,* **14,** 611 (1973).

470. McCLUER, R. H. and EVANS, J. E., *J. Lipid Res.,* **17,** 412 (1976).

471. McCREARY, D. K., KOSSA, W. C., RAMACHANDRAN, S. and KURTZ, R. R., *J. Chromat. Sci.,* **16,** 329 (1978).

472. MACFARLANE, M. G., *Adv. Lipid Res.,* **2,** 91 (1964).

473. McKIBBIN, J. M., in *Lipid Chromatographic Analysis,* Vol. 1, pp 497–507 (1967) (edited by G. V. Marinetti, Edward Arnold Limited, London).

474. McLAFFERTY, F. W., *Analyt. Chem.,* **31,** 82 (1959).

475. MAHADEVAN, V., *Prog. Chem. Fats,* **11,** 83 (1971).

476. MAHADEVAN, V., *Prog. Chem. Fats,* **15,** 255 (1978).

477. MAHADEVAN, V., in *Fatty Acids,* pp 527–542 (1979) (edited by E. H. Pryde, American Oil Chemists' Society, Champaign, Illinois).

478. MAHADEVAN, V., PHILLIPS, F. and LUNDBERG, W. O., *J. Lipid Res.,* **6,** 434 (1965).

479. MAHADEVAN, V., PHILLIPS, F. and LUNDBERG, W. O., *Lipids,* **1,** 183 (1966).

480. MAHADEVAN, V., VISWANATHAN, C. V. and LUNDBERG, W. O., *J. Chromat.,* **24,** 357 (1966).

481. MALINS, D. C., *Prog. Chem. Fats,* **8,** 303 (1966).

482. MALINS, D. C. and MANGOLD, H. K., *J. Am. Oil Chem. Soc.,* **37,** 576 (1960).

483. MANGOLD, H. K., *Thin Layer Chromatography,* pp 137–186 (1965) (edited by E. Stahl, Springer-Verlag, Germany).

484. MANGOLD, H. K. and MUKHERJEE, K. D., *J. Chromat. Sci.,* **13,** 398 (1975).

485. MANGOLD, H. K. and SAND, D. M., *Biochim. biophys. Acta,* **164,** 124 (1968).

486. MARES, P., TVRZICKA, E. and SKOREPA, J., *J. Chromat.,* **164,** 331 (1979).

487. MARES, P., TVRZICKA, E. and TAMCHYNA, V. *J. Chromat.,* **146,** 241 (1978).

488. MARSH, J. B. and WEINSTEIN, D. B., *J. Lipid Res.,* **7,** 574 (1966).

489. MATSUBARA, T. and HAYASHI, A., *Biochim. Biophys. Acta,* **296,** 171 (1973).

489a. MATUCHA, M. and SMOLKOVA, E., *J. Chromat.,* **127,** 163 (1976).

490. MEAD, J. F., *Prog. Chem. Fats,* **9,** 159 (1968).

491. MEAKINS, G. D. and SWINDELLS, R., *J. Chem. Soc.,* 1044 (1959).

492. MECHAM, D. K. and MOHAMMAD, A., *Cereal Chem.,* **32,** 405 (1955).

493. METCALFE, L. D., *J. Gas Chromat.,* **1,** 7 (1963).

494. MIKOLAJCZAK, K. L. and BAGBY, M. O., *J. Am. Oil Chem. Soc.,* **41,** 391 (1964).

495. MILLER, R. A., BUSSELL, N. E. and RICKETTS, C., *J. Liquid Chromat.,* **1,** 291 (1978).

496. MINNIKIN, D. E., *Lipids,* **7,** 398 (1972).

497. MINNIKIN, D. E., *Lipids,* **10,** 55 (1975).

498. MINNIKIN, D. E., ABLEY, P., McQUILLIN, F. J., KUSAMRAN, K., MASKENS, K. and POLGAR, N., *Lipids,* **9,** 135 (1974).

499. MINNIKIN, D. E. and SMITH, S., *J. Chromat.,* **103,** 205 (1975).

500. MIURA, Y. and FULCO, A. J., *J. biol. Chem.,* **249,** 1880 (1974).

501. MIWA, T. K., KWOLEK, W. F. and WOLFF, I. A., *Lipids,* **1,** 152 (1966).

502. MIWA, T. K., MIKOLAJCZAK, K. L., EARLE, F. R. and WOLFF, I. A., *Analyt. Chem.*, **32**, 1739 (1960).
503. MIZUNO, K., TOYOSATO, M., YABUMOTO, S., TANIMIZU, I. and HIRAKAWA, H., *Analyt. Biochem.*, **108**, (1980).
504. MOLD, J. D., MEANS, R. E., STEVENS, R. K. and RUTH, J. M., *Biochemistry*, **3**, 1293 (1964).
505. MOMSEN, W. E. and BROCKMAN, H. L., *J. Lipid Res.*, **19**, 1032 (1978).
506. MONTHONY, J. F., WALLACE, E. G. and ALLEN, D. M., *Clin. Chem.*, **24**, 1825 (1978).
507. MORRIS, L. J., *J. Chromat.*, **12**, 321 (1963).
508. MORRIS, L. J., *J. Lipid Res.*, **7.**, 717 (1966).
509. MORRIS, L. J. and HALL, S. W., *Lipids*, **1**, 188 (1966).
510. MORRIS, L. J., HOLMAN, R. T. and FONTELL, K., *J. Lipid Res.*, **1**, 412 (1960).
511. MORRIS, L. J. and MARSHALL, M. O., *Chem. Ind.* (London), 460 (1966).
512. MORRIS, L. J., MARSHALL, M. O. and KELLY, W., *Tetrahedron Letters*, 4249 (1966).
513. MORRIS, L. J. and WHARRY, D. M., *J. Chromat.*, **20**, 27 (1965).
514. MORRIS, L. J., WHARRY, D. M. and HAMMOND, E. W., *J. Chromat.*, **31**, 69 (1967).
515. MORRIS, L. J., WHARRY, D. M. and HAMMOND, E. W., *J. Chromat.*, **33**, 471 (1968).
516. MORRISON, W. R., *Biochim. biophys. Acta*, **176**, 537 (1969).
517. MORRISON, W. R., *Biochim. biophys. Acta*, **316**, 98 (1973).
518. MORRISON, W. R. and HAY, J. D., *Biochim. biophys. Acta*, **202**, 460 (1970).
519. MORRISON, W. R., LAWRIE, T. D. V. and BLADES, J., *Chem. Ind.* (London), 1534 (1961).
520. MORRISON, W. R. and SMITH, L. M., *J. Lipid Res.*, **5**, 600 (1964).
521. MORRISON, W. R., TAN, S. L. and HARGIN, K. D., *J. Sci. Food Agric.*, **31**, 329 (1980).
522. MUKHERJEE, K. D. and MANGOLD, H. K., *J. Labelled Compounds*, **9**, 779 (1973).
523. MURATA, T., ARIGA, T., OSHIMA, M. and MIYATAKE, T., *J. Lipid Res.*, **19**, 370 (1978).
524. MURATA, T. and TAKAHASHI, S., *Analyt. Chem.*, **45**, 1816 (1973).
525. MURATA, T. and TAKAHASHI, S., *Analyt. Chem.*, **49**, 728 (1977).
526. MURRAY, K. E., *Aust. J. Chem.*, **12**, 657 (1959).
527. MURRAY, R. K., LEVINE, M. and KORNBLATT, M. J., in *Glycolipid Methodology*, pp 305–327 (1976) (edited by A. L. Witting, American Oil Chemists' Society, Champaign, Illinois).
528. MYHER, J. J., in *Handbook of Lipid Research*, Vol. 1, pp 123–196 (1978) (edited by D. J. Hanahan & A. Kuksis, Plenum Press, New York).
529. MYHER, J. J. and KUKSIS, A., *Lipids*, **9**, 382 (1974).
530. MYHER, J. J. and KUKSIS, A., *J. Chromat. Sci.*, **13**, 138 (1975).
531. MYHER, J. J. and KUKSIS, A., *Can. J. Biochem.*, **57**, 117 (1979).
532. MYHER, J. J., KUKSIS, A., MARAI, L. and YEUNG, S. K. F., *Analyt. Chem.*, **50**, 557 (1978).
533. MYHER, J. J., MARAI, L. and KUKSIS, A., *Aanalyt. Biochem.*, **62**, 188 (1974).
534. NADENICEK, J. D. and PRIVETT, O. S., *Chem. Phys. Lipids*, **2**, 409 (1968).
535. NARAYAN, K. A., in *Analysis of Lipids and Lipoproteins*, pp 225–249 (1975) (edited by E. G. Perkins, American Oil Chemists' Society, Champaign, Illinois).

536. NELSON, G. J., *Lipids*, **3**, 104 (1968).
537. NETTING, A. G., *Analyt. Biochem.*, **86**, 580 (1978).
538. NEUDOERFFER, T. S. and LEA, C. H., *J. Chromat.*, **21**, 138 (1966).
539. NICHOLS, B. W., *Biochim. biophys. Acta*, **70**, 417 (1963).
540. NICHOLS, B. W., *New Biochemical Separations*, pp 321–337 (1964) (edited by A. T. James and L. J. Morris, Van Norstrand, New York).
541. NICHOLS, B. W. and JAMES, A. T., *Fette Seifen Anstrich.*, **66**, 1003 (1964).
542. NICHOLS, B. W. and MOOREHOUSE, R., *Lipids*, **4**, 311 (1969).
543. NICHOLLS, R. G. and HARWOOD, J. L., *Phytochemistry*, **18**, 1151 (1979).
544. NICKELL, E. C. and PRIVETT, O. S., *Separation Sci.*, **2**, 307 (1967).
545. NICOLAIDES, N., *J. Chromat. Sci.*, **8**, 717 (1970).
546. NICOLAIDES, N. and FU, H. C., *Lipids*, **4**, 83 (1969).
547. NICOLOSI, R. J., SMITH, S. C. and SANTERRE, R. F., *J. Chromat.*, **60**, 111 (1971).
548. NIEHAUS, W. G. and RYHAGE, R., *Analyt. Chem.*, **40**, 1840 (1968).
549. NOBLE, R. C. and SHAND, J. H., *Lipids*, **15**, 269 (1980).
550. NOBLE, R. C., SHAND, J. H. and WAGSTAFF, H., *Biochem. Soc. Trans.*, **10**, 34 (1982).
551. NODA, M. and FUJIWARA, N., *Biochim. biophys. Acta*, **137**, 199 (1967).
552. NONAKA, G. and KISHIMOTO, Y., *Biochim. biophys. Acta*, **572**, 423 (1979).
553. NUTTER, L. J. and PRIVETT, O. S., *Lipids*, **1**, 234 (1966).
554. NUTTER, L. J. and PRIVETT, O. S., *Lipids*, **1**, 258 (1966).
555. NUTTER, L. J. and PRIVETT, O. S., *J. Dairy Sci.*, **50**, 298 (1967).
556. NUTTER, L. J. and PRIVETT, O. S., *J. Dairy Sci.*, **50**, 1194 (1967).
557. NYSTROM, E. and SJÖVALL, J., *Methods in Enzymology*, **35**, 378 (1975).
558. O'BRIEN, J. S. and BENSON, A. A., *J. Lipid Res.*, **5**, 432 (1964).
559. OKUYAMA, H. and NOJIMA, S., *J. Biochem* (Tokyo), **57**, 529 (1965).
560. OPLIGER, C. E., HEINRICH, P. C. and OLSON, R. E., *J. Lipid Res.*, **15**, 281 (1974).
561. ORD, W. O. and BAMFORD, P. C., *Chem. Ind. (London)*, 1681 (1966).
562. OSHIMA, M., ARIGA, T. and MURATA, T., *Chem. Phys. Lipids*, **19**, 289 (1977).
563. OTNAESS, A.-B., *FEBS Lett.*, **114**, 202 (1980).
564. OTTENSTEIN, D. M., BARTLEY, D. A. and SUPINA, W. R., *J. Chromat.*, **119**, 401 (1976).
565. OTTOLENGHI, A. C., *Methods in Enzymology*, **14**, 188 (1969).
566. OULEVEY, J., BODDEN, E. and THIELE, O. W., *Eur. J. Biochem.*, **79**, 265 (1977).
567. OULEVEY, J. and THIELE, O. W., *Lipids*, **15**, 194 (1980).
568. OWENS, K., *Biochem. J.*, **100**, 354 (1966).
569. PALMER, F. B. ST. C., *Biochim. Biophys. Acta*, **231**, 134 (1971).
570. PALTAUF, F., *Biochim. biophys. Acta*, **239**, 38 (1971).
571. PALTAUF, F., *Lipids*, **13**, 165 (1978).
572. PARKER, F. and PFETERSON, N. F., *J. Lipid Res.*, **6**, 455 (1965).
573. PARRIS, N. A., *J. Chromat.*, **149**, 615 (1978).
574. PEI, P. T. S., HENLY, R. S. and RAMACHANDRAN, S., *Lipids*, **10**, 152 (1975).
575. PEIFER, J. J., *Mikrochim. Acta*, 529 (1962).

576. PERKINS, E. G., McCARTHY, T. P., O'BRIEN, M. A. and KUMMEROW, F. A., *J. Am. Oil Chem. Soc.*, **54**, 279 (1977).

577. PHILLIPS, F. C. and PRIVETT, O. S., *Lipids*, **14**, 590 (1979).

578. PHILLIPS, F. C. and PRIVETT, O. S., *Lipids*, **14**, 949 (1979).

579. PITT, G. A. J. and MORTON, R. A., *Prog. Chem. Fats*, **4**, 228 (1957).

580. PLATTNER, R. D., SPENCER, G. F. and KLEIMAN, R., *Lipids*, **11**, 222 (1976).

581. PLATTNER, R. D., SPENCER, G. F. and KLEIMAN, R., *J. Am. Oil Chem. Soc.*, **54**, 511 (1977).

582. POHL, P., GLASL, H. and WAGNER, H., *J. Chromat.*, **49**, 488 (1970).

583. POLGAR, N., in *Topics in Lipid Chemistry*, Vol. 2, pp 207–246 (1971) (edited by F. D. Gunstone, Logos Press, London).

584. POLITO, A. J., AKITA, T. and SWEELEY, C. C., *Biochemistry*, **7**, 2609 (1968).

585. POLITO, A. J., NAWORAL, J. and SWEELEY, C. C., *Biochemistry*, **8**, 1811 (1969).

586. POLLET, S., ERMIDOU, S., LE SAUX, F., MONGE, M. and BAUMANN, N., *J. Lipid Res.*, **19**, 916 (1978).

587. POOLE, C. F., in *Handbook of Derivatives for Chromatography* pp 152–200 (1978) (edited by K. Blau & G. S. King. Heyden & Sons, London).

588. POOLE, C. F. and ZLATKIS, A., *J. Chromat. Sci.*, **17**, 115 (1979).

589. POORTHUIS, B. J. H. M., YAZAKI, P. J. and HOSTETLER, K. Y., *J. Lipid Res.*, **17**, 433 (1976).

590. PORTNER, N. A., WOLFF, R. A. and NIXON, J. R., *Lipids*, **14**, 20 (1979).

591. POSSMAYER, F., SCHERPHOF, G. L., DUBBELMAN, T. M. A. R., VAN GOLDE, L. M. G. and VAN DEENEN, L. L. M., *Biochim. biophys. Acta*, **176**, 95 (1969).

592. POWELL, R. G. and SMITH, C. R., *Biochemistry*, **5**, 625 (1966).

593. POWELL, R. G., SMITH, C. R., GLASS, C. A. and WOLFF, I. A., *J. org. Chem.*, **30**, 610 (1965).

594. PRIVETT, O. S., *Prog. Chem. Fats*, **9**, 91 (1966).

595. PRIVETT, O. S., *Prog. Chem. Fats*, **9**, 407 (1966).

596. PRIVETT, O. S., DOUGHERTY, K. A., ERDAHL, W. L. and STOLYHWO, A., *J. Am. Oil Chemists' Soc.*, **50**, 516 (1973).

597. PRIVETT, O. S., NADENICEK, J. D., WEBER, R. P. and PUSCH, F., *J. Am. Oil Chem. Soc.*, **40**, 28 (1963).

598. PRIVETT, O. S. and NICKELL, E. C., *J. Am. Oil Chem. Soc.*, **40**, 189 (1963).

599. PRIVETT, O. S. and NICKELL, E. C., *Lipids*, **1**, 98 (1966).

600. PRIVETT, O. S. and NUTTER, L. J., *Lipids*, **2**, 149 (1967).

601. PURCELL, J. E. and ETTRE, L. S., *J. Gas Chromat.*, **4**, 23 (1966).

602. QURESHI, N., TAKAYAMA, K. and SCHNOES, H. K., *J. biol. Chem.*, **255**, 182 (1980).

603. RADIN, N. S., *J. Am. Oil Chem. Soc.*, **42**, 569 (1965).

604. RADIN, N. S., *Adv. Neurochem.*, **1**, 51 (1975).

605. RAHN, C. H. and SCHLENK, H., *Lipids*, **8**, 612 (1973).

606. RAJU, P. K. and REISER, R., *Lipids*, **1**, 10 (1966).

607. RAJU, P. K. and REISER, R., *Lipids*, **2**, 197 (1967).

608. RAKOFF, H., WEISLEDER, D. and EMKEN, E. A., *Lipids*, **14**, 81 (1979).

609. RAMACHANDRAN, S., PANGANAMALA, R. V. and CORNWELL, D. G., *J. Lipid Res.*, **10**, 465 (1969).

610. RAMACHANDRAN, S., SPRECHER, H. W. and CORNWELL, D. G., *Lipids*, **3**, 511 (1968).

611. RAMACHANDRAN, S., VENKATA RAO, P. and CORNWELL, D. G., *J. Lipid Res.*, **9**, 137 (1968).

612. RAMWELL, P. W. (editor), *The Prostaglandins*, Vols 1–4 (1976–1979) (Plenum Press, New York).

613. RAPPORT, M. M. and ALONZO, N., *J. biol. Chem.*, **217**, 199 (1955).

614. RATNAYAKE, W. N. and ACKMAN, R. G., *Lipids*, **14**, 580 (1979).

615. RECOURT, J. H., JURRIENS, G. and SCHMITZ, M., *J. Chromat.*, **30**, 35 (1967).

616. REDGRAVE, T. G., ROBERTS, D. C. K. and WEST, C. E., *Analyt. Biochem.*, **65**, 42 (1975).

617. RENKONEN, O., *J. Am. Oil Chem. Soc.*, **42**, 298 (1965).

618. RENKONEN, O., *Lipids*, **1**, 160 (1966).

619. RENKONEN, O., *Biochim. biophys. Acta*, **125**, 288 (1966).

620. RENKONEN, O., *Adv. Lipid Res.*, **5**, 329 (1967).

621. RENKONEN, O., *Biochim. biophys. Acta*, **137**, 575 (1967).

622. RENKONEN, O., *J. Lipid Res.*, **9**, 34 (1968).

623. RENKONEN, O., *Lipids*, **3**, 191 (1968).

624. RENKONEN, O., *Biochim. biophys. Acta*, **152**, 114 (1968).

625. RENKONEN, O., LIUSVAARA, S. and MIETTINEN, E., *Annls Med. exp. Biol. Fenn.*, **43**, 200 (1965).

626. RENKONEN, O. and LUUKKONEN, A., in *Lipid Chromatographic Analysis*, 2nd Edition Vol. 1, pp 1–58 (1976) (edited by G. V. Marinetti, Marcel Dekker, N.Y.).

627. RENKONEN, O. and VARO, P., in *Lipid Chromatographic Analysis*, Vol. 1, pp 41–98 (1967) (edited by G. V. Marinetti, Edward Arnold Ltd., London).

628. RICKERT, S. J. and SWEELEY, C. C., *J. Chromat.*, **147**, 317 (1978).

629. ROBERTS, R. N., *Lipid Chromatographic Analysis*, Vol. 1, pp 447–463 (1967) (edited by G. V. Marinetti, Edward Arnold Ltd., London).

630. ROBERTSON, A. F. and LANDS, W. E. M., *Biochemistry*, **1**, 804 (1962).

631. ROCH, L. A. and GROSSBERG, S. E., *Analyt. Biochem.*, **41**, 105 (1971.

632. RODRIGUEZ DE TURCO, E. B. and AVELDANO DE CALDIRONI, M. I., *Analyt. Biochem.*, **104**, 62 (1980).

633. RODRIGUEZ DE TURCO, E. B. and BAZAN, N. G., *J. Chromat.*, **137**, 194 (1977).

634. ROEHM, J. N. and PRIVETT, O. S., *Lipids*, **5**, 353 (1970).

635. ROUSER, G., *J. Chromat. Sci.*, **11**, 60 (1973).

636. ROUSER, G., BAUMAN, A. J., KRITCHEVSKY, G., HELLER, D. and O'BRIEN, J., *J. Am. Oil. Chem. Soc.*, **38**, 544 (1961).

637. ROUSER, G., GALLI, C., LIEBER, E., BLANK, M. L. and PRIVETT, O. S., *J. Am. Oil Chem. Soc.*, **41**, 836 (1964).

638. ROUSER, G., KRITCHEVSKY, G., HELLER, D. and LIEBER, E., *J. Am. Oil Chem. Soc.*, **40**, 425 (1963).

639. ROUSER, G., KRITCHEVSKY, G., SIMON, G. and NELSON, G. J., *Lipids*, **2**, 37 (1967).

640. ROUSER, G., KRITCHEVSKY, G. and YAMAMOTO, A., in *Lipid Chromatographic Analysis* Vol. 1, pp 99–162 (1967) (edited by G. V. Marinetti, Edward Arnold Ltd., London).

641. ROUSER, G., KRITCHEVSKY, G. and YAMAMOTO, A., in *Lipid Chromatographic Analysis*, Second Edition, Vol. 3, pp 713–776 (1976) (edited by G. V. Marinetti, Marcel Dekker, N.Y.).

642. ROUSER, G., KRITCHEVSKY, G., YAMAMOTO, A., SIMON, G., GALLI, C. and BAUMAN, A. J., *Methods in Enzymology*, **14**, 272 (1969).

643. ROUSER, G., O'BRIEN, J. and HELLER, D., *J. Am. Oil Chem. Soc.*, **38**, 14 (1961).

644. RYHAGE, R. and STENHAGEN, E., *Arkiv. Kemi.*, **13**, 523 (1959).

645. RYHAGE, R. and STENHAGEN, E., *Arkiv. Kemi.*, **15**, 291 (1960).

646. RYHAGE, R. and STENHAGEN, E., *Arkiv. Kemi.,* **15**, 332 (1960).

647. RYHAGE, R. and STENHAGEN, E., *Arkiv. Kemi.,* **15**, 545 (1960).

648. RYU, E. K. and MACCOSS, M., *J. Lipid Res.,* **20**, 561 (1979).

649. SAFFORD, R. and NICHOLS, B. W., *Biochim. biophys. Acta,* **210**, 57 (1970).

650. SAITO, M. and MUKOYAMA, K., *J. Biochem.* (Tokyo), **69**, 83 (1971).

651. SAITO, T. and HAKOMORI S-I., *J. Lipid Res.,* **12**, 257 (1971).

652. SAITO, Y., SILVIUS, J. R. and MCELHANEY, R. N., *Arch. Biochem. Biophys.,* **182**, 443 (1977).

653. SALEM, N., ABOOD, L. G. and HOSS, W., *Analyt. Biochem.,* **76**, 407 (1976).

654. SAMUELSSON, B. and SAMUELSSON, K., *Biochim. biophys. Acta,* **164**, 421 (1968).

655. SAMUELSSON, B. and SAMUELSSON, K., *J. Lipid Res.,* **10**, 41 (1969).

656. SANDER, L. C., STURGEON, R. L. and FIELD, L. R., *J. Liquid Chromat.,* **4**, 61 (1981).

657. SANDERS, H., *Biochim. biophys. Acta,* **144**, 485 (1967).

658. SASTRY, P. S., *Adv. Lipid Res.,* **12**, 251 (1974).

659. SATOUCHI, K. and SAITO, K., *Biomed. Mass Spectrom.,* **6**, 396 (1979).

660. SCHACHT, J., *J. Lipid Res.,* **19**, 1063 (1978).

661. SCHIEFER, H. G. and NEUHOFF, V., *Hoppe Seyler's Z. physiol. Chem.,* **352**, 913 (1971).

662. SCHLENK, H., *Prog. Chem. Fats,* **2**, 243 (1954).

663. SCHLENK, H. and GELLERMAN, J. L., *Analyt. Chem.,* **32**, 1412 (1960).

664. SCHLENK, H. and SAND, D. M., *Analyt. Chem.,* **34**, 1676 (1962).

665. SCHMID, H. H. O., BANDI, P. C. and SU, K. L., *J. Chromat. Sci.,* **13**, 478 (1975).

666. SCHMID, H. H. O., BAUMANN, W. J. and MANGOLD, H. K., *J. Am. Chem. Soc.,* **89**, 4797 (1967).

667. SCHMID, H. H. O., JONES, L. L. and MANGOLD, H. K., *J. Lipid Res.,* **8**, 692 (1967).

668. SCHMID, H. H. O. and MANGOLD, H. K., *Biochim. biophys. Acta,* **125**, 182 (1966).

669. SCHMID, H. H. O., MANGOLD, H. K. and LUNDBERG, W. O., *J. Am. Oil Chem. Soc.,* **42**, 372 (1965).

670. SCHMID, H. H. O. and TAKAHASHI, T., *Hoppe-Seyler's Z. physiol. Chem.,* **349**, 1673 (1968).

671. SCHMID, P., *Physiol. Chem. Phys.,* **5**, 141 (1973).

672. SCHMID, P., CALVERT, J. and STEINER, R., *Physiol. Chem. Phys.,* **5**, 157 (1973).

673. SCHMID, P., HUNTER, E. and CALVERT, J., *Physiol. Chem. Phys.,* **5**, 151 (1973).

674. SCHMITZ, B. and EGGE, H., *Chem. Phys. Lipids,* **25**, 287 (1979).

675. SCHOGT, J. C. M. and HAVERKAMP-BEGEMANN, P., *J. Lipid Res.,* **6**, 466 (1965).

676. SCHOLFIELD, C. R., *J. Am. Oil Chem. Soc.,* **52**, 36 (1975).

677. SCHOLFIELD, C. R., *Analyt. Chem.,* **47**, 1417 (1975).

678. SCHOLFIELD, C. R. and MOUNTS, T. L., *J. Am. Oil Chem. Soc.,* **54**, 319 (1977).

679. SCHOMBURG, G., DIELMANN, R., HUSMANN, H. and WEEKE, F., *J. Chromat.,* **122**, 55 (1976).

680. SCHULTE, K. E. and RUCKER, G., *J. Chromat.,* **49**, 317 (1970).

681. SCHWARTZ, D. P., *Analyt. Biochem.,* **71**, 24 (1976).

682. SELVAM, R. and RADIN, N. S., *Analyt. Biochem.,* **112**, 338 (1981).

683. SÉMÉRIVA, M. and DUFOUR, C., *Biochim. biophys. Acta,* **260**, 393 (1972).

684. SERDAREVICH, B. and CARROLL, K. K., *J. Lipid Res.,* **7**, 277 (1966).

685. SHAND, J. H. and NOBLE, R. C., *Analyt. Biochem.,* **101**, 427 (1980).

686. SHAPIRO, I. L. and KRITCHEVSKY, D., *J. Chromat.,* **18**, 599 (1965).

687. SHAW, N., *Biochim. biophys Acta,* **164**, 435 (1968).

688. SHAW, N., *Bacteriol. Revs.,* **34**, 365 (1970).

689. SHAW, W. A., HARLAN, W. R. and BENNET, A., *Analyt. Biochem.,* **43**, 119 (1971).

690. SHEPPARD, A. J. and IVERSON, J. L., *J. Chromat. Sci.,* **13**, 448 (1975).

691. SHEPPARD, A. J., WALTKING, A. E., ZMACHINSKI, H. and JONES, S. T., *J. Assoc. off. Analyt. Chem.,* **61**, 1419 (1978).

692. SHIMIZU, S., INOUE, K., TANI, Y. and YAMADA, H., *Analyt. Biochem.,* **98**, 341 (1979).

693. SHIOIRI, T., NINOMIYA, K. and YAMADE, S., *J. Am. Chem. Soc.,* **94**, 6203 (1972).

694. SHUKLA, V. K. S., ABDEL-MOETY, E. M., LARSEN, E. and EGSGAARD, H., *Chem. Phys. Lipids,* **23**, 285 (1979).

695. SIAKOTOS, A. N., *Lipids,* **2**, 87 (1967).

696. SIAKOTOS, A. N. and ROUSER, G., *J. Am. Oil Chem. Soc.,* **42**, 913 (1965).

697. SIMPSON, T. H., *J. Chromat.,* **38**, 24 (1968).

698. SINGH, H. and PRIVETT, O. S., *Lipids,* **5**, 692 (1970).

699. SINGLETON, W. S., GRAY, M. S., BROWN, M. L. and WHITE, J. L., *J. Am. Oil Chem. Soc.,* **42**, 53 (1965).

700. SINNHUBER, R. O., CASTELL, J. D. and LEE, D. J., *Fedn. Proc. Fedn. Am. Soc. exp. Biol.,* **31**, 1436 (1972).

701. SIPOS, J. C. and ACKMAN, R. G., *J. Chromat. Sci.,* **16**, 443 (1978).

702. SKIPSKI, V. P. and BARCLAY, M., *Methods in Enzymology,* **14**, 530 (1969).

703. SKIPSKI, V. P., BARCLAY, M., REICHMAN, E. S. and GOOD, J. J., *Biochim. biophys. Acta,* **137**, 80 (1967).

704. SKIPSKI, V. P., PETERSON, R. F. and BARCLAY, M., *Biochem. J.,* **90**, 374 (1964).

705. SLAKEY, P. M. and LANDS, W. E. M., *Lipids,* **3**, 30 (1968).

706. SLAWSON, V., ADAMSON, A. W. and MEAD, J. F., *Lipids,* **8**, 129 (1973).

707. SLAWSON, V and MEAD, J. F., *J. Lipid Res.,* **13**, 143 (1972).

708. SLAWSON, V. and STEIN, R. A., *Lipids,* **5**, 713 (1970).

709. SLOTBOOM, A. J., DE HAAS, G. H., BONSEN, P. P. M., BURBACH-WESTERHUIS, G. J. and VAN DEENEN, L. L. M., *Chem. Phys. Lipids,* **4**, 15 (1970).

710. SLOTBOOM, A. J., DE HAAS, G. H., BURBACH-WESTERHUIS, G. J. and VAN DEENEN, L. L. M., *Chem. Phys. Lipids,* **4**, 30 (1970).

711. SLOVER, H. T. and LANZA, E., *J. Am. Oil Chem. Soc.,* **56**, 933 (1979).

712. SMID, F. and REINISOVA, J., *J. Chromat.,* **86**, 200 (1973).

713. SMITH, A. and LOUGH, A. K., *J. Chromat. Sci.,* **13**, 486 (1975).

714. SMITH, C. R., *Lipids,* **1**, 268 (1966).

715. SMITH, C. R., *Prog. Chem. Fats,* **11**, 137 (1970).

716. SMITH, C. R., in *Topics in Lipid Chemistry,* Vol. 1, pp 277–368 (1970) (edited by F. D. Gunstone, Logos Press, London).

717. SMITH, C. R., in *Topics in Lipid Chemistry,* Vol. 3, pp 89–124 (1972) (edited by F. D. Gunstone, Logos Press, London).

718. SMITH, C. R., *J. Chromat. Sci.,* **14**, 36 (1976).

719. SMITH, L. C., POWNALL, H. J. and GOTTO, A. M., *Ann. Rev. Biochem.*, **47**, 751 (1978).
720. SNYDER, F., *Prog. Chem. Fats*, **10**, 287 (1970).
721. SNYDER, F. (editor), *Ether Lipids: Chemistry and Biology*, (1972) (Academic Press, New York and London).
722. SNYDER, F., in *Lipid Chromatographic Analysis*, Second Edition, Vol. 1, pp 111–148 (1976) (edited by G. V. Marinetti, Marcel Dekker Inc., N.Y.).
723. SNYDER, F., BLANK, M. L. and WYKLE, R. L., *J. biol. Chem.*, **246**, 3639 (1971).
724. SNYDER, F. and PIANTADOSI, C., *Biochim. biophys. Acta*, **152**, 794 (1968).
725. SNYDER, L. R. and KIRKLAND, J. J., *Introduction to Modern Liquid Chromatography*, (1974) (Plenum Press, New York).
726. SNYDER, W. R. and LAW, J. H., *Lipids*, **5**, 800 (1970).
727. SORM, F., WOLLRAB, V., JAROLIMEK, P and STREIBL, M., *Chem. Ind. (London)*, 1833 (1964).
728. SPENCE, M. W., *Can. J. Biochem.*, **47**, 735 (1969).
729. SPENCER, B., *Biochim. J.*, **75**, 435 (1960).
730. STEARNS, E. M., WHITE, H. B. and QUACKENBUSH, F. W., *J. Am. Oil Chem. Soc.*, **39**, 61 (1962).
731. STEIN, R. A. and NICOLAIDES, N., *J. Lipid Res.*, **3**, 476 (1962).
732. STEIN, R. A. and SLAWSON, V., *Prog. Chem. Fats*, **8**, 373 (1966).
733. STEINBERG, G., SLATON, W. H., HOWTON, D. R. and MEAD, J. F., *J. biol. Chem.*, **220**, 257 (1956).
734. STEWART, J. C. M., *Analyt. Biochem.*, **104**, 10 (1980).
735. STILLWAY, L. W. and HARMON, S. J., *J. Lipid Res.*, **21**, 1141 (1980).
736. STODOLA, F. H., DEINEMA, M. H. and SPENCER, J. F. T., *Bacteriol. Revs.*, **31**, 194 (1967).
737. STOLYHWO, A. and PRIVETT, O. S., *J. Chromat. Sci.*, **11**, 20 (1973).
738. STOLYHWO, A., PRIVETT, O. S. and ERDAHL, W. L., *J. Chromat. Sci.*, **11**, 263 (1973).
739. STORRY, J. E. and TUCKLEY, B., *Lipids*, **2**, 501 (1967).
740. STRICKLAND, K. P., in *Form and Function of Phospholipids*, pp 9–42 (1973) (edited by G. B. Ansell, J. N. Hawthorne and R. M. C. Dawson, Elsevier Publishing Co., Amsterdam).
741. STROCCHI, A. and BONAGA, G., *Chem. Phys. Lipids*, **15**, 87 (1975).
742. SU, K. L. and SCHMID, H. H. O., *Lipids*, **9**, 208 (1974).
743. SUN, G. Y. and HORROCKS, L. A., *J. Lipid Res.*, **10**, 153 (1969).
744. SUNDLER, R. and AKESSON, B., *J. Chromat.*, **80**, 233 (1973).
745. SUYAMA, K., HORI, K. and ADACHI, S., *J. Chromat.*, **174**, 234 (1979).
746. SUZUKI, A., HANDA, S., ISHIZUKA, I. and YAMAKAWA, T., *J. Biochem.* (Tokyo), **81**, 127 (1977).
747. SUZUKI, A., HANDA, S. and YAMAKAWA, T., *J. Biochem.* (Tokyo), **80**, 1181 (1976).
748. SUZUKI, A., HANDA, S. and YAMAKAWA, T., *J. Biochem.* (Tokyo), **82**, 1185 (1977).
749. SUZUKI, A., KUNDU, S. K. and MARCUS, D. M., *J. Lipid Res.*, **21**, 473 (1980).
750. SVENNERHOLM, L., *J. Neurochem.*, **1**, 42 (1956).
751. SVENNERHOLM, L., *Biochim. biophys. Acta*, **24**, 604 (1957).
752. SVENNERHOLM, L., *Acta Soc. Med. Upsalien*, **61**, 287 (1957).
753. SVENNERHOLM, L. and FREDMAN, P., *Biochim. biophys. Acta*, **617**, 97 (1980).
754. SVENNERHOLM, E. and SVENNERHOLM, L., *Biochim. biophys. Acta*, **70**, 432 (1963).
755. SWEELEY, C. C., *Methods in Enzymology*, **14**, 255 (1969).
756. SWEELEY, C. C. and MOSCATELLI, E. A., *J. Lipid Res.*, **1**, 40 (1960).
757. SWEELEY, C. C. and SIDDIQUI, B., in *The Glycoconjugates*, Vol. 1, pp 459–540 (1977) (edited by M. Horowitz and W. Pigman, Academic Press, New York and London).
758. SWEELEY, C. C. and WALKER, B., *Analyt. Chem.*, **36**, 1461 (1964).
759. TAGUCHI, R., ASAHI, Y. and IKEZAWA, H., *Biochim. biophys. Acta*, **619**, 48 (1980).
760. TAKAGI, T. and ITABASHI, Y., *J. Chromat. Sci.*, **15**, 121 (1977).
761. TAKAGI, T. and ITABASHI, Y., *Lipids*, **12**, 1062 (1977).
762. TAKAGI, T., ITABASHI, Y., OTA, T. and HAYASHI, K., *Lipids*, **11**, 354 (1976).
763. TAKAHASHI, T. and SCHMID, H. H. O., *Chem. Phys. Lipids*, **2**, 220 (1968).
764. TALLENT, W. H. and KLEIMAN, R., *J. Lipid Res.*, **9**, 146 (1968).
765. Tentative Method Cd 14–61, in *Official and Tentative Methods of the American Oil Chemists' Society* (1946) (A.O.C.S. Chicago, Illinois).
766. THENOT, J-P., HORNING, E. C., STAFFORD, M. and HORNING, M. G., *Analyt. Lett.*, **5**, 217 (1972).
767. THOMAS, A. E., SHAROUN, J. E. and RALSTON, H., *J. Am. Oil Chem. Soc.*, **42**, 789 (1965).
768. THOMAS, P. J. and DUTTON, H. J., *Analyt. Chem.*, **41**, 657 (1969).
769. THOMPSON, W., *Biochim. biophys. Acta*, **187**, 150 (1969).
770. THOMPSON, W. and MACDONALD, G., *Eur. J. Biochem.*, **65**, 107 (1976).
771. THOMSON, G. A. and LEE, P., *Biochim. biophys. Acta*, **98**, 151 (1965).
772. THORPE, C. W., POHLAND, L. and FIRESTONE, D., *J. Ass. off Analyt. Chem.*, **52**, 774 (1969).
773. TIMMS, R. E., *Aust. J. Dairy Technol.*, **33**, 4 (1978).
774. TORELLO, L. A., YATES, A. J. and THOMPSON, D. K., *J. Chromat.*, **202**, 195 (1980).
775. TRENNER, N. R., SPETH, O. C., GRUBER, V. B. and VANDENHEUVEL, W. J. A., *J. Chromat.*, **71**, 415 (1972).
776. TRINDER, P., *J. clin. Pathol.*, **22**, 158 (1969).
777. TULLOCH, A. P., *J. Am. Oil Chem. Soc.*, **41**, 833 (1964).
778. TULLOCH, A. P., *J. Am. Oil Chem. Soc.*, **43**, 670 (1966).
779. TULLOCH, A. P., *J. Chromat. Sci.*, **13**, 403 (1975).
780. TULLOCH, A. P., in *Glycolipid Methodology*, pp 329–344 (1976) (edited by L. A. Witting, American Oil Chemists' Society, Champaign, Illinois).
781. TULLOCH, A. P. and MAZUREK, M., *Lipids*, **11**, 228 (1976).
782. TURNER, J. D. and ROUSER, G., *Analyt. Biochem.*, **38**, 423 (1970).
783. TURNER, J. D. and ROUSER, G., *Analyt. Biochem.*, **38**, 437 (1970).
784. UDENFRIEND, S., STEIN, S., BOHLEN, P., DAIRMAN, W., LEIMGRUBER, W. and WEIGELE, M., *Science*, **178**, 871 (1972).
785. ULLMAN, M. D. and McCLUER, R. H., *J. Lipid Res.*, **18**, 371 (1977).
786. ULLMAN, M. D. and McCLUER, R. H., *J. Lipid Res.*, **19**, 910 (1978).
787. VALICENTI, A. J., HEIMERMANN, W. H. and HOLMAN, R. T., *J. org. Chem.*, **44**, 1068 (1979).
788. VAN BEERS, G. J., DE JONGH, H. and BOLDINGH, J., in *Essential Fatty Acids*, pp 43–47 (1958) (edited by H. Sinclair, Butterworth Press, London).
789. VANDAMME, D., BLATON, V. and PEETERS, H., *J. Chromat.*, **145**, 151 (1978).

790. VAN DEENEN, L. L. M. and DE HAAS, G. H., *Adv. Lipid Res.*, **2**, 167 (1964).
791. VAN DER HORST, D. J., VAN GENNIP, A. H. and VOOGT, P. A., *Lipids*, **4**, 300 (1969).
792. VAN GENT, C. M. and ROSELEUR, O. J., *Clin. Chim. Acta*, **57**, 197 (1974).
793. VAN GOLDE, L. M. G. and VAN DEENEN, L. L. M., *Biochim. biophys. Acta*, **125**, 496 (1966).
794. VAN GOLDE, L. M. G. and VAN DEENEN, L. L. M., *Chem. Phys. Lipids*, **1**, 157 (1967).
795. VAN GORKOM, M. and HALL, G. E., *Spectrochim. Acta*, **22**, 990 (1966).
796. VASKOVSKY, V. E. and DEMBITZKY, V. M., *J. Chromat.*, **115**, 645 (1975).
797. VASKOVSKY, V. E. and SUPPES, Z. S., *J. Chromat.*, **63**, 455 (1971).
798. VENKATA RAO, P., RAMACHANDRAN, S. and CORNWELL, D. G., *J. Lipid Res.*, **8**, 380 (1967).
799. VERESHCHAGIN, A. G., *J. Chromat.*, **14**, 184 (1964).
800. VINSON, J. A. and HOOYMAN, J. E., *J. Chromat.*, **135**, 226 (1977).
801. VIOQUE, E. and HOLMAN, R. T., *Analyt. Chem.*, **33**, 1444 (1961).
802. VIOQUE, E. and HOLMAN, R. T., *J. Am. Oil Chem. Soc.*, **39**, 63 (1962).
803. VISWANATHAN, C. V., *J. Chromat.*, **98**, 129 (1974).
804. VISWANATHAN, C. V., HOEVET, S. P., LUNDBERG, W. O., WHITE, J. M. and MUCCINI, G. A., *J. Chromat.*, **40**, 225 (1969).
805. VISWANATHAN, C. V. and NAGABHUSHANAM, A., *J. Chromat.*, **75**, 227 (1973).
806. VITIELLO, F. and ZANETTA, J.-P., *J. Chromat.*, **166**, 637 (1978).
807. VOGEL, A. I., *Practical Organic Chemistry* (3rd edition) (1956) (Longmans-Green, London).
808. VON RUDLOFF, E., *J. Am. Oil Chem. Soc.*, **33**, 126 (1956).
809. VON RUDLOFF, E., *Can. J. Chem.*, **34**, 1413 (1956).
810. VORBECK, M. L. and MARINETTI, G. V., *J. Lipid Res.*, **6**, 3 (1965).
811. WAGNER, H., HORHAMMER, L. and WOLFF, P., *Biochem. Z.*, **334**, 175 (1961).
812. WAKU, K. and NAKAZAWA, Y., *J. Biochem.* (Tokyo), **72**, 149 (1972).
813. WANG, C. S. and SMITH, R. L., *Analyt. Biochem.*, **63**, 414 (1975).
814. WATHELET, J-P., CLAUSTRIAUX, J-J. and SEVERIN, M., *J. Chromat.*, **110**, 157 (1975).
815. WATSON, G. R. and WILLIAMS, J. P., *J. Chromat.*, **67**, 221 (1972).
816. WATTS, R. and DILS, R., *Chem. Phys. Lipids*, **3**, 168 (1969).
817. WAYS, P. and HANAHAN, D. J., *J. Lipid Res.*, **5**, 318 (1964).
818. WEBB, R. A. and METTRICK, D. F., *J. Chromat.*, **67**, 75 (1972).
819. WEISS, B., in *Lipid Chromatographic Analysis Second Edition*, Vol. 2, pp 701–712 (1976) (edited by G. V. Marinetti, Marcel Dekker Inc., N.Y.).
820. WELLS, M. A. and DITTMER, J. C., *Biochemistry*, **2**, 1259 (1963).
821. WELLS, M. A. and DITTMER, J. C., *Biochemistry*, **4**, 2459 (1965),
822. WELLS, M. A. and DITTMER, J. C., *J. Chromat.*, **18**, 503 (1965).
823. WELLS, W. W., PITTMAN, T. A. and WELLS, H. J., *Analyt. Biochem.*, **10**, 450 (1965).
824. WHITE, H. B., *J. Chromat.*, **21**, 213 (1966).
825. WHITE, H. B. and QUACKENBUSH, F. W., *J. Am. Oil Chem. Soc.*, **39**, 517 (1962).
826. WHITTAKER, V. P. and WIJESUNDERA, S., *Biochem. J.*, **51**, 348 (1952).
827. WHITTLE, K. J., DUNPHY, P. J. and PENNOCK, J. F., *Chem. Ind. (London)*, 1303 (1966).
828. WIEGANDT, H., *Adv. Lipid Res.*, **9**, 249 (1971).
829. WILLIAMS, J. P., WATSON, G. R., KHAN, M., LEUNG, S., KUKSIS, A., STACHNYK, O. and MYHER, J. J., *Analyt. Biochem.*, **66**, 110 (1975).
830. WILLNER, D., *Chem. Ind. (London)*, 1839 (1965).
831. WINEBERG, J. P. and SWERN, D., *J. Am. Oil Chem. Soc.*, **50**, 142 (1973).
832. WINEBERG, J. P. and SWERN, D., *J. Am. Oil Chem. Soc.*, **51**, 528 (1974).
833. WINTERFELD, M. and DEBUCH, H., *Hoppe Seyler's Z. physiol. Chem.*, **345**, 11 (1966).
834. WITTING, L. A. (editor), *Glycolipid Methodology* (1976) (American Oil Chemists' Society, Champaign, Illinois).
835. WOELK, H., DEBUCH, H. and PORCELLATI, G., *Hoppe-Seyler's Z. physiol. Chem.*, **354**, 1265 (1973).
836. WOLFF, I. A. and MIWA, T. K., *J. Am. Oil Chem. Soc.*, **42**, 208 (1965).
837. WOOD, P., ENGLISH, J., CHAKRABORTY, J. and HINTON, R., *Lab. Practice*, **24**, 739 (1975).
838. WOOD, R., *Lipids*, **2**, 199 (1967).
839. WOOD, R., BEVER, E. L. and SNYDER, F., *Lipids*, **1**, 399 (1966).
840. WOOD, R. and HARLOW, R. D., *Arch. Biochem. Biophys.*, **131**, 495 (1969).
841. WOOD, R. and HARLOW, R. D., *Arch. Biochem. Biophys.*, **135**, 272 (1969).
842. WOOD, R. and HARLOW, R. D., *J. Lipid Res.*, **10**, 463 (1969).
843. WOOD, R. and HEALY, K., *Lipids*, **5**, 661 (1970).
844. WOOD, R., RAJU, P. K. and REISER, R., *J. Am. Oil Chem. Soc.*, **42**, 81 (1965).
845. WOOD, R., RAJU, P. K. and REISER, R., *J. Am. Oil Chem. Soc.*, **42**, 161 (1965).
846. WOOD, R. and SNYDER, F., *Lipids*, **1**, 62 (1966).
847. WOOD, R. and SNYDER, F., *J. Am. Oil Chem. Soc.*, **43**, 53 (1966).
848. WOOD, R. and SNYDER, F., *Lipids*, **2**, 161 (1967).
849. WOOD, R. and SNYDER, F., *Lipids*, **3**, 129 (1968).
850. WOOD, R. and SNYDER, F., *Arch. Biochem. Biophys.*, **131**, 478 (1969).
851. WOODFORD, F. P. and VAN GENT, C. M., *J. Lipid Res.*, **1**, 188 (1960).
852. WRENN, J. J., *J. Chromat.*, **4**, 173 (1960).
853. WREN, J. J. and SZCZEPANOWSKA, A. D., *J. Chromat.*, **14**, 404 (1964).
854. WURSTER, C. F. and COPENHAVER, J. H., *Lipids*, **1**, 422 (1966).
855. WUTHIER, R. E., *J. Lipid Res.*, **7**, 558 (1966).
856. WYBENGA, D. R., PILEGGI, V. J., DIRSTINE, P. H. and DI GIORGIO, J., *Clin. Chem.*, **16**, 980 (1970).
857. YABUUCHI, H. and O'BRIEN, J. S., *J. Lipid Res.*, **9**, 65 (1968).
858. YAMADA, K., IMURA, K., TANIGUCHI, M. and SAKAGAMI, T., *J. Biochem.* (Tokyo), **79**, 809 (1976).
859. YAMAMOTO, A. and ROUSER, G., *Lipids*, **5**, 442 (1970).
860. YAMAZAKI, T., SUZUKI, A., HANDA, S. and YAMAKAWA, T., *J. Biochem.* (Tokyo), **86**, 803 (1979).
861. YANG, H. and HAKOMORI, S., *J. biol. Chem.*, **246**, 1192 (1971).
862. YATES, A. J. and THOMPSON, D., *J. Lipid Res.*, **18**, 660 (1977).

863. YEUNG, S. K. F., KUKSIS, A., MARAI, L. and MYHER, J. J., *Lipids,* **12,** 529 (1977).

864. YOUNG, O. M. and KANFER, J. N., *J. Chromat.,* **19,** 611 (1965).

865. YU, R. K. and LEDEEN, R. W., *J. Lipid Res.,* **13,** 680 (1972).

866. YURKOWSKI, M. and BROCKERHOFF, H., *Biochim. biophys. Acta,* **125,** 55 (1966).

867. ZAHLER, P. and NIGGLI, V., in *Methods in Membrane Biology,* Vol. 8, pp 1–50 (1970) (edited by E. D. Korn, Plenum Press, New York).

868. ZAK, B., *Lipids,* **15,** 698 (1980).

869. ZANETTA, J-P., VITIELLO, F. and ROBERT, J., *J. Chromat.,* **137,** 481 (1977).

870. ZEMAN, I. and POKORNY, J., *J. Chromat.,* **10,** 15 (1963).

871. ZEMAN, A. and SCHARMANN, H., *Fette Seifen Anstrich.,* **74,** 509 (1972); **75,** 32 (1973); **75,** 170 (1973).

872. ZIMINSKI, T. and BOROWSKI, E., *J. Chromat.,* **23,** 480 (1966).

873. ZINKEL, D. F. and ROWE, J. W., *Analyt. Chem.,* **36,** 1160 (1964).

874. ZLATKIS, A. and KAISER, R. E., *HPTLC High Performance Thin-Layer Chromatography* (1977) (Elsevier, Amsterdam).

Index